Professor Stewart's Cabinet of Mathematical Curiosities

By the Same Author

Concepts of Modern Mathematics
Game, Set, and Math
Does God Play Dice?
Another Fine Math You've Got Me Into
Fearful Symmetry (with Martin Golubitsky)
Nature's Numbers
From Here to Infinity
The Magical Maze
Life's Other Secret
Flatterland
What Shape is a Snowflake?
The Annotated Flatland (with Edwin A. Abbott)
Math Hysteria
The Mayor of Uglyville's Dilemma
Letters to a Young Mathematician
How to Cut a Cake
Why Beauty is Truth
Taming the Infinite

with Jack Cohen
The Collapse of Chaos
Figments of Reality
What Does a Martian Look Like?
Wheelers (science fiction)
Heaven (science fiction)

with Terry Pratchett and Jack Cohen
The Science of Discworld
The Science of Discworld II: The Globe
The Science of Discworld III: Darwin's Watch

Professor Stewart's Cabinet of Mathematical Curiosities

Ian Stewart

BASIC
BOOKS

A Member of the Perseus Books Group
New York

Published in the United States in 2008 by Basic
Books, A Member of the Perseus Books Group
Published in Great Britain in 2008 by Profile
Books LTD

Designed by Sue Lamble

A CIP catalog record for this book is available
from the Library of Congress.
ISBN: 978-0-465-01302-9
British ISBN: 978 1 84668 064 9

10 9 8

Contents

Start Here

●●● There are three kinds of people
in the world:
those who can count,
and those who can't.

When I was fourteen years old, I started a notebook. A *math*
notebook. Before you write me off as a sad case, I hasten to
add that it wasn't a notebook of school math. It was a notebook of every interesting thing I could find out about the
math that *wasn't* taught at school. Which, I discovered, was
a lot, because I soon had to buy another notebook.

OK, *now* you can write me off. But before you do, have
you spotted the messages in this sad little tale? *The math
you did at school is not all of it.* Better still: *the math you
didn't do at school is interesting.* In fact, a lot of it is fun—especially when you don't have to pass a test or get the sums
right.

My notebook grew to a set of six, which I still have, and
then spilled over into a filing cabinet when I discovered
the virtues of the photocopier. *Curiosities* is a sample from
my cabinet, a miscellany of intriguing mathematical
games, puzzles, stories, and factoids. Most items stand by
themselves, so you can dip in at almost any point. A few
form short mini-series. I incline to the view that a miscellany should be miscellaneous, and this one is.

The games and puzzles include some old favorites, which tend to reappear from time to time and often cause renewed excitement when they do—the car and the goats, and the 12-ball weighing puzzle, both caused a huge stir in the media: one in the USA, the other in the UK. A lot of the material is new, specially designed for this book. I've striven for variety, so there are logic puzzles, geometric puzzles, numerical puzzles, probability puzzles, odd items of mathematical culture, things to do, and things to make.

One of the virtues of knowing a bit of math is that you can impress the hell out of your friends. (Be modest about it, though, that's my advice. You can also annoy the hell out of your friends.) A good way to achieve this desirable goal is to be up to speed on the latest buzzwords. So I've scattered some short "essays" here and there, written in an informal, nontechnical style. The essays explain some of the recent breakthroughs that have featured prominently in the media. Things like Fermat's Last Theorem—remember the TV program? And the Four-Color Theorem, the Poincaré Conjecture, Chaos Theory, Fractals, Complexity Science, Penrose Patterns. Oh, and there are also some unsolved questions, just to show that math isn't all *done*. Some are recreational, some serious—like the P = NP? problem, for which a million-dollar prize is on offer. You may not have heard of the problem, but you need to know about the prize.

Shorter, snappy sections reveal interesting facts and discoveries about familiar but fascinating topics: π, prime numbers, Pythagoras' Theorem, permutations, tilings. Amusing anecdotes about famous mathematicians add a historical dimension and give us all a chance to chuckle sympathetically at their endearing foibles ...

Now, I *did* say you could dip in anywhere—and you can, believe me—but to be brutally honest, it's probably better to start at the beginning and dip in following much the same order as the pages. A few of the early items help with later ones, you see. And the early ones tend to be a bit easier, while some of the later ones are, well, a bit … *challenging*. I've made sure that a lot of easy stuff is mixed in everywhere, though, to avoid wearing your brain out too quickly.

What I'm trying to do is to excite your imagination by showing you lots of amusing and intriguing pieces of mathematics. I want you to have fun, but I'd also be overjoyed if *Curiosities* encouraged you to *engage* with mathematics, experience the thrill of discovery, and keep yourself informed about important developments—be they from four thousand years ago, last week—or tomorrow.

Ian Stewart
Coventry, January 2008

Alien Encounter

The starship *Indefensible* was in orbit around the planet
Noncomposmentis, and Captain Quirk and Mr Crock had
beamed down to the surface.

'According to the *Good Galaxy Guide*, there are two species of
intelligent aliens on this planet,' said Quirk.

'Correct, Captain – Veracitors and Gibberish. They all speak
Galaxic, and they can be distinguished by how they answer
questions. The Veracitors always reply truthfully, and the
Gibberish always lie.'

'But physically—'

'—they are indistinguishable, Captain.'

Quirk heard a sound, and turned to find three aliens creeping
up on them. They looked identical.

'Welcome to Noncomposmentis,' said one of the aliens.

'I thank you. My name is Quirk. Now, you are . . .' Quirk
paused. 'No point in asking their names,' he muttered. 'For all we
know, they'll be wrong.'

'That is logical, Captain,' said Crock.

'Because we are poor speakers of Galaxic,' Quirk improvised,
'I hope you will not mind if I call you Alfy, Betty and Gemma.' As
he spoke, he pointed to each of them in turn. Then he turned to
Crock and whispered, 'Not that we know what sex they are,
either.'

'They are all hermandrofemigynes,' said Crock.

'Whatever. Now, Alfy: to which species does Betty belong?'

'Gibberish.'

'Ah. Betty: do Alfy and Gemma belong to different species?'

'No.'

'Right . . . Talkative lot, aren't they? Um . . . Gemma: to
which species does Betty belong?'

'Veracitor.'

Quirk nodded knowledgeably. 'Right, that's settled it, then!'

'Settled what, Captain?'

'Which species each belongs to.'

'I see. And those species are—?'

'Haven't the foggiest idea, Crock. *You're* the one who's supposed to be logical!'

Answer on page 252

● ●

Tap-an-Animal

This is a great mathematical party trick for children. They take turns to choose an animal. Then they spell out its name while you, or another child, tap successive points of the ten-pointed star. You must start at the point labelled 'Rhinoceros', and move in a clockwise direction along the lines. Miraculously, as they say the final letter, you tap the correct animal.

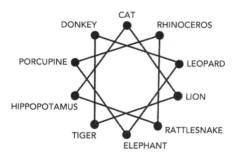

Spell the name to find the animal.

How does it work? Well, the third word along the star is 'Cat', which has three letters, the fourth is 'Lion', with four, and so on. To help conceal the trick, the animals in positions 0, 1 and 2 have 10, 11 and 12 letters. Since 10 taps takes you back to where you started, everything works out perfectly.

To conceal the trick, use *pictures* of the animals – in the diagram I've used their names for clarity.

● ●

Curious Calculations

Your calculator can do tricks.

(1) Try these multiplications. What do you notice?

1×1
11×11
111×111
1111×1111
$11,111 \times 11,111$

Does the pattern continue if you use longer strings of 1's?

(2) Enter the number

142,857

(preferably into the memory) and multiply it by 2, 3, 4, 5, 6 and 7. What do you notice?

Answers on page 252

. . . .

Triangle of Cards

I have 15 cards, numbered consecutively from 1 to 15. I want to lay them out in a triangle. I've put numbers on the top three for later reference:

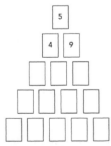

Triangle of cards.

However, I don't want any old arrangement. I want each card to be the difference between the two cards immediately below it, to left and right. For example, 5 is the difference between 4 and 9. (The differences are always calculated so that they are positive.) This condition does not apply to the cards in the bottom row, you appreciate.

The top three cards are already in place – and correct. Can you find how to place the remaining twelve cards?

Mathematicians have found 'difference triangles' like this with two, three or four rows of cards, bearing consecutive whole numbers starting from 1. It has been proved that no difference triangle can have six or more rows.

Answer on page 253

Pop-up Dodecahedron

The *dodecahedron* is a solid made from twelve pentagons, and is one of the five regular solids (page 174).

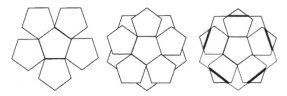

Three stages in making a pop-up dodecahedron.

Cut out two identical copies of the left-hand diagram – 10 cm across – from thickish card. Crease heavily along the joins so that the five pentagonal flaps are nice and floppy. Place one copy on top of the other, like the centre diagram. Lace an elastic band alternately over and under, as in the right-hand diagram (thick solid lines show where the band is on top) – while holding the pieces down with your finger.

Now let go.

If you've got the right size and strength of elastic band, the whole thing will pop up to form a three-dimensional dodecahedron.

Popped-up
dodecahedron.

• •

Sliced Fingers

Here's how to wrap a loop of string around somebody's fingers –
your own or those of a 'volunteer' – so that when it is pulled tight
it seems to slice through the fingers. The trick is striking because
we know from experience that if the string is genuinely linked
with the fingers then it shouldn't slip off. More precisely,
imagine that your fingers all touch a fixed surface – thereby
preventing the string from sliding off their tips. The trick is
equivalent to removing the loop from the holes created by your
fingers and the surface. If the loop were really linked through
those holes, you couldn't remove it at all, so it has to appear to be
linked without actually being linked.

If by mistake it *is* linked, it really would have to slice through
your fingers, so be careful.

How (not) to slice
your fingers off.

Why is this a mathematical trick? The connection is
topology, a branch of mathematics that emerged over the past
150 years, and is now central to the subject. Topology is about
properties such as being knotted or linked – geometrical features
that survive fairly drastic transformations. Knots remain knotted
even if the string is bent or stretched, for instance.

Make a loop from a 1-metre length of string. Hook one end over the little finger of the left hand, twist, loop it over the next finger, twist in the same direction, and keep going until it passes behind the thumb (left picture). Now bring it round in front of the thumb, and twist it over the fingers in reverse order (right picture). Make sure that when coming back, all the twists are in the *opposite* direction to what they were the first time.

Fold the thumb down to the palm of the hand, releasing the string. Pull hard on the free loop hanging from the little finger ... and you can *hear* it slice through those fingers. Yet, miraculously, no damage is done.

Unless you get a twist in the wrong direction somewhere.

Turnip for the Books

'It's been a good year for turnips,' farmer Hogswill remarked to his neighbour, Farmer Suticle.

'Yup, that it has,' the other replied. 'How many did you grow?'

'Well ... I don't exactly recall, but I do remember that when I took the turnips to market, I sold six-sevenths of them, plus one-seventh of a turnip, in the first hour.'

'Must've been tricky cuttin' 'em up.'

'No, it was a whole number that I sold. I never cuts 'em.'

'If'n you say so, Hogswill. Then what?'

'I sold six-sevenths of what was left, plus one-seventh of a turnip, in the second hour. Then I sold six-sevenths of what was left, plus one-seventh of a turnip, in the third hour. And finally I sold six-sevenths of what was left, plus one-seventh of a turnip, in the fourth hour. Then I went home.'

'Why?'

''Cos I'd sold the lot.'

How many turnips did Hogswill take to market?

Answer on page 253

The Four-Colour Theorem

Problems that are easy to state can sometimes be very hard to answer. The four-colour theorem is a notorious example. It all began in 1852 with Francis Guthrie, a graduate student at University College, London. Guthrie wrote a letter to his younger brother Frederick, containing what he thought would be a simple little puzzle. He had been trying to colour a map of the English counties, and had discovered that he could do it using four colours, so that no two adjacent counties were the same colour. He wondered whether this fact was special to the map of England, or more general. 'Can every map drawn on the plane be coloured with four (or fewer) colours so that no two regions having a common border have the same colour?' he wrote.

It took 124 years to answer him, and even now, the answer relies on extensive computer assistance. No simple conceptual proof of the four-colour theorem – one that can be checked step by step by a human being in less than a lifetime – is known.

Colouring England's counties with four colours – one solution out of many.

Frederick Guthrie couldn't answer his brother's question, but he 'knew a man who could' – the famous mathematician Augustus De Morgan. However, it quickly transpired that De Morgan *couldn't*, as he confessed in October of the same year in a letter to his even more famous Irish colleague, Sir William Rowan Hamilton.

It is easy to prove that *at least* four colours are necessary for some maps, because there are maps with four regions, each adjacent to all the others. Four counties in the map of England (shown here slightly simplified) form such an arrangement, which proves that at least four colours are necessary in this case. Can you find them on the map?

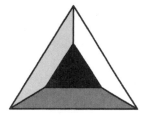

A simple map needing four colours.

De Morgan did make some progress: he proved that it is not possible to find an analogous map with *five* regions, each adjacent to all four of the others. However, this does not prove the four-colour theorem. All it does is prove that the simplest way in which it might go wrong doesn't happen. For all we know, there might be a very complicated map with, say, a hundred regions, which can't be coloured using only four colours because of the way long chains of regions connect to their neighbours. There's no reason to suppose that a 'bad' map has only five regions.

The first printed reference to the problem dates from 1878, when Arthur Cayley wrote a letter to the *Proceedings of the London Mathematical Society* (a society founded by De Morgan) to ask whether anyone had solved the problem yet. They had not, but in the following year Arthur Kempe, a barrister, published a proof, and that seemed to be that.

Kempe's proof was clever. First he proved that any map contains at least one region with five or fewer neighbours. If a region has three neighbours, you can shrink it away, getting a simpler map, and if the simpler map can be 4-coloured, so can the original one. You just give the region that you shrunk whichever colour differs from those of its three neighbours. Kempe had a more elaborate method for getting rid of a region with four or five neighbours. Having established this key fact, the rest of the proof was straightforward: to 4-colour a map, keep shrinking it, region by region, until it has four regions or fewer. Colour those regions with different colours, and then reverse the procedure, restoring regions one by one and colouring them according to Kempe's rules. Easy!

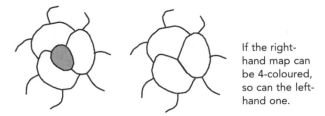

If the right-hand map can be 4-coloured, so can the left-hand one.

It looked too good to be true – and it was. In 1890 Percy Heawood discovered that Kempe's rules didn't always work. If you shrunk a region with five neighbours, and then tried to put it back, you could run into terminal trouble. In 1891 Peter Guthrie Tait thought he had fixed this error, but Julius Petersen found a mistake in Tait's method, too.

Heawood did observe that Kempe's method can be adapted to prove that five colours are always sufficient for any map. But no one could find a map that *needed* more than four. The gap was tantalising, and quickly became a disgrace. When you know that a mathematical problem has either 4 or 5 as its answer, surely you ought to be able to decide which!

But … no one could.

The usual kind of partial progress then took place. In 1922 Philip Franklin proved that all maps with 26 or fewer regions can be 4-coloured. This wasn't terribly edifying in itself, but Franklin's method paved the way for the eventual solution by introducing the idea of a *reducible configuration*. A configuration is any connected set of regions within the map, plus some information about how many regions are adjacent to those in the configuration. Given some configuration, you can remove it from the map to get a simpler map – one with fewer regions. The configuration is reducible if there is a way to 4-colour the original map, provided you can 4-colour the simpler map. In effect, there has to be a way to 'fill in' colours in that configuration, once everything else has been 4-coloured.

A single region with only three neighbours forms a reducible configuration, for instance. Remove it, and 4-colour what's left – if you can. Then put that region back, and give it a colour that has *not* been used for its three neighbours. Kempe's failed proof does establish that a region with four neighbours forms a reducible configuration. Where he went wrong was to claim the same thing for a region with five neighbours.

Franklin discovered that configurations containing several regions can sometimes work when single regions don't. Lots of multi-region configurations turn out to be reducible.

Kempe's proof would have worked if every region with five neighbours were reducible, and the reason why it would have worked is instructive. Basically, Kempe thought he had proved two things. First, every map contains a region with either three, four or five adjacent ones. Second, each of the associated configurations is reducible. Now these two facts together imply that *every* map contains a reducible configuration. In particular, when you remove a reducible configuration, the resulting simpler map also contains a reducible configuration. Remove that one, and the same thing happens. So, step by step, you can get rid of reducible configurations until the result is so simple that it has at most four regions. Colour those however you wish – at most four colours will be needed. Then restore the previously removed configuration; since this was reducible, the resulting map can also be 4-coloured ... and so on. Working backwards, we eventually 4-colour the original map.

This argument works because *every* map contains one of our irreducible configurations: they form an 'unavoidable set'.

Kempe's attempted proof failed because one of his configurations, a region with five neighbours, isn't reducible. But the message from Franklin's investigation is: don't worry. Try a bigger list, using lots of more complicated configurations. Dump the region with five neighbours; replace it by several configurations with two or three regions. Make the list as big as you need. If you can find *some* unavoidable set of reducible configurations, however big and messy, you're done.

In fact – and this matters in the final proof – you can get away with a weaker notion of unavoidability, applying only to 'minimal criminals': hypothetical maps that require five colours, with the nice feature that any smaller map needs only four colours. This condition makes it easier to prove that a given set is unavoidable. Ironically, once you prove the theorem, it turns out that no minimal criminals exist. No matter: that's the proof strategy.

In 1950 Heinrich Heesch, who had invented a clever method for proving that many configurations are reducible, said that he believed the four-colour theorem could be proved by finding an unavoidable set of reducible configurations. The only difficulty was to find one – and it wouldn't be easy, because some rule-of-thumb calculations suggested that such a set would have to include about 10,000 configurations.

By 1970 Wolfgang Haken had found some improvements to Heesch's method for proving configurations to be reducible, and began to feel that a computer-assisted proof was within reach. It should be possible to write a computer program to check that each configuration in some proposed set is reducible. You could write down several thousand configurations by hand, if you really had to. Proving them unavoidable would be time-consuming, but not necessarily out of reach. But with the computers then available, it would have taken about a century to deal with an unavoidable set of 10,000 configurations. Modern computers can do the job in a few hours, but Haken had to work with what was available, which meant that he had to improve the theoretical methods, and cut the calculation down to a feasible size.

Working with Kenneth Appel, Haken began a 'dialogue' with his computer. He would think of potential new methods for attacking the problem; the computer would then do lots of sums designed to tell him whether these methods were likely to succeed. By 1975, the size of an unavoidable set was down to only 2,000, and the two mathematicians had found much faster tests for irreducibility. Now there was a serious prospect that a human–machine collaboration could do the trick. In 1976 Appel and Haken embarked on the final phase: working out a suitable unavoidable set. They would tell the computer what set they had in mind, and it would then test each configuration to see whether it was reducible. If a configuration failed this test, it was removed and replaced by one or more alternatives, and the computer would repeat the test for irreducibility. It was a delicate process, and there was no guarantee that it would stop – but if it ever did, they would have found an unavoidable set of irreducible configurations.

In June 1976 the process stopped. The computer reported that the current set of configurations – which at that stage contained 1,936 of them, a figure they later reduced to 1,405 – was unavoidable, and every single one of those 1,936 configurations was irreducible. The proof was complete.

The computation took about 1,000 hours in those days, and the test for reducibility involved 487 different rules. Today, with faster computers, we can repeat the whole thing in about an hour. Other mathematicians have found smaller unavoidable sets and improved the tests for reducibility. But no one has yet managed to cut down the unavoidable set to something so small that an unaided human can verify that it does the job. And even if somebody did do that, this type of proof doesn't provide a very satisfactory explanation of why the theorem is true. It just says 'do a lot of sums, and the end result works'. The sums are clever, and there are some neat ideas involved, but most mathematicians would like to get a bit more insight into what's really going on. One possible approach is to invent some notion of 'curvature' for maps, and interpret reducibility as a kind of 'flattening out' process. But no one has yet found a suitable way to do this.

Nevertheless, we now know that the four-colour theorem is true, answering Francis Guthrie's innocent-looking question. Which is an amazing achievement, even if it does depend on a little bit of help from a computer.

Answer on page 254

• •

Shaggy Dog Story

Brave Sir Lunchalot was travelling through foreign parts. Suddenly there was a flash of lighting and a deafening crack of thunder, and the rain started bucketing down. Fearing rust, he headed for the nearest shelter, Duke Ethelfred's castle. He arrived to find the Duke's wife, Lady Gingerbere, weeping piteously.

Sir Lunchalot liked attractive young ladies, and for a brief moment he noticed a distinct glint through Gingerbere's tears. Ethelfred was very old and frail, he observed ... Only one thing, he vowed, would deter him from a secret tryst with the Lady – the one thing in all the world that he could not stand.

Puns.

Having greeted the Duke, Lunchalot enquired why Gingerbere was so sad.

'It is my uncle, Lord Elpus,' she explained. 'He died yesterday.'

'Permit me to offer my sincerest condolences,' said Lunchalot.

'That is not why I weep so ... so piteously, sir knight,' replied Gingerbere. 'My cousins Gord, Evan and Liddell are unable to fulfil the terms of uncle's will.'

'Why ever not?'

'It seems that Lord Elpus invested the entire family fortune in a rare breed of giant riding-dogs. He owned 17 of them.'

Lunchalot had never heard of a riding-dog, but he did not

wish to display his ignorance in front of such a lithesome lady. But this fear, it appeared, could be set to rest, for she said, 'Although I have heard much of these animals, I myself have never set eyes on one.'

'They are no fit sight for a fair lady,' said Ethelfred firmly.

'And the terms of the will—?' Lunchalot asked, to divert the direction of the conversation.

'Ah. Lord Elpus left everything to his three sons. He decreed that Gord should receive half the dogs, Evan should receive one-third, and Liddell one-ninth.'

'Mmm. Could be messy.'

'No dog is to be subdivided, good knight.'

Lunchalot stiffened at the phrase 'good knight', but decided it had been uttered innocently and was not a pathetic attempt at humour.

'Well—' Lunchalot began.

'Pah, 'tis a puzzle as ancient as yonder hills!' said Ethelfred scathingly. 'All you have to do is take one of our own riding-dogs over to the castle. Then there are 18 of the damn' things!'

'Yes, my husband, I understand the numerology, but—'

'So the first son gets half that, which is 9; the second gets one-third, which is 6; the third son gets one-ninth, which is 2. That makes 17 altogether, and our own dog can be ridden back here!'

'Yes, my husband, but we have no one here who is manly enough to ride such a dog.'

Sir Lunchalot seized his opportunity. 'Sire, *I* will ride your dog!' The look of admiration in Gingerbere's eye showed him how shrewd his gallant gesture had been.

'Very well,' said Ethelfred. 'I will summon my houndsman and he will bring the animal to the courtyard. Where we shall meet them.'

They waited in an archway as the rain continued to fall. When the dog was led into the courtyard, Lunchalot's jaw dropped so far that it was a good job he had his helmet on. The animal was twice the size of an elephant, with thick striped fur, claws like broadswords, blazing red eyes the size of Lunchalot's shield, huge floppy ears dangling to the ground, and a tail like a pig's – only with more twists and covered in sharp spines. Rain cascaded off its coat in waterfalls. The smell was indescribable.

Perched improbably on its back was a saddle.

Gingerbere seemed even more shocked than he by the sight of this terrible monstrosity. However, Sir Lunchalot was undaunted. *Nothing* could daunt his confidence. *Nothing* could prevent a secret tryst with the Lady, once he returned astride the giant hound, the will executed in full. Except ...

Well, as it happened, Sir Lunchalot did *not* ride the monstrous dog to Lord Elpus's castle, and for all he knows the will has still not been executed. Instead, he leaped on his horse and rode off angrily into the stormy darkness, mortally offended, leaving Gingerbere to suffer the pangs of unrequited lust.

It wasn't Ethelfred's dodgy arithmetic – it was what the Lady had said to her husband in a stage whisper.

What did she say?

Answer on page 254

• •

Shaggy Cat Story

No cat has eight tails.
One cat has one tail.
Adding: one cat has nine tails.

• •

Rabbits in the Hat

The Great Whodunni, a stage magician, placed his top hat on the table.

'In this hat are two rabbits,' he announced. 'Each of them is either black or white, with equal probability. I am now going to convince you, with the aid of my lovely assistant Grumpelina, that I can deduce their colours without looking inside the hat!'

He turned to his assistant, and extracted a black rabbit from her costume. 'Please place this rabbit in the hat.' She did.

Pop him in the hat and deduce what's already there.

Whodunni now turned to the audience. 'Before Grumpelina added the third rabbit, there were four equally likely combinations of rabbits.' He wrote a list on a small blackboard: BB, BW, WB and WW. 'Each combination is equally likely – the probability is $\frac{1}{4}$.

'But then I added a black rabbit. So the possibilities are BBB, BWB, BBW and BWW – again, each with probability $\frac{1}{4}$.

'Suppose – I won't do it, this is hypothetical – *suppose* I were to pull a rabbit from the hat. What is the probability that it is black? If the rabbits are BBB, that probability is 1. If BWB or BBW, it is $\frac{2}{3}$. If BWW, it is $\frac{1}{3}$. So the overall probability of pulling out a black rabbit is

$$\frac{1}{4} \times 1 + \frac{1}{4} \times \frac{2}{3} + \frac{1}{4} \times \frac{2}{3} + \frac{1}{4} \times \frac{1}{3}$$

which is exactly $\frac{2}{3}$.

'*But* – if there are three rabbits in a hat, of which exactly r are black and the rest white, the probability of extracting a black rabbit is $r/3$. Therefore $r = 2$, so there are two black rabbits in the hat.' He reached into the hat and pulled out a black rabbit. 'Since I added this black rabbit, the original pair must have been one black and one white!'

The Great Whodunni bowed, to tumultuous applause. Then

he pulled two rabbits from the hat – one pale lilac and the other shocking pink.

It seems evident that you can't deduce the contents of a hat without finding out what's inside. Adding the extra rabbit and then removing it again (was it the *same* black rabbit? Do we care?) is a clever piece of misdirection. *But why is the calculation wrong?*

Answer on page 255

River Crossing 1 – Farm Produce

Alcuin of Northumbria, aka Flaccus Albinus Alcuinus or Ealhwine, was a scholar, a clergyman and a poet. He lived in the eighth century and rose to be a leading figure at the court of the emperor Charlemagne. He included this puzzle in a letter to the emperor, as an example of 'subtlety in Arithmetick, for your enjoyment'. It still has mathematical significance, as I'll eventually explain. It goes like this.

A farmer is taking a wolf, a goat and a basket of cabbages to market, and he comes to a river where there is a small boat. He can fit only one item of the three into the boat with him at any time. He can't leave the wolf with the goat, or the goat with the cabbages, for reasons that should be obvious. Fortunately the wolf detests cabbage. How does the farmer transport all three items across the river?

Answer on page 256

More Curious Calculations

The next few calculator curiosities are variations on one basic theme.

(1) Enter a three-digit number – say 471. Repeat it to get 471,471.

Now divide that number by 7, divide the result by 11, and divide the result by 13. Here we get

$$471,471/7 = 67,353$$
$$67,353/11 = 6,123$$
$$6,123/13 = 471$$

which is the number you first thought of.

Try it with other three-digit numbers – you'll find that exactly the same trick works.

Now, mathematics isn't just about noticing curious things – it's also important to find out *why* they happen. Here we can do that by reversing the entire calculation. The reverse of division is multiplication, so – as you can check – the reverse procedure starts with the three-digit result 471, and gives

$$471 \times 13 = 6,123$$
$$6,123 \times 11 = 67,353$$
$$67,353 \times 7 = 471,471$$

Not terribly helpful as it stands ... but what this is telling us is that

$$471 \times 13 \times 11 \times 7 = 471,471$$

So it could be a good idea to see what $13 \times 11 \times 7$ is. Get your calculator and work this out. What do you notice? Does it explain the trick?

(2) Another thing mathematicians like to do is 'generalise'. That is, they try to find related ideas that work in similar ways. Suppose we start with a four-digit number, say 4,715. What should we multiply it by to get 47,154,715? Can we achieve that in several stages, multiplying by a series of smaller numbers?

To get started, divide 47,154,715 by 4,715.

(3) If your calculator runs to ten digits (nowadays a lot of them do), what would the corresponding trick be with five-digit numbers?

(4) If your calculator handles numbers with at least 12 digits, go back to a three-digit number, say 471 again. This time, instead of multiplying it by 7, 11 and 13, try multiplying it by 7, then 11, then 13, then 101, then 9,901. What happens? Why?

(5) Think of a three-digit number, such as 128. Now multiply repeatedly by 3, 3, 3, 7, 11, 13 and 37. (Yes, *three* multiplications by 3.) The result is 127,999,872 – nothing special here. So add the number your first thought of: 128. *Now* what do you get?

Answer on page 257

Extracting the Cherry

This puzzle is a golden oldie, with a simple but elusive answer.

The cocktail cherry is inside the glass, which is formed from four matches. Your task is to move at most two of the matches, so that the cherry is then outside the glass. You can turn the glass sideways or upside down if you wish, but the shape must remain the same.

Move two matches
to extract the cherry.

Answer on page 258

Make Me a Pentagon

You have a long, thin rectangular strip of paper. Your task is to make from it a regular pentagon (a five-sided figure with all edges the same length and all angles the same size).

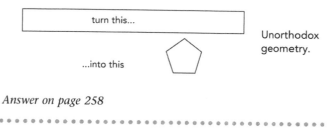

turn this...

...into this

Unorthodox
geometry.

Answer on page 258

What is π?

The number π, which is approximately 3.141 59, is the length of the circumference of a circle whose diameter is exactly 1. More generally, a circle of diameter *d* has a circumference of π*d*. A simple approximation to π is $3\frac{1}{7}$ or 22/7, but this is not exact. $3\frac{1}{7}$ is approximately 3.14 285, which is wrong by the third decimal place. A better approximation is 355/113, which is 3.141 592 9 to seven places, whereas π is 3.141 592 6 to seven places.

How do we know that π is not an exact fraction? However much you continue to improve the approximation *x/y* by using ever larger numbers, you can never get to π itself, only better and better approximations. A number that cannot be written exactly as a fraction is said to be *irrational*. The simplest proof that π is irrational uses calculus, and it was found by Johann Lambert in 1770. Although we can't write down an exact numerical representation of π, we can write down various formulas that define it precisely, and Lambert's proof uses one of those formulas.

More strongly, π is *transcendental* – it does not satisfy any algebraic equation that relates it to rational numbers. This was proved by Ferdinand Lindemann in 1882, also using calculus.

The fact that π is transcendental implies that the classical geometric problem of 'squaring the circle' is impossible. This problem asks for a Euclidean construction of a square whose area is equal to that of a given circle (which turns out to be equivalent to constructing a line whose length is the circumference of the circle). A construction is called Euclidean if it can be performed using an unmarked ruler and a compass. Well, to be pedantic, a 'pair of compasses', which means a single instrument, much as a 'pair of scissors' comes as one gadget.

● ●

Legislating the Value of π

There is a persistent myth that the State Legislature of Indiana (some say Iowa, others Idaho) once passed a law declaring the correct value of π to be – well, sometimes people say 3, sometimes $3\frac{1}{6}$ ...

Anyway, the myth is false.

However, something uncomfortably close nearly happened. The actual value concerned is unclear: the document in question seems to imply at least nine different values, all of them wrong. The law was not passed: it was 'indefinitely postponed', and apparently still is. The law concerned was House Bill 246 of the Indiana State Legislature for 1897, and it empowered the State of Indiana to make sole use of a 'new mathematical truth' at no cost. This Bill *was* passed – there was no reason to do otherwise, since it did not oblige the State to do anything. In fact, the vote was unanimous.

The new truth, however, was a rather complicated, and incorrect, attempt to 'square the circle' – that is, to construct π geometrically. An Indianapolis newspaper published an article pointing out that squaring the circle is impossible. By the time the Bill went to the Senate for confirmation, the politicians – even though most of them knew nothing about π – had sensed that there were difficulties. (The efforts of Professor C.A. Waldo of the Indiana Academy of Science, a mathematician who happened to be visiting the House when the Bill was debated, probably helped concentrate their minds.) They did not debate the validity of the mathematics; they decided that the matter was not suitable for legislation. So they postponed the bill ... and as I write, 111 years later, it remains that way.

The mathematics involved was almost certainly the brainchild of Edwin J. Goodwin, a doctor who dabbled in mathematics. He lived in the village of Solitude in Posey County, Indiana, and at various times claimed to have trisected the angle and duplicated the cube – two other famous and equally impossible feats – as well as squaring the circle. At any rate, the legislature of Indiana did not *consciously* attempt to give π an incorrect value by law – although there is a persuasive argument that passing the Bill would have 'enacted' Goodwin's approach, implying its accuracy in law, though perhaps not in mathematics. It's a delicate legal point.

• •

If They *Had* Passed It ...

If the Indiana State Legislature had passed Bill 246, and if the worst-case scenario had proved legally valid, namely that the value of π in law was different from its mathematical value, the consequences would have been distinctly interesting. Suppose that the legal value is $p \neq \pi$, but the legislation states that $p = \pi$. Then

$$\frac{p - \pi}{p - \pi} = 1 \text{ mathematically}$$

but

$$\frac{p - \pi}{p - \pi} = 0 \text{ legally}$$

Since mathematical truths are legally valid, the law would then be maintaining that $1 = 0$. Therefore all murderers have a cast-iron defence: admit to one murder, then argue that legally it is zero murders. And that's not the last of it. Multiply by one billion, to deduce that one billion equals zero. Now any citizen apprehended in possession of no drugs is in possession of drugs to a street value of $1 billion.

In fact, any statement whatsoever would become legally provable.

It seems likely that the Law would not be *quite* logical enough for this kind of argument to stand up in court. But sillier legal arguments, often based on abuse of statistics, have done just that, causing innocent people to be locked away for long periods. So Indiana's legislators might have opened up Pandora's box.

Empty Glasses

I have five glasses in a row. The first three are full and the other two empty. How can I arrange them so that they are alternately full and empty, by moving only *one* glass?

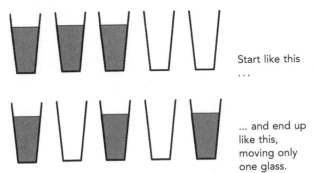

Start like this . . .

. . . and end up like this, moving only one glass.

Answer on page 258

How Many—

Ways are there to rearrange the letters of the (English) alphabet?

$403,291,461,126,605,635,584,000,000$

Ways are there to shuffle a pack of cards?

$80,658,175,170,943,878,571,660,636,856,403,766,975,$
$289,505,440,883,277,824,000,000,000,000$

Different positions are there for a Rubik cube?

$43,252,003,274,489,856,000$

Different sudoku puzzles are there?

$6,670,903,752,021,072,936,960$

(Calculated by Bertram Felgenhauer and Frazer Jarvis in 2005.)

Different sequences of 100 zeros and ones are there?

$1,267,650,600,228,229,401,496,703,205,376$

• •

Three Quickies

(1) After four bridge hands have been dealt, which is the more likely: that you and your partner hold all the spades, or that you and your partner hold no spades?

(2) If you took three bananas from a dish holding thirteen bananas, how many bananas would you have?

(3) A secretary prints out six letters from the computer and addresses six envelopes to their intended recipients. Her boss, in a hurry, interferes and stuffs the letters into the envelopes at random, one letter in each envelope. What is the probability that exactly five letters are in the right envelope?

Answers on page 258

• •

Knight's Tours

The knight in chess has an unusual move. It can jump two squares horizontally or vertically, followed by a single square at right angles, and it hops over any intermediate pieces. The geometry of the knight's move has given rise to many mathematical recreations, of which the simplest is the *knight's tour*. The knight is required to make a series of moves, visiting each square on a chessboard (or any other grid of squares) exactly once. The diagram shows a tour on a 5 × 5 board, and also shows what the possible moves look like. This tour is not 'closed'– that is, the start and finish squares are not one knight's move apart. Can you find a closed tour on the 5 × 5 board?

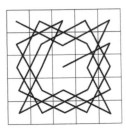 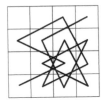

(Left) A 5 × 5 knight's tour, and (right) a partial 4 × 4 knight's tour.

I tried to find a 4 × 4 knight's tour, but I got stuck after visiting 13 squares. Can you find a knight's tour that visits all 16 squares? If not, what is the largest number of squares that the knight can visit?

There is a vast literature on this topic. Good websites include:
www.ktn.freeuk.com
mathworld.wolfram.com/KnightsTour.html

Answers on page 259

Much Undo About Knotting

A mathematician's knot is like an ordinary knot in a piece of string, but the ends of the string are glued together so that the knot can't escape. More precisely, a knot is a closed loop in space. The simplest such loop is a circle, which is called the *unknot*. The next simplest is the *trefoil knot*.

Unknot and trefoil.

Mathematicians consider two knots to be 'the same' – the jargon is *topologically equivalent* – if one can be continuously transformed into the other. 'Continuously' means you have to keep the string in one piece – no cutting – and it can't pass through itself. Knot theory becomes interesting when you discover that a really complicated knot, such as *Haken's Gordian knot*, is in fact just the unknot in disguise.

Haken's Gordian knot.

The trefoil knot is genuine – it can't be unknotted. The first proof of this apparently obvious fact was found in the 1920s.

Knots can be listed according to their complexity, which is measured by the *crossing number* – the number of apparent crossings that occur in a picture of the knot when you draw it using as few crossings as possible. The crossing number of the trefoil knot is 3.

The number of topologically distinct knots with a given number of crossings grows rapidly. Up to 16 crossings, the numbers are:

No. of crossings	3	4	5	6	7	8	9	10
No. of knots	1	1	2	3	7	21	49	165

No. of crossings	11	12	13	14	15	16
No. of knots	552	2176	9,988	46,972	253,293	1,388,705

(For pedants: these numbers refer to *prime knots*, which can't be transformed into two separate knots tied one after the other, and mirror images are ignored.)

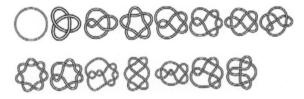

The knots with 7 or fewer crossings.

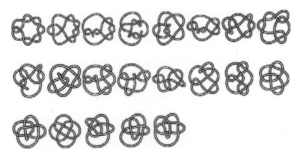

The knots with 8 crossings.

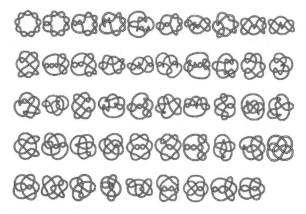

The knots with 9 crossings.

Knot theory is used in molecular biology, to understand knots in DNA, and in quantum physics. Just in case you thought knots were good only for tying up parcels.

For further information see
katlas.math.toronto.edu/wiki/The_Rolfsen_Knot_Table

• •

White-Tailed Cats

'I see you've got a cat,' said Ms Jones to Ms Smith. 'I *do* like its cute white tail! How many cats do you have?'

'Not a lot,' said Ms Smith. 'Ms Brown next door has twenty, which is a lot more than I've got.'

'You still haven't told me how many cats you have!'

'Well … let me put it like this. If you chose two distinct cats of mine at random, the probability that both of them have white tails is exactly one-half.'

'That doesn't tell me how many you've got!'

'Oh yes it does.'

How many cats does Ms Smith have – and how many have white tails?

Answer on page 259

• •

To Find Fake Coin

In February 2003 Harold Hopwood of Gravesend wrote a short letter to the *Daily Telegraph*, saying that he had solved the newspaper's crossword every day since 1937, but one conundrum had been nagging away at the back of his mind since his schooldays, and at the age of 82 he had finally decided to enlist some help.

The puzzle was this. You are given 12 coins. They all have the same weight, except for one, which may be either lighter or heavier than the rest. You have to find out which coin is different, and whether it is light or heavy, using at most three weighings on a pair of scales. The scales have no graduations for weights; they just have two pans, and you can tell whether they are in balance, or the heavier one has gone down and the lighter one has gone up.

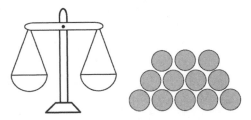

Exactly one coin is either light or heavy: find out which in three weighings.

Before reading on, you should have a go. It's quite addictive.

Within days, the paper's letters desk had received 362 letters and calls about the puzzle, nearly all asking for the answer, and they phoned me. I recognised the problem as one of the classic puzzles, typical of the 'weights and scales' genre, but I'd forgotten the answer. But my friend Marty, who happened to be in the room when I answered the phone, also recognised the problem. The same puzzle had inspired him as a teenager, and his successful solution had led him to become a mathematician.

Of course he had forgotten how the solution went, but we

came up with a method in which we weighed various sets of coins against various others, and faxed it to the newspaper.

In fact, there are many answers, including a very clever one which I finally remembered on the day that the *Telegraph* printed our less elegant method. I had seen it twenty years earlier in *New Scientist* magazine, and it had been reproduced in Thomas H. O'Beirne's *Puzzles and Paradoxes*, which I had on my bookshelf.

Puzzles like this seem to come round every twenty years or so, presumably when a new generation is exposed to them, a bit like an epidemic that gets a new lease of life when the population loses all immunity. O'Beirne traced it back to Howard Grossman in 1945, but it is almost certainly much older, going back to the seventeenth century. It wouldn't surprise me if one day we find it on a Babylonian cuneiform tablet.

O'Beirne offered a 'decision tree' solution, along the lines that Marty and I had concocted. He also recalled the elegant 1950 solution published by 'Blanche Descartes' in *Eureka*, the journal of the Archimedeans, Cambridge University's undergraduate mathematics society. Ms Descartes was in actuality Cedric A.B. Smith, and his solution is a masterpiece of ingenuity. It is presented as a poem about a certain Professor Felix Fiddlesticks, and the main idea goes like this:

> F set the coins out in a row
>   And chalked on each a letter, so,
> To form the words 'F AM NOT LICKED'
>   (An idea in his brain had clicked.)
> And now his mother he'll enjoin:
>   MA DO LIKE
>   ME TO FIND
>   FAKE COIN

This cryptic list of three weighings, one set of four against another, solves the problem, as *Eureka* explains, also in verse. To convince you, I'm going to list all the outcomes of the weighings, according to which coin is heavy or light. Here

L means that the left pan goes down, R that the right pan goes down, and – means they stay balanced.

False coin	1st weighing	2nd weighing	3rd weighing
F heavy	—	R	L
F light	—	L	R
A heavy	L	—	L
A light	R	—	R
M heavy	L	L	—
M light	R	R	—
N heavy	—	R	R
N light	—	L	L
O heavy	L	L	R
O light	R	R	L
T heavy	—	L	—
T light	—	R	—
L heavy	R	—	—
L light	L	—	—
I heavy	R	R	R
I light	L	L	L
C heavy	—	—	R
C light	—	—	L
K heavy	R	—	L
K light	L	—	R
E heavy	R	L	L
E light	L	R	R
D heavy	L	R	—
D light	R	L	—

You can check that no two possibilities give the same results.

The *Telegraph*'s publication of a valid solution did not end the matter. Readers wrote in to object to our answer, on spurious grounds. They wrote to improve it, not always by valid methods. They e-mailed to point out Ms Descartes's solution or similar ones. They told us about other weighing puzzles. They thanked us for setting their minds at rest. They cursed us for reopening an old wound. It was as if some vast, secret reservoir of folk wisdom had suddenly been breached. One correspondent remembered that the puzzle had featured on BBC television in the 1960s, with the solution being given the following night. Ominously, the letter continued, 'I do not recall why it was raised in the first place, or whether that was my first acquaintance with it; *I have a feeling that it was not.*'

Perpetual Calendar

In 1957 John Singleton patented a desk calendar that could represent any date from 01 to 31 using two cubes, but he let the patent lapse in 1965. Each cube bears six digits, one on each face.

A two-cube calendar, and two of the days it can represent.

The picture shows how such a calendar represents the 5th and the 25th day of the month. I have intentionally omitted any other numbers from the faces. You are allowed to place the cubes with any of the six faces showing, and you can also put the grey one on the left and the white one on the right.

What are the numbers on the two cubes?

Answer on page 260

Mathematical Jokes 1*

A biologist, a statistician and a mathematician are sitting outside a cafe watching the world go by. A man and a woman enter a building across the road. Ten minutes later, they come out accompanied by a child.

'They've reproduced,' says the biologist.

* The purpose of these jokes is not primarily to make you laugh. It is to show you what makes *mathematicians* laugh, and to provide you with a glimpse into an obscure corner of the world's mathematical subculture.

'No,' says the statistician. 'It's an observational error. On average, two and a half people went each way.'

'No, no, no,' says the mathematician. 'It's perfectly obvious. If someone goes in now, the building will be empty.'

• •

Deceptive Dice

The Terrible Twins, Innumeratus and Mathophila, were bored.

'I know,' said Mathophila brightly. 'Let's play dice!'

'Don't like dice.'

'Ah, but these are *special* dice,' said Mathophila, digging them out of an old chocolate box. One was red, one yellow and one blue.

Innumeratus picked up the red dice.* 'There's something funny about this one,' he said. 'It's got two 3's, two 4's and two 8's.'

'They're all like that,' said Mathophila carelessly. 'The yellow one has two 1's, two 5's and two 9's – and the blue one has two 2's, two 6's and two 7's.'

'They look rigged to me,' said Innumeratus, deeply suspicious.

'No, they're perfectly fair. Each face has an equal chance of turning up.'

'How do we play, anyway?'

'We each choose a different one. We roll them simultaneously, and the highest number wins. We can play for pocket money.' Innumeratus looked sceptical, so his sister quickly added: 'Just to be fair, I'll let you choose first! Then you can choose the best dice!'

'Weeelll ... ' said Innumeratus, hesitating.

Should he play? If not, why not?

Answer on page 260

• •

---

* Strictly speaking, 'dice' is the plural, and I should have used 'die' – but I've given up fighting that particular battle. I mention this to stop people writing in to tell me I've got it wrong. Anyway, the proverb tells us 'never say die'.

An Age-Old Old-Age Problem

The Emperor Scrumptius was born in 35 BC, and died on his birthday in AD 35. What was his age when he died?

Answer on page 262

• •

Why Does Minus Times Minus Make Plus?

When we first meet negative numbers, we are told that multiplying two negative numbers together makes a positive number, so that, for example, $(-2) \times (-3) = +6$. This often seems rather puzzling.

The first point to appreciate is that starting from the usual conventions for arithmetic with positive numbers, we are free to define $(-2) \times (-3)$ to be anything we want. It could be -99, or 127π, if we wished. So the main question is not what is the true value, but what is the sensible value. Several different lines of thought all converge on the same result – namely, that $(-2) \times (-3) = +6$. I include the $+$ sign for emphasis.

But *why* is this sensible? I rather like the interpretation of a negative number as a debt. If my bank account contains £–3, then I owe the bank £3. Suppose that my debt is multiplied by 2 (positive): then it surely becomes a debt of £6. So it makes sense to insist that $(+2) \times (-3) = -6$, and most of us are happy with that. What, though, should $(-2) \times (-3)$ be? Well, if the bank kindly writes off (takes away) two debts of £3 each, I am £6 better off – my account has changed exactly as it would if I had deposited £+6. So in banking terms, we want $(-2) \times (-3)$ to equal $+6$.

The second argument is that we can't have both $(+2) \times (-3)$ and $(-2) \times (-3)$ equal to $+6$. If that were the case, then we could cancel the -3 and deduce that $+2 = -2$, which is silly.

The third argument begins by pointing out an unstated assumption in the second one: that the usual laws of arithmetic should remain valid for negative numbers. It proceeds by adding that this is a reasonable thing to aim for, if only for mathematical elegance. If we require the usual laws to be valid, then

$$(+2) \times (-3) + (-2) \times (-3) = (2 - 2) \times (-3) = 0 \times (-3) = 0$$

so

$$-6 + (-2) \times (-3) = 0$$

Adding 6 to both sides, we find that

$$(-2) \times (-3) = +6$$

In fact a similar argument justifies taking $(+2) \times (-3) = -6$, as well.

Putting all this together: mathematical elegance leads us to *define* minus times minus to be plus. In applications such as finance, this choice turns out to match reality in a straightforward manner. So as well as keeping arithmetic simple, we end up with a good model for important aspects of the real world.

We could do it differently. But we'd end up by complicating arithmetic, and reducing its applicability. Basically, there's no contest. Even so, 'minus times minus makes plus' is a conscious human convention, not an inescapable fact of nature.

● ●

Heron Suit

No cat that wears a heron suit is unsociable.
No cat without a tail will play with a gorilla.
Cats with whiskers always wear heron suits.
No sociable cat has blunt claws.
No cats have tails unless they have whiskers.
> *Therefore*:
No cat with blunt claws will play with a gorilla.
> *Is the deduction logically correct?*

Answer on page 262

● ●

How to Unmake a Greek Cross

To paraphrase the old music-hall joke, almost any insult will make a Greek cross. But what I want you to do is *unmake* a Greek cross. In this region of Puzzledom, a Greek cross is five equal squares joined to make a + shape. I want you to convert it to a square, by cutting it into pieces and reassembling them. Here's one solution, using five pieces. But can you find an alternative, using four pieces, *all the same shape*?

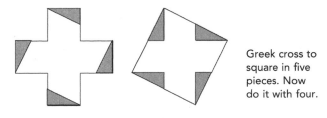

Greek cross to square in five pieces. Now do it with four.

Answer on page 263

How to Remember a Round Number

A traditional French rhyme goes like this:*

> Que j'aime a faire apprendre
> Un nombre utile aux sages!
> Glorieux Archimède, artiste ingenieux,
> Toi, de qui Syracuse loue encore le mérite!

But to which 'number useful to the sages' does it refer? Counting the letters in each word, treating 'j' as a word with one letter and placing a decimal point after the first digit, we get

3.141 592 653 589 793 238 462 6

---

* A loose translation is:
   How I like to make
   The sages learn a useful number!
   Glorious Archimedes, ingenious artist,
   You whose merit Syracuse still praises.

which is π to the first 22 decimal places. Many similar mnemonics for π exist in many languages. In English, one of the best known is

> How I want a drink, alcoholic of course, after the heavy chapters involving quantum mechanics. One is, yes, adequate even enough to induce some fun and pleasure for an instant, miserably brief.

It probably stopped there because the next digit is a 0, and it's not entirely clear how best to represent a word with no letters. (For one convention, see later.) Another is:

> Sir, I bear a rhyme excelling
> In mystic force, and magic spelling
> Celestial sprites elucidate
> All my own striving can't relate.

An ambitious π-mnemonic featured in *The Mathematical Intelligencer* in 1986 (volume 8, page 56). This is an informal 'house journal' for professional mathematicians. The mnemonic is a self-referential story encoding the first 402 decimals of π. It uses punctuation marks (ignoring full stops) to represents the digit zero, and words with more than 9 letters represent two consecutive digits – for instance, a word with 13 letters represents the digits 13 in that order. Oh, and any actual digit represents itself. The story begins like this:

> For a time I stood pondering on circle sizes. The large computer mainframe quietly processed all of its assembly code. Inside my entire hope lay for figuring out an elusive expansion. Value: pi. Decimals expected soon. I nervously entered a format procedure. The mainframe processed the request. Error. I, again entering it, carefully retyped. This iteration gave zero error printouts in all – success.

For the rest of the story, and many other π-mnemonics in various languages, see
www.geocities.com/capecanaveral/lab/3550/pimnem.htm

• •

The Bridges of Königsberg

Occasionally, a simple puzzle starts a whole new area of mathematics. Such occurrences are rare, but I can think of at least three. The most famous such puzzle is known as the Bridges of Königsberg, which led Leonhard Euler[*] to invent a branch of graph theory in 1735. Königsberg, which was in Prussia in those days, straddled the river Pregel. There were two islands, connected to the banks and each other by seven bridges. The puzzle was: is it possible to find a path that crosses each bridge exactly once?

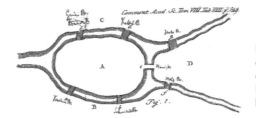

Euler's
diagram
of the
Königsberg
bridges.

Euler solved the puzzle by proving that no solution exists. More generally, he gave a criterion for any problem of this kind to have a solution, and observed that it did not apply to this particular example. He pointed out that the exact geometry is irrelevant – what matters is how everything is connected. So the puzzle reduces to a simple network of dots, joined by lines, here shown superimposed on the map. Each dot corresponds to one connected piece of land, and two dots are joined by lines if there is a bridge linking the corresponding pieces of land.

---

[*] It is mandatory to point out that his name is pronounced 'oiler', not 'yooler'. Numerous oil-based puns then become equally mandatory, but I won't mention any.

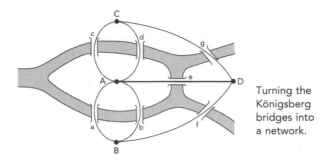

Turning the
Königsberg
bridges into
a network.

So we get four dots, A, B, C and D, and seven edges, a, b, c, d, e, f and g, one for each bridge. The puzzle now simplifies to this: is it possible to find a path through the network that includes each line exactly once? You might like to experiment before reading on.

To work out when a solution exists, Euler distinguished two kinds of path. An *open tour* starts and ends at different dots; a *closed tour* starts and ends at the same dot. He proved that for this particular network, neither kind of tour exists. The main theoretical idea is the *valency* of each dot: how many lines meet there. For instance, 5 lines meet at dot A, so the valency of A is 5.

Suppose that a closed tour exists on some network. Whenever one of the lines in the tour enters a dot, then the next line must exit from that dot. So, if a closed tour is possible, the number of lines at any given dot must be even: every dot must have even valency. This already rules out any closed tour of the Königsberg bridges, because that network has three dots of valency 3 and one of valency 5 – all odd numbers.

A similar criterion works for open tours, but now there must be exactly two dots of odd valency: one at the start of the tour, the other at its end. The Königsberg diagram has four vertices with odd valency, so there is no open tour either.

Euler also proved that these conditions are sufficient for a tour to exist, provided the diagram is connected – any two dots must be linked by *some* path. Euler's proof of this is quite lengthy. Nowadays a proof takes just a few lines, thanks to new discoveries inspired by his pioneering efforts.

• •

How to do Lots of Mathematics

Leonhard Euler was the most prolific mathematician of all time. He was born in 1707 in Basel, Switzerland, and died in 1783 in St Petersburg, Russia. He wrote more than 800 research papers, and a long list of books. Euler had 13 children, and often worked on his mathematics while one of them sat on his knee. He lost the sight of one eye in 1735, probably because of a cataract; the other eye failed in 1766. Going blind seems to have had no effect on his productivity. His family took notes, and he had an astonishing mental powers – once doing a mental calculation to fifty decimal places to decide which of two students had the right answer.

Leonhard Euler.

Euler spent many years at the court of Queen Catherine the Great. It has been suggested that to avoid becoming embroiled in court politics – which could easily prove fatal – Euler spent nearly all of his time working on mathematics, except when he was asleep. That way it was obvious that he had no time for intrigue.

Which reminds me of a mathematical joke: Why should a mathematician keep a mistress as well as a wife? (For gender-equality reasons, feel free to change to 'a lover as well as a

husband'.) Answer: when the wife thinks you're with the
mistress, and the mistress thinks you're with the wife, you have
time to get on with your mathematics.

Euler's Pentagonal Holiday

Here's your chance to put Euler's discoveries about tours on
networks to the test. (a) Find an open tour of this network.
(b) Find one that looks the same when you reflect the figure to
interchange left and right.

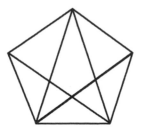

A network with
an open tour.

Answer on page 263

Ouroborean Rings

Around 1960 the American mathematician Sherman K. Stein
discovered a curious pattern in the Sanskrit nonsense word
yamátárájabhánasalagám. The composer George Perle told Stein
that the stressed (*á*) and unstressed (*a*) syllables form a
mnemonic for rhythms, and correspond to long and short
beats. Thus the first three syllables, *ya má tá*, have the rhythm
short, long, long. The second to fourth are *má tá rá*, long, long,
long – and so on. There are eight possible triplets of long or
short rhythms, and each occurs in the nonsense word exactly
once.

Stein rewrote the word using 0 for short and 1 for long,

getting 0111010001. Then he noticed that the first two digits are the same as the last two, so the string of digits can be bent into a loop, swallowing its own tail. Now you can generate all possible sequences of three digits 0 and 1 by moving along the loop one space at a time:

I call such sequences *ouroborean rings*, after the mythical serpent Ouroboros, which eats its own tail.

There is an ouroborean ring for pairs: 0011. It is unique except for rotations. Your task is to find one for quadruplets. That is, arrange eight 0's and eight 1's in a ring so that every possible string of four digits, from 0000 to 1111, appears as a series of consecutive symbols. (Each string of four must then occur exactly once.)

Answer on page 263

● ●

The Ourotorus

Are there higher-dimensional analogues of ouroborean rings?

For example, there are sixteen 2×2 squares with entries 0 or 1. Is it possible to write 0's and 1's in a 4×4 square so that each possibility occurs exactly once as a subsquare? You must pretend that opposite edges of the square are joined together, so that it wraps round into an *ourotorus*.

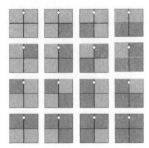

16 pieces for the
ourotorus puzzle.

You can turn this puzzle into a game. Cut out the sixteen pieces shown – the small dot near the top tells you which way up they go. Can you arrange them in a 4×4 grid, keeping the dot at the top, so that adjacent squares have the same colours along common edges? This rule also applies to squares that become adjacent if the top and bottom, or the left and right sides, of the grid are 'wrapped round' so that they join.

Answer on page 264

Who Was Pythagoras?

We recognise the name 'Pythagoras' because it is attached to a theorem, one that most of us have grappled with at school. 'The square on the hypotenuse of a right-angled triangle is equal to the sum of the squares on the other two sides.' That is, if you take any right-angled triangle, then the square of the longest side is equal to the sum of the squares of the other two sides. Well known as his theorem may be, the actual person has proved rather elusive, although we know more about him as a historical figure than we do for, say, Euclid. What we *don't* know is whether he proved his eponymous theorem, and there are good reasons to suppose that, even if he did, he wasn't the first to do so.

But more of that story later.

Pythagoras was Greek, born around 569 BC on the island of Samos in the north-eastern Aegean. (The exact date is disputed, but this one is wrong by at most 20 years.) His father,

Mnesarchus, was a merchant from Tyre; his mother, Pythais, was from Samos. They may have met when Mnesarchus brought corn to Samos during a famine, and was publicly thanked by being made a citizen.

Pythagoras studied philosophy under Pherekydes. He probably visited another philosopher, Thales of Miletus. He attended lectures given by Anaximander, a pupil of Thales, and absorbed many of his ideas on cosmology and geometry. He visited Egypt, was captured by Cambyses II, the King of Persia, and taken to Babylon as a prisoner. There he learned Babylonian mathematics and musical theory. Later he founded the school of Pythagoreans in the Italian city of Croton (now Crotone), and it is for this that he is best remembered. The Pythagoreans were a mystical cult. They believed that the universe is mathematical, and that various symbols and numbers have a deep spiritual meaning.

Various ancient writers attributed various mathematical theorems to the Pythagoreans, and by extension to Pythagoras – notably his famous theorem about right-angled triangles. But we have no idea what mathematics Pythagoras himself originated. We don't know whether the Pythagoreans could prove the theorem, or just believed it to be true. And there is evidence from the inscribed clay tablet known as Plimpton 322 that the ancient Babylonians may have understood the theorem 1200 years earlier – though they probably didn't possess a proof, because Babylonians didn't go much for proofs anyway.

Proofs of Pythagoras

Euclid's method for proving Pythagoras's Theorem is fairly complicated, involving a diagram known to Victorian schoolboys as 'Pythagoras's pants' because it looked like underwear hung on a washing line. This particular proof fitted into Euclid's development of geometry, which is why he chose it. But there are many other proofs, some of which make the theorem much more obvious.

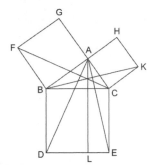

Pythagoras's
pants.

One of the simplest proofs is a kind of mathematician's
jigsaw puzzle. Take any right-angled triangle, make four copies,
and assemble them inside a carefully chosen square. In one
arrangement we see the square on the hypotenuse; in the other,
we see the squares on the other two sides. Clearly, the areas
concerned are equal, since they are the difference between the
area of the surrounding square and the areas of the four copies of
the triangle.

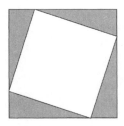
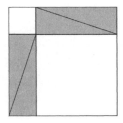

(Left) The square on the hypotenuse (plus four triangles). (Right)
The sum of the squares on the other two sides (plus four
triangles). Take away the triangles ... and Pythagoras's
Theorem is proved.

Then there's a cunning tiling pattern. Here the slanting grid
is formed by copies of the square on the hypotenuse, and the
other grid involves both of the smaller squares. If you look at
how one slanting square overlaps the other two, you can see how
to cut the big square into pieces that can be reassembled to make
the two smaller squares.

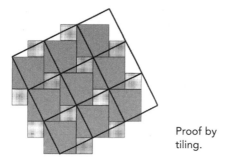

Proof by
tiling.

Another proof is a kind of geometric 'movie', showing how to split the square on the hypotenuse into two parallelograms, which then slide apart – without changing area – to make the two smaller squares.

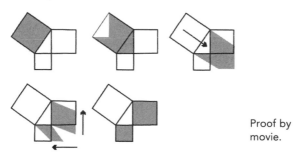

Proof by
movie.

A Constant Bore

'Now, *this* component is a solid copper sphere with a cylindrical hole bored exactly through its centre,' said Rusty Nail, the construction manager. He opened a blueprint on his laptop's screen:

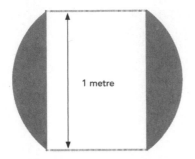

1 metre

Cross-section of sphere
with cylindrical hole.

'Looks straightforward,' said the foreman, Lewis Bolt. 'That's quite a lot of copper.'

'Coincidentally, that's what I want you to work out,' said Rusty. 'What volume of copper do we need to cast it?'

Lewis stared at the blueprint. 'It doesn't say how big the sphere is.' He paused. 'I can't find the answer unless you tell me the radius of the sphere.'

'Hmmm,' said Rusty. 'They must have forgotten to put that in. But I'm sure you can work something out. I need the answer by lunchtime.'

What is the volume of copper required? Does it depend on the size of the sphere?

Answer on page 264

• •

Fermat's Last Theorem

The great virtue of Fermat's Last Theorem is that it's easy to understand what it means. What made the theorem famous is that proving it turned out to be amazingly hard. So hard, in fact, that it took about 350 years of effort, by many of the world's leading mathematicians, to polish it off. And to do that, they had to invent entire new mathematical theories, and prove things that looked much harder.

Pierre de Fermat.

It all started around 1650, when Pierre de Fermat wrote an enigmatic note in the margin of his copy of Diophantus's book *Arithmetica*: 'of which fact I have found a remarkable proof, 'but this margin is too small to contain it.' Proof of what? Let me back up a bit.

Diophantus was probably Greek, and he lived in ancient Alexandria. Some time around AD 250 he wrote a book about solving algebraic equations – with a slight twist: the solutions were required to be fractions, or better still, whole numbers. Such equations are called *Diophantine equations* to this day. A typical Diophantine problem is: find two squares whose sum is square (using only whole numbers). One possible answer is 9 and 16, which add up to 25. Here 9 is the square of 3, while 16 is the square of 4, and 25 is the square of 5. Another answer is 25 (the square of 5) and 144 (the square of 12), which add up to 169 (the square of 13). These are the tip of an iceberg.

This particular problem is linked to Pythagoras's Theorem, and Diophantus was following a long tradition of looking for *Pythagorean triples* – whole numbers that can form the sides of a right-angled triangle. Diophantus wrote down a general rule for finding all Pythagorean triples. He wasn't the first to discover it, but it belonged very naturally in his book. Now Fermat wasn't a professional mathematician – he never held an academic position. In his day job he was a legal advisor. But his passion was mathematics, especially what we now call *number theory*, the properties of ordinary whole numbers. This area uses the simplest ingredients anywhere in mathematics, but paradoxically it is one of the most difficult areas to make progress in. The simpler the ingredients, the harder it is to make things with them.

Fermat pretty much created number theory. He took over where Diophantus had left off, and by the time he had finished, the subject was virtually unrecognisable. And some time around 1650 – we don't even know the exact date – he must have been thinking about Pythagorean triples, and wondered 'can we do it with *cubes*?'

Just as the square of a number is what you get by multiplying two copies of the same number, the cube is what you get by multiplying three copies. That is, the square of 5, say, is $5 \times 5 = 25$, and the cube of 5 is $5 \times 5 \times 5 = 125$. These are written more compactly as 5^2 and 5^3, respectively. No doubt Fermat tried a few possibilities. Is the sum of the cubes of 1 and 2 a cube, for instance? The cubes here are 1 and 8, so their sum is 9. That's a square, but not a cube: no banana.

He surely noticed that you can get pretty close. The cube of 9 is 729; the cube of 10 is 1,000; their sum is 1,729. That's *very nearly* the cube of 12, which is 1,728. Missed by one! Still no banana.

Like any mathematician, Fermat would have tried bigger numbers, and used any short cuts he could think of. Nothing worked. Eventually he gave up: he hadn't found any solutions, and by now he suspected that there weren't any. Except for the cube of 0 (which is also 0) and any cube whatsoever, which add up to the whatsoever – but we all know that adding zero makes no difference to anything, so that's 'trivial', and he wasn't interested in trivialities.

OK, so cubes don't get us anywhere. What about the next such type of number, fourth powers? You get those by multiplying four copies of the same number, for example $3 \times 3 \times 3 \times 3 = 81$ is the fourth power of 3, written as 3^4. Still no joy. In fact, for fourth powers Fermat found a logical proof that no solutions exist except trivial ones. Very few of Fermat's proofs have survived, and few of them were written down, but we know how this one went, and it's both cunning and correct. It takes some hints from Diophantus's method of finding Pythagorean triples.

Fifth powers? Sixth powers? Still nothing. By now Fermat was ready to make a bold statement: 'To resolve a cube into the sum of two cubes, a fourth power into two fourth powers, or in general any power higher than the second into two powers of the same kind, is impossible.' That is: the only way for two nth powers to add up to an nth power is when n is 2 and we are looking at Pythagorean triples. This is what he wrote in his margin, and it's what caused so much fuss for the next 350 years.

We don't actually have Fermat's copy of the *Arithmetica* with its marginal notes. What we have is a printed edition of the book prepared later by his son, which has the notes printed in it.

Fermat included various other unproved but fascinating bits of number theory in his letters and the marginal notes published by his son, and the world's mathematicians rose to the challenge. Soon all but one of Fermat's statements had been proved – apart from one that was disproved, but in that case Fermat never claimed he had a proof anyway. The sole remaining 'last theorem' – not the last one he wrote down, but the last one that no one else could prove or disprove – was the marginal note about sums of like powers.

Fermat's Last Theorem became notorious. Euler proved that there is no solution in cubes. Fermat himself had done fourth powers. Peter Lejeune Dirichlet dealt with fifth powers in 1828, and 14th powers in 1832. Gabriel Lamé published a proof for 7th powers, but it had a mistake in it. Carl Friedrich Gauss, one of the best mathematicians who has ever lived and an expert in number theory, tried to patch up Lamé's attempt, but failed, and gave up on the whole problem. He wrote to a scientific friend that the problem 'has little interest for me, since a multitude of such propositions, which one can neither prove nor refute, can easily be formulated'. But for once Gauss's instincts let him down: the problem *is* interesting, and his remark seems to have been a case of sour grapes.

In 1874 Lamé had a new idea, linking Fermat's Last Theorem to a special type of complex number – involving the square root of minus one (see page 184). There was nothing wrong with complex numbers, but there was a hidden assumption in Lamé's argument, and Ernst Kummer wrote to him to inform him that it went wrong for 23rd powers. Kummer managed to fix Lamé's idea, eventually proving Fermat's Last Theorem for all powers up to the 100th, except for 37, 59 and 67. Later mathematicians polished off these powers too, and extended the list, until by 1980 Fermat's Last Theorem had been proved for all powers up to the 125,000th.

You might think that this would be good enough, but mathematicians are made of sterner stuff. It has to be *all* powers, or nothing. The first 125,000 whole numbers are minuscule compared with the infinity of numbers that remain. But Kummer's methods needed special arguments for each power, and they weren't really up to the job. What was needed was *a new idea*. Unfortunately, nobody knew where to look for one.

So number theorists abandoned Fermat's Last Theorem and headed off into areas where they could still make progress. One such area, the theory of *elliptic curves*, started to get really exciting, but also very technical. An elliptic curve is not an ellipse – if it were, we wouldn't need a different name for it. It is a curve in the plane whose y-coordinate, squared, is a cubic formula in its x-coordinate. These curves in turn are connected with some remarkable expressions involving complex numbers, called *elliptic functions*, which were in vogue in the late nineteenth century. The theory of elliptic curves, and their associated elliptic functions, became very deep and powerful.

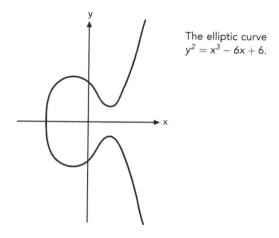

The elliptic curve
$y^2 = x^3 - 6x + 6$.

Starting around 1970, a series of mathematicians started to get glimpses of a strange connection between elliptic curves and Fermat's Last Theorem. Roughly speaking, if Fermat was wrong, and two nth powers do add up to another nth power, then those three numbers determine an elliptic curve. And because the powers add like that, it is a very strange elliptic curve, with a surprising combination of properties. So surprising, in fact, that it looks wildly unlikely that it can exist at all, as Gerhard Frey pointed out in 1985.

This observation opens the way to a 'proof by contradiction', what Euclid called '*reductio ad absurdum*' (reduction to the absurd). To prove that some statement is true, you begin by assuming that, on the contrary, it is false. Then you deduce the logical consequences of this falsity. If the consequences contradict each other or known facts, then your assumption must have been wrong – so the statement must be true after all. In 1986 Kenneth Ribet pinned this idea down by proving that if Fermat's Last Theorem is false, then the associated elliptic curve violates a conjecture (that is, a plausible but unproved theorem) introduced by the Japanese mathematicians Yutaka Taniyama and Goro Shimura. This Taniyama–Shimura conjecture, which dates from 1955, says that every elliptic curve is associated with a special class of elliptic functions, called *modular functions*.

Ribet's discovery implies that any proof of the Taniyama–Shimura conjecture automatically proves – by contradiction – Fermat's Last Theorem as well. Because the assumed falsity of Fermat's Last Theorem tells us that Frey's elliptic curve exists, but the Taniyama–Shimura conjecture tells us that it doesn't.

Unfortunately, the Taniyama–Shimura conjecture was just that – a conjecture.

Enter Andrew Wiles. When Wiles was a child he heard about Fermat's Last Theorem, and decided that when he grew up he would become a mathematician and prove it. He did become a mathematician, but by then he had decided that Fermat's Last Theorem was much as Gauss had complained – an isolated question of no particular interest for the mainstream of mathematics. But Frey's discovery changed all that. It meant that Wiles could work on the Taniyama–Shimura conjecture, an important mainstream problem, and polish off Fermat's Last Theorem too.

Now, the Taniyama–Shimura conjecture is very difficult – that's why it remained a conjecture for some forty years. But it has good links to many areas of mathematics, and sits firmly in the middle of an area where the techniques are very powerful: elliptic curves. For seven years Wiles worked away in his study, trying every technique he could think of, striving to prove the Taniyama–Shimura conjecture. Hardly anybody knew that he was working on that problem; he wanted to keep it secret.

In June 1993 Wiles gave a series of three lectures at the Isaac Newton Institute in Cambridge, one of the world's top mathematical research centres. Their title was 'Modular forms, elliptic curves and Galois representations', but the experts knew it must really be about the Taniyama–Shimura conjecture – and, just possibly, Fermat's Last Theorem. On day three, Wiles announced that he had proved the Taniyama–Shimura conjecture, not for all elliptic curves, but for a special kind called 'semistable'.

Frey's elliptic curve, if it exists, is semistable. Wiles was telling his audience that he had proved Fermat's Last Theorem.

But it wasn't quite that straightforward. In mathematics you don't get credit for solving a big problem by giving a few lectures in which you say you've got the answer. You have to publish your ideas in full, so that everyone else can check that they are right. And when Wiles started that process – which involves getting experts to go over the work in detail before it gets into print – some logical gaps emerged. He quickly filled most of them, but one seemed much harder, and it wouldn't go away. As rumours spread that the proposed proof had collapsed, Wiles made one final attempt to shore up his increasingly rickety proof – and, contrary to most expectations, he succeeded. One final technical point was supplied by his former student, Richard Taylor, and by the end of October 1994 the proof was complete. The rest, as they say, is history.

By developing Wiles's new methods, the Taniyama–Shimura conjecture has now been proved for all elliptic curves, not just semistable ones. And although the *result* of Fermat's Last Theorem is still just a minor curiosity – nothing important rests on it being true or false – the *methods* used to prove it have become a permanent and important addition to the mathematical armoury.

One question remains. Did Fermat really have a valid proof, as he claimed in his margin? If he did, it certainly wasn't the one that Wiles found, because the necessary ideas and methods simply did not exist in Fermat's day. As an analogy: today we could erect the pyramids using huge cranes, but we can be confident that however the ancient Egyptians built their pyramids, they didn't use modern machinery. Not just because there is no evidence of such machines, but because the necessary infrastructure could not have existed. If it had done, the whole culture would have been different. So the consensus among mathematicians is that what Fermat thought was a proof probably had a logical gap that he missed. There are some plausible but incorrect attempts that would have been feasible in his day. But we don't know whether his proof – if one ever existed – followed those lines. Maybe – just maybe – there is a much simpler proof lurking out there in some unexplored realm of mathematical imagination, waiting for somebody to stumble into it.* Stranger things have happened.

● ●

Pythagorean Triples

I can't really get away without telling you Diophantus's method for finding all Pythagorean triples, can I?

OK, here it is. Take any two whole numbers, and form:

- twice their product
- the difference between their squares
- the sum of their squares

Then the resulting three numbers are the sides of a Pythagorean triangle.

For instance, take the numbers 2 and 1. Then

- twice their product $= 2 \times 2 \times 1 = 4$
- the difference between their squares $= 2^2 - 1^2 = 3$
- the sum of their squares $= 2^2 + 1^2 = 5$

and we obtain the famous 3–4–5 triangle. If instead we take numbers 3 and 2, then

- twice their product $= 2 \times 3 \times 2 = 12$
- the difference between their squares $= 3^2 - 2^2 = 5$
- the sum of their squares $= 3^2 + 2^2 = 13$

and we get the next-most-famous 5–12–13 triangle. Taking numbers 42 and 23, on the other hand, leads to

- twice their product $= 2 \times 42 \times 23 = 1,932$

---

* If you think you've found it, *please don't send it to me*. I get too many attempted proofs as it is, and so far – well, just don't get me started, OK?

- the difference between their squares $= 42^2 - 23^2 = 1,235$
- the sum of their squares $= 42^2 + 23^2 = 2,293$

and no one has ever heard of the 1,235–1,932–2,293 triangle. But these numbers do work:

$$1,235^2 + 1,932^2 = 1,525,225 + 3,732,624 = 5,257,849$$
$$= 2,293^2$$

There's a final twist to Diophantus's rule. Having worked out the three numbers, we can choose any other number we like and multiply them all by that. So the 3–4–5 triangle can be converted to a 9–12–15 triangle by multiplying all three numbers by 3, or to an 18–24–30 triangle by multiplying all three numbers by 6. We can't get these two triples from the above prescription using whole numbers. Diophantus knew that.

• •

Prime Factoids

Prime numbers are among the most fascinating in the whole of mathematics. Here's a Prime Primer.

A whole number bigger than 1 is *prime* if it is not the product of two smaller numbers. The sequence of primes begins

2, 3, 5, 7, 11, 13, 17, 19, 23, 29, 31, 37, ...

Note that 1 is excluded, by convention. Prime numbers are of fundamental importance in mathematics because every whole number is a product of primes – for instance,

$$2,007 = 3 \times 3 \times 223$$
$$2,008 = 2 \times 2 \times 2 \times 251$$
$$2,009 = 7 \times 7 \times 41$$

Moreover (only mathematicians worry about this sort of thing, but actually it's kind of important and surprisingly difficult to prove), there is only one way to achieve this, apart from rearranging the order of the prime numbers concerned. For instance, $2,008 = 251 \times 2 \times 2 \times 2$, but that doesn't count as different. This property is called 'unique prime factorisation'.

If you're worried about 1 here, mathematicians consider it to be the product of *no* primes. Sorry, mathematics is like that sometimes.

The primes seem to be scattered fairly unpredictably. Apart from 2, they are all odd – because an even number is divisible by 2, so it can't be prime unless it is equal to 2. Similarly, 3 is the only prime that is a multiple of 3, and so on.

Euclid proved that there is no largest prime. In other words, there exist infinitely many primes. Given any prime p, you can always find a bigger one. In fact, any prime divisor of $p! + 1$ will do the job. Here $p! = p \times (p - 1) \times (p - 2) \times \cdots \times 3 \times 2 \times 1$, a product called the *factorial* of p. For instance,

$$7! = 7 \times 6 \times 5 \times 4 \times 3 \times 2 \times 1 = 5,040.$$

The largest *known* prime is another matter, because Euclid's method isn't a practical way to generate new primes explicitly. As I write, the largest known prime is

$$2^{43,112,609} - 1$$

which has 12,978,189 digits when written out in decimal notation (see page 153).

Twin primes are pairs of primes that differ by 2. Examples are (3, 5), (5, 7), (11, 13), (17, 19), and so on. The twin primes conjecture states that there are infinitely many twin primes. This is widely believed to be true, but has never been proved. Or disproved. The largest known twin primes, to date, are:

$$2,003,663,613 \times 2^{195,000} - 1 \text{ and } 2,003,663,613 \times 2^{195,000} + 1$$

with 58,711 digits each.

Nicely does it ... In 1994 Thomas Nicely was investigating twin primes by computer, and noticed that his results disagreed with previous computations. After spending weeks searching for errors in his program, he traced the problem to a previously unknown bug in the Intel™ Pentium™ microprocessor chip. At

that time, the Pentium was the central processing unit of most of the world's computers. See
www.trnicely.net/pentbug/bugmail1.html

● ●

A Little-Known Pythagorean Curiosity

It is well known that any two Pythagorean triples can be combined to yield another one. In fact, if

$$a^2 + b^2 = c^2$$

and

$$A^2 + B^2 = C^2$$

then

$$(aA - bB)^2 + (aB + bA)^2 = (cC)^2$$

However, there is a lesser-known feature of this method for combining Pythagorean triples. If you think of it as a kind of 'multiplication' for triples, then we can define a triple to be *prime* if it is not the product of two smaller triples. Then every Pythagorean triple is a product of distinct prime Pythagorean triples; moreover, this 'prime factorisation' of triples is essentially unique, except for some trivial distinctions which I won't go into here.

It turns out that the prime triples are those for which the hypotenuse is a prime number of the form $4k + 1$ and the other two sides are both non-zero, *or* the hypotenuse is 2 or a prime of the form $4k - 1$ and one of the other sides *is* zero (a 'degenerate' triple).

For instance, the 3–4–5 triple is prime, and so is the 5–12–13 triple, because their hypotenuses are both $4k + 1$ primes. The 0–7–7 triple is also prime. The 33–56–65 triple is not prime – it is the 'product' of the 3–4–5 and 5–12–13 triples.

Just thought you'd like to know.

● ●

Digital Century

Place *exactly* three common mathematical symbols between the digits

123456789

so that the result equals 100. The same symbol can be repeated if you wish, but each repeat counts towards your limit of three. Rearranging the digits is not permitted.

Answer on page 266

Squaring the Square

We all know that a rectangular floor can be tiled with square tiles of equal size – provided its edges are integer (whole-number) multiples of the size of the tile. But what happens if we are required to use square tiles which all have *different* sizes?

The first 'squared rectangle' was published in 1925 by Zbigniew Morón, using ten square tiles of sizes 3, 5, 6, 11, 17, 19, 22, 23, 24 and 25.

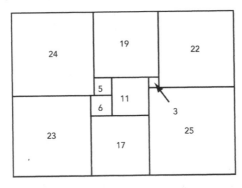

Morón's first squared rectangle.

Not long after, he found a squared rectangle using nine square tiles with sizes 1, 4, 7, 8, 9, 10, 14, 15 and 18. *Can you arrange these tiles to make a rectangle?* As a hint, it has size 32×33. What about making a *square* out of different square tiles? For a long time this was thought to be impossible, but in 1939 Roland Sprague found 55 distinct square tiles that fit together to make a square. In 1940 four mathematicians (Leonard Brooks, Cedric Smith, Arthur Stone and William Tutte, then undergraduates at Trinity College, Cambridge) published a paper relating the problem to electrical networks – the network encodes what size the squares are, and how they fit together. This method led to more solutions.

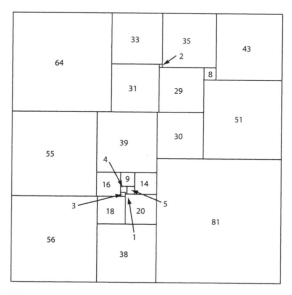

Willcocks's squared square with 24 tiles.

In 1948 Theophilus Willcocks found 24 squares that fit together to make a square. For a while it was thought that no smaller set would do the job, but in 1962 Adrianus Duijvestijn used a computer to show that only 21 square tiles are needed, and this is the minimum number. Their sizes are 2, 4, 6, 7, 8, 9, 11, 15, 16, 17, 18, 19, 24, 25, 27, 29, 33, 35, 37, 42 and 50. *Can you arrange Duijvestijn's 21 tiles to make a square?* As a hint, it has size 112 × 112.

Finally, a really hard one: can you tile the infinite plane, leaving no gaps, using exactly one tile of each whole number size: 1, 2, 3, 4, and so on? This problem remained open until 2008, when Frederick and James Henle proved that you can. See their article 'Squaring the plane', *American Mathematical Monthly*, volume 115 (2008), pages 3–12.

For further information, see **www.squaring.net**

Answers on page 267

Magic Squares

I'm on a bit of a square kick here, so let me mention the most ancient 'square' mathematical recreation of them all. According to a Chinese myth, the Emperor Yu, who lived in the third millennium BC, came across a sacred turtle in a tributary of the Yellow River, with strange markings on its shell. These markings are now known as the *Lo shu* ('Lo river writing').

The *Lo Shu*.

The markings are numbers, and they form a square pattern:

4 9 2
3 5 7
8 1 6

Here every row, every column and every diagonal adds to the same number, 15. A number square with these properties is said to be *magic*, and the number concerned is its *magic constant*. Usually the square is made from successive whole numbers, 1, 2, 3, 4, and so on, but sometimes this condition is relaxed.

Dürer's *Melancholia* and its magic square.

In 1514 the artist Albrecht Dürer produced an engraving, 'Melancholia', containing a 4×4 magic square (top right corner). The middle numbers in the bottom row are 15–14, the date of the work. This square contains the numbers

16	3	2	13
5	10	11	8
9	6	7	12
4	15	14	1

and has magic constant 34.

Using consecutive whole numbers 1, 2, 3, ..., and counting rotations and reflections of a given square as being the same, there are precisely:

- 1 magic square of size 3×3
- 880 magic squares of size 4×4
- 27,5305,224 magic squares of size 5×5

The number of 6×6 magic squares is unknown, but has been estimated by statistical methods to be about 1.77×10^{19}.

The literature on magic squares is gigantic, including many variations such as magic cubes. The website mathworld.wolfram.com/MagicSquare.html is a good place to look, but there are plenty of others.

● ●

Squares of Squares

Magic squares are so well known that I'm not going to say a lot about the common ones, but some of the variants are more interesting. For instance, is it possible to make a magic square whose entries are all distinct perfect squares? Call this a *square of squares*. (Clearly the condition of using *consecutive* whole numbers must be ignored!)

We still have no idea whether a 3×3 square of squares exists. Near misses include Lee Sallows's

$$127^2 \quad 46^2 \quad 58^2$$
$$2^2 \quad 113^2 \quad 94^2$$
$$74^2 \quad 82^2 \quad 97^2$$

for which all rows, columns, and *one* diagonal have the same sum. Another near miss is magic:

$$373^2 \quad 289^2 \quad \quad 565^2$$
$$\mathbf{360,721} \quad 425^2 \quad \quad 23^2$$
$$205^2 \quad 527^2 \quad \mathbf{222,121}$$

However, only seven entries are square – I've marked the exceptions in bold. It was found by Sallows and (independently) Andrew Bremner.

In 1770 Euler sent the first 4×4 square of squares to Joseph-Louis Lagrange:

$$68^2 \quad 29^2 \quad 41^2 \quad 37^2$$
$$17^2 \quad 31^2 \quad 79^2 \quad 32^2$$
$$59^2 \quad 28^2 \quad 23^2 \quad 61^2$$
$$11^2 \quad 77^2 \quad 8^2 \quad 49^2$$

It has magic constant, 8515.

Christian Boyer has found $5 \times 5, 6 \times 6$ and 7×7 squares of squares. The 7×7 square uses squares of consecutive integers from 0^2 to 48^2:

$$25^2 \quad 45^2 \quad 15^2 \quad 14^2 \quad 44^2 \quad 5^2 \quad 20^2$$
$$16^2 \quad 10^2 \quad 22^2 \quad 6^2 \quad 46^2 \quad 26^2 \quad 42^2$$
$$48^2 \quad 9^2 \quad 18^2 \quad 41^2 \quad 27^2 \quad 13^2 \quad 12^2$$
$$34^2 \quad 37^2 \quad 31^2 \quad 33^2 \quad 0^2 \quad 29^2 \quad 4^2$$
$$19^2 \quad 7^2 \quad 35^2 \quad 30^2 \quad 1^2 \quad 36^2 \quad 40^2$$
$$21^2 \quad 32^2 \quad 2^2 \quad 39^2 \quad 23^2 \quad 43^2 \quad 8^2$$
$$17^2 \quad 28^2 \quad 47^2 \quad 3^2 \quad 11^2 \quad 24^2 \quad 38^2$$

• •

Ring a-Ring a-Ringroad

The M25 motorway completely encircles London, and in Britain we drive on the left. So if you travel clockwise round the M25 you stay on the outside carriageway, whereas travelling anti-clockwise keeps you on the inside carriageway, which is shorter. But how much shorter? The total length of the M25 is 188 km (117 miles), so the advantage of being on the inside carriageway ought to be quite a lot – shouldn't it?

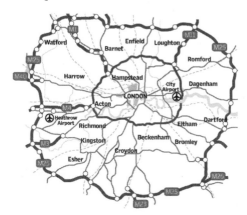

The M25
motorway.

Suppose that two cars travel right round the M25, staying in the outside lane—no, make that two *white vans* travelling right round the M25, staying in the outside lane, as they tend to do. Assume that one is going clockwise and the other anticlockwise, and suppose (which is not entirely true but makes the problem specific) that the distance between these two lanes is always 10 metres. How much further does the clockwise van travel than the anticlockwise one? You may assume that the roads all lie in a flat plane (which also isn't entirely true).

Answer on page 267

Pure v. Applied

Relations between pure and applied mathematicians are based on trust and understanding. Pure mathematicians do not trust applied mathematicians, and applied mathematicians do not understand pure mathematicians.

Magic Hexagon

Magic hexagons are like magic squares, but using a hexagon-shaped arrangement of hexagons, like a chunk of a honeycomb:

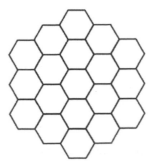

Grid for magic hexagon.

Your task is to place the numbers from 1 to 19 in the hexagons so that any straight line of three, four or five cells, in any of the three directions, add up to the same magic constant, which I can reveal must be 38.

Answer on page 269

● ●

Pentalpha

This ancient geometrical puzzle is easy if you look at it the right way, and baffling if you don't.

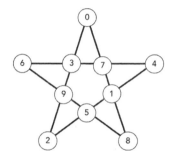

Follow the rules to place nine counters.

You have nine counters, to be placed on the circles in a five-pointed star. Here I've numbered the circles to help explain the solution. In the real game, there aren't any numbers. Successive counters must be positioned by placing them in an empty circle, jumping over an adjacent circle (which may be empty or full) to land on an empty circle adjacent to the one jumped over, so that all three circles involved in the move are on the same straight line. For instance, if circles 7 and 8 are empty, you can place a counter on 7 and jump over 1 to land on 8. Here 1 can be empty or full – it doesn't matter. But you are not allowed to jump 7 over 1 to land on 4 or 5 because now the three circles involved are not in a straight line.

If you try placing counters at random, you usually run out of suitable pairs of empty circles before finishing the puzzle.

Answer on page 270

● ●

Wallpaper Patterns

A *wallpaper pattern* repeats the same image in two directions: down the wall and across the wall (or on a slant). The repetition down the wall comes from the paper being printed in a continuous roll, using a revolving cylinder to create the pattern. The repetition across the wall makes it possible to continue the pattern sideways, across adjacent strips of paper, to cover the entire wall. A 'drop' from one panel to the next causes no problems, and can actually make it easier to hang the paper.

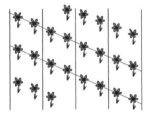

Wallpaper patterns repeat in two directions.

The number of possible *designs* for wallpaper is effectively infinite. But different designs can have the same underlying pattern, it's just that the basic image that gets repeated is different. For instance, the flower in the design above could be replaced by a butterfly, or a bird, or an abstract shape. So mathematicians distinguish *essentially* different patterns by their symmetries. What are the different ways in which we can slide the basic image, or rotate it, or even flip it over (like reflecting it in a mirror), so that the end result is the same as the start?

For my pattern of flowers, the only symmetries are slides along the two directions in which the basic image repeats, or several such slides performed in turn. This is the simplest type of symmetry, but there are more elaborate ones involving rotations and reflections as well. In 1924 George Pólya and Paul Niggli proved that there are exactly 17 different symmetry types of wallpaper pattern – surprisingly few.

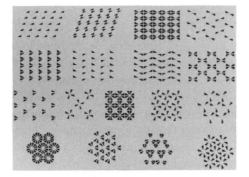

The 17 types of wallpaper pattern.

In three dimensions the corresponding problem is to list all possible symmetry types of atomic lattices of crystals. Here there are 230 types. Curiously, that answer was discovered before anyone solved the much easier two-dimensional version for wallpaper.

How Old Was Diophantus?

A few pages back, in the section on Fermat's Last Theorem, I mentioned Diophantus of Alexandria, who lived around AD 250 and wrote a famous book on equations, the *Arithmetica*. That is virtually all we know about him, except that a later source tells us his age – assuming it is authentic. That source says this:

Diophantus's childhood lasted one-sixth of his life. His beard grew after one-twelfth more. He married after one-seventh more. His son was born five years later. The son lived to half his father's age. Diophantus died four years after his son. How old was Diophantus when he died?

Answer on page 271

If You Thought Mathematicians were Good at Arithmetic ...

Ernst Kummer was a German algebraist, who did some of the best work on Fermat's Last Theorem before the modern era. However, he was poor at arithmetic, so he always asked his students to do the calculations for him. On one occasion he needed to work out 9×7. 'Umm ... nine times seven is ... nine times ... seven ... is ...'

'Sixty-one,' suggested one student. Kummer wrote this on the blackboard.

'No, Professor! It should be sixty-seven!' said another.

'Come, come, gentlemen,' said Kummer. 'It can't be both. It must be one or the other.'

The Sphinx is a Reptile

Well, a rep-tile, which isn't quite the same thing. Short for 'replicating tile', this word refers to a shape that appears – magnified – when several copies of it are put together. The most obvious rep-tile is a square.

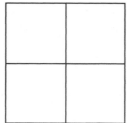

Four square tiles make a bigger square.

However, there are many others, such as these:

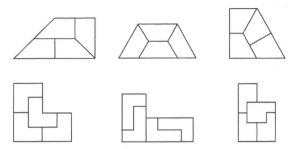

More interesting rep-tiles.

A famous rep-tile is the *sphinx*. Can you assemble four copies of a sphinx to make a bigger sphinx? You can turn some of the tiles over if you want.

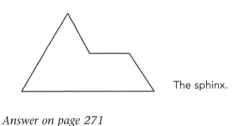

The sphinx.

Answer on page 271

Six Degrees of Separation

In 1998 Duncan Watts and Steven Strogatz published a research paper in the science journal *Nature* about 'small-world networks'. These are networks in which certain individuals are unusually well-connected. The paper triggered a mass of research in which the ideas were applied to real networks such as the Internet and the transmission of disease epidemics.

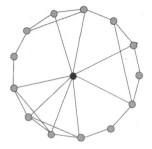

A small-world network. The black individual in the middle is connected to many others, unlike the grey individuals.

The story began in 1967 when the psychologist Stanley Milgram prepared 160 letters with the name of his stockbroker on the envelope – but no address. Then he 'lost' the letters so that random members of the populace could find them, and, he hoped, send them on. Many of the letters duly arrived at the stockbroker's office, and when they did, they had done so in at most six steps. This led Milgram to the idea that we are connected to every other person on the planet by at most five intermediaries – six degrees of separation.

I was explaining the *Nature* paper and its background to my friend Jack Cohen, in the maths common room. Our head of department walked past, stopped, and said, 'Nonsense! Jack, how many steps are there between you and a Mongolian yak herder?' Jack's instant response was: 'One!' He then explained that the person in the office next to his was an ecologist who had worked in Mongolia. This kind of thing happens to Jack, because he is one of those unusually well-connected people who makes small-world networks hang together. For example, he causes both me and my head of department to be only two steps away from a Mongolian yak herder.

You can explore the small-world phenomenon using the Oracle of Bacon, at oracleofbacon.org. Kevin Bacon is an actor who has appeared in a lot of movies. Anyone who has appeared in the same movie as Kevin has a *Bacon number* of 1. Anyone who has appeared in the same movie as anyone with a Bacon number of 1 has a Bacon number of 2, and so on. If Milgram is right, virtually every actor (movies being the relevant 'world') has a

Bacon number of 6 or less. At the Oracle, when you type in an actor's name it tells you the Bacon number, and which movies form the links. For instance,

- Michelle Pfeiffer appeared in *Amazon Women on the Moon* in 1987 with:
- David Alan Grier, who appeared in *The Woodsman* in 2004 with:
- Kevin Bacon

so Michelle's Bacon number is 2.

It's not easy to find anyone with Bacon number bigger than 2! One of them is

- Hayley Mooy, who appeared in *Star Wars: Episode III – Revenge of the Sith* in 2005 with:
- Samuel L. Jackson, who appeared in *Snakes on a Plane* in 2006 with:
- Rachel Blanchard, who appeared in *Where the Truth Lies* in 2005 with:
- Kevin Bacon

so Hayley has Bacon number 3.

Mathematicians have their own version of the Oracle of Bacon, centred on the late Paul Erdős. Erdős wrote more joint research papers than any other mathematician, so the game goes the same way but the links are joint papers. My Erdős number is 3, because

- I wrote a joint paper with:
- Marty Golubitsky, who wrote a joint paper with:
- Bruce Rothschild, who wrote a joint paper with:
- Paul Erdős

and no shorter chain exists. One of my former students, who has written a joint paper with me but with no one else, has Erdős number 4.

Usually, more people take part in the same movie than co-author the same mathematics research paper – though I can't say the same about some areas of biology or physics. So you'd expect bigger Erdős numbers than Bacon numbers, on the whole. All mathematicians with Erdős number 1 or 2 are listed at www.oakland.edu/enp

● ●

Trisectors Beware!

Euclid tells us how to bisect an angle – divide it into two equal parts – and by repeating this method we can divide any given angle into $4, 8, 16, \ldots, 2^n$ equal parts. But Euclid doesn't explain how to *trisect* an angle – divide it into three equal parts. (Or *quinquisect*, five equal parts, or ....)

Traditionally, Euclidean constructions are carried out using only two instruments – an idealised ruler, with no markings along its edge, to draw a straight line, and an idealised (pair of) compass(es), which can draw a circle. It turns out that these instruments are inadequate for trisecting angles, but the proof had to wait until 1837, when Pierre Wantzel used algebraic methods to show that no ruler-and-compass trisection of the angle 60° is possible. Undeterred, many amateurs continue to seek trisections. So it may be worth explaining why they don't exist.

Any point can be constructed approximately, and the approximation can be as accurate as we wish. To trisect an angle to an accuracy of, say one-trillionth of a degree is easy – in principle. The mathematical problem is not about practical solutions: it is about the existence, or not, of ideal, infinitely accurate ones. It is also about *finite* sequences of applications of ruler and compass: if you allow infinitely many applications, again any point can be constructed – this time exactly.

The key feature of Euclidean constructions is their ability to form square roots. Repeating an operation leads to complicated combinations of square roots of quantities involving square roots of ... well, you get the idea. But that's all you can do with the traditional instruments.

Turning now to algebra, we find the coordinates of such points by starting with rational numbers and repeatedly taking square roots. Any such number satisfies an algebraic equation of a very specific kind. The highest power of the unknown that appears in the equation (called the *degree*) must be the square, or the fourth power, or the eighth power ... That is, the degree must be a power of 2.

An angle of 60° can be formed from three constructible points: (0, 0), (1, 0), and $(\frac{1}{2}, \frac{\sqrt{3}}{2})$, which lie on the unit circle (radius 1, with its centre at the origin of the coordinate system). Trisecting this angle is equivalent to constructing a point (x, y) where the line at 20° to the horizontal axis crosses this circle. Using trigonometry and algebra, the coordinate x of this point is a solution of a *cubic* equation with rational coefficients. In fact x satisfies the equation $8x^3 - 6x - 1 = 0$. But the degree of a cubic is 3, which is not a power of 2. Contradiction – so no trisection is possible. Yes, you can get as close as you wish, but not spot on.

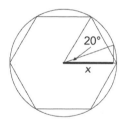

Trisecting 60° is equivalent to constructing x.

Trisectors often look for the impossible method even though they've heard of Wantzel's proof. They say things like 'I know it's impossible algebraically, but what about *geometrically*?' But Wantzel's proof shows that there is no geometric solution. It uses algebraic *methods* to do that, but algebra and geometry are mutually consistent parts of mathematics.

I always tell would-be trisectors that if they think they've found a trisection, a direct consequence is that 3 is an even number. Do they really want to go down in history as making that claim?

If the conditions of the problem are relaxed, many trisections exist. Archimedes knew of one that used a ruler with just two marks along its edge. The Greeks called this kind of technique a *neusis* construction. It involves sliding the rule so that the marks fall on two given curves – here a line and a circle:

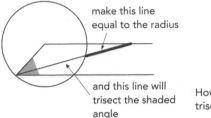

make this line
equal to the radius

and this line will
trisect the shaded
angle

How Archimedes
trisected an angle.

Langford's Cubes

The Scottish mathematician C. Dudley Langford was watching his young son playing with six coloured blocks – two of each colour. He noticed that the boy had arranged them so that the two yellow blocks (say) were separated by one block, the two blue blocks were separated by two blocks, and the two red blocks were separated by three blocks. Here I've used white for yellow, grey for blue and black for red to show you what I mean:

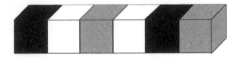

Langford's
cubes.

In between the white blocks we find just one block (which happens to be grey). Between the grey blocks are two blocks (one black, one white). And between the black blocks are three blocks (two white and one grey). Langford thought about this and was able to prove that this is the only such arrangement, except for its left–right reversal.

He wondered if you could do the same with more colours – such as four. And he found that again there is only one arrangement, plus its reversal. Can you find it? The simplest way to work on the puzzle is to use playing cards instead of blocks. Take two aces, two 2's, two 3's and two 4's. Can you lay the cards in a row to get exactly one card between the two aces, two cards between the two 2's, three cards between the two 3's and four cards between the two 4's?

There are no such arrangements with five or six pairs of cards, but there are 26 of them with seven pairs. In general, solutions exist if and only if the number of pairs is a multiple of 4 or one less than a multiple of 4. No formula is known for how many solutions there are, but in 2005 Michaël Krajecki, Christophe Jaillet and Alain Bui ran a computer for three months and found that there are precisely 46,845,158,056,515,936 arrangements with 24 pairs.

Answer on page 271

Duplicating the Cube

I'll briefly mention another cube problem: the third famous 'geometric problem of antiquity'. It's nowhere near as well known as the other two – trisecting the angle and squaring the circle. The traditional story is that an altar in the shape of a perfect cube must be doubled in volume. This is equivalent to constructing a line of length $\sqrt[3]{2}$ starting from the rational points of the plane. The desired length satisfies another cubic equation, this time the obvious one, $x^3 - 2 = 0$. For the same reason that trisecting the angle is impossible, so is duplicating the cube, as Pierre Wantzel pointed out in his 1837 paper. Cube-duplicators are so rare that you hardly ever come across one.* Trisectors are ten a penny.

---

* Though we shouldn't forget Edwin J. Goodwin, whose work on squaring the circle nearly caused a rumpus in Indiana (page 25).

Magic Stars

Here is a five-pointed star. It is a *magic* star because the numbers on any line of four circles add up to the same total, 24. But it's not a very good magic pentacle, because it doesn't use the numbers from 1 to 10. Instead, it uses the numbers from 1 to 12 with 7 and 11 missing.

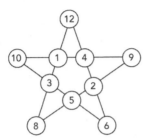

Five-pointed magic star.
Numbers are not consecutive.

It turns out that this is the best you can do with a five-pointed star. But if you use a six-pointed star, it is possible to place the numbers 1 to 12 in the circles, using one of each, so that each line of four numbers has the same total. (As a hint, the total has to be 26.) And, just to make the puzzle harder, I want you to make the six outermost numbers add up to 26 as well.

Where do the numbers go?

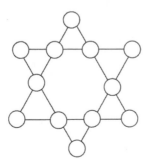

Write the numbers 1 to 12 in the circles to make this star magic.

Answer on page 271

Curves of Constant Width

The circle has the same width in any orientation. If you place it between two parallel lines, it can turn into any position. This is one reason why wheels are circular, and it's why circular logs make excellent rollers.

Is it the *only* curve like that?

turn to any angle

Is the circle the only curve like this?

Answer on page 272

Connecting Cables

Can you connect the fridge, cooker and dishwasher to the three corresponding electrical sockets, using cables that don't go through the kitchen walls or any of the three appliances, so that no two cables cross?

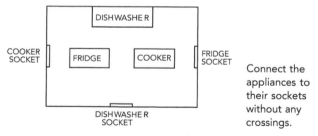

Connect the appliances to their sockets without any crossings.

In ordinary three-dimensional space this puzzle is a bit artificial, but in two dimensions it's a genuine problem, as any inhabitant of Flatland will tell you. A kitchen with no doors is an even bigger problem, but there you go.

Answer on page 272

Coin Swap

The first diagram shows six silver coins, A, C, E, G, I and K, and six gold coins, B, D, F, H, J and L. Your job is to move the coins into the arrangement shown in the second diagram. Each move must swap one silver coin with one adjacent gold coin; two coins are adjacent if there is a straight line joining them. The smallest number of moves that solves this puzzle is known to be 17. Can you find a 17-move solution?

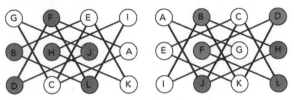

Move the coins from the first position to the second.

Answer on page 273

The Stolen Car

Nigel Fenderbender bought a secondhand car for £900 and advertised it in the local paper for £2,900. A respectable-looking elderly gentleman dressed as a clergyman turned up at the doorstep and enquired about the car, and bought it at the asking price. However, he mistakenly made his cheque out for £3,000, and it was the last cheque in his chequebook.

Now, Fenderbender had no cash in the house, so he nipped next door to the local newsagent, Maggie Zine, who was a friend of his, and got her to change the cheque. He paid the clergyman £100 change. However, when Maggie tried to pay the cheque in at the bank, it bounced. In order to pay back the newsagent, she was forced to borrow £3,000 from another friend, Honest Harry.

After Fenderbender had repaid this debt as well, he complained vociferously: 'I lost £2,000 profit on the car, £100 in

change, £3,000 repaying the newsagent and another £3,000 repaying Honest Harry. That's £8,100 altogether!'
How much money had he actually lost?

Answer on page 273

••

Space-Filling Curves

We ordinarily think of a curve as being much 'thinner' than, say, the interior of a square. For a long time, mathematicians thought that since a curve is one-dimensional, and a square is two-dimensional, it must be impossible for a curve to pass through every point inside a square.

Not so. In 1890 the Italian mathematician Giuseppe Peano discovered just such a *space-filling curve*. It was infinitely long and infinitely wiggly, but it still fitted the mathematical concept of a curve – which basically is some kind of bent line. In this case, *very* bent. A year later the German mathematician David Hilbert found another one. These curves are too complicated to draw – and if you could, you might as well just draw a solid black square like the left-hand picture. Mathematicians define space-filling curves using a step-by-step process that introduces more and more wiggles. At each step, the new wiggles are finer than the previous ones. The right-hand picture shows the fifth stage of this process for Hilbert's curve.

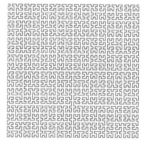

Hilbert's space-filling curve and an approximation.

There is an excellent animation showing successive stages of the construction of the Hilbert curve at en.wikipedia.org/wiki/ Hilbert_curve. Similar curves can fill a solid cube, indeed the analogue of a cube in any number of dimensions. So examples like this forced mathematicians to rethink basic concepts, such as 'dimension'. Space-filling curves have been proposed as the basis of an efficient method for searching databases by computer.

Compensating Errors

The class had been given a sum to do, involving three positive whole numbers ('positive' here means 'greater than zero'). During the break, two classmates compared notes.

'Oops. I added the three numbers instead of multiplying them,' said George.

'You're lucky, then,' said Henrietta. 'It's the same answer either way.'

What were the three numbers? What would they have been if there had been only two of them, or four of them, again with their sum equal to their product?

Answer on page 273

The Square Wheel

We seldom see a square wheel, but that's not because such a wheel can't roll without creating a bumpy ride. Circular wheels are great on flat roads. For square wheels, you just need a different shape of road:

A square-wheeled bicycle stays level on this bumpy road.

In fact, the correct shape is a series of inverted arcs of a catenary. A catenary is a U-shaped curve formed by a hanging chain. If the arcs meet at a right angle, a square of the right size will fit snugly, and its centre will stay level as it moves. It turns out that almost any shape of wheel will work provided you build the right kind of road for it to run on. Reinventing the wheel is easy. What counts is reinventing the road.

Why Can't I Divide by Zero?

In general, any number can be divided by any other number – except when the number we are dividing by is 0. 'Division by zero' is forbidden; even our calculators put up error messages if we try it. Why is zero a pariah in the division stakes?

The difficulty is not that we can't *define* division by zero. We could for instance, insist that any number divided by zero gives 42. What we can't do is make that kind of definition and still have all the usual arithmetical rules working properly. With this admittedly very silly definition, we could start from $1/0 = 42$ and apply standard arithmetical rules to deduce that $1 = 42 \times 0 = 0$.

Before we worry about division by zero, we have to agree on the rules that we want division to obey. Division is normally introduced as a kind of opposite to multiplication. What is 6 divided by 2? It is whichever number, multiplied by 2, gives 6. Namely, 3. So the two statements

$$6/2 = 3 \quad \text{and} \quad 6 = 2 \times 3$$

are logically equivalent. And 3 is the only number that works here, so 6/2 is unambiguous.

Unfortunately, this approach runs into major problems when we try to define division by zero. What is 6 divided by 0? It is whichever number, multiplied by 0, gives 6. Namely … uh. *Any* number multiplied by 0 makes 0; *you can't get 6.*

So 6/0 is out, then. So is any other number divided by 0, except – perhaps – 0 itself. What about 0/0?

Usually, if you divide a number by itself, you get 1. So we could define 0/0 to be 1. Now $0 = 1 \times 0$, so the relation with multiplication works. Nevertheless, mathematicians insist that 0/0 doesn't make sense. What worries them is a different arithmetical rule. Suppose that $0/0 = 1$. Then

$$2 = 2 \times 1 = 2 \times (0/0) = (2 \times 0)/0 = 0/0 = 1$$

Oops.

The main problem here is that since any number multiplied by 0 makes 0, we deduce that 0/0 can be any number whatsoever. If the rules of arithmetic work, and division is the opposite of multiplication, then 0/0 can take any numerical value. It's not unique. Best avoided, then.

Hang on – if you divide by zero, don't you get *infinity*?

Sometimes mathematicians use that convention, yes. But when they do, they have to check their logic rather carefully, because 'infinity' is a slippery concept. Its meaning depends on the context, and in particular you can't assume that it behaves like an ordinary number.

And even when infinity makes sense, 0/0 still causes headaches.

● ●

River Crossing 2 – Marital Mistrust

Remember Alcuin's letter to Charlemagne (page 20) and the wolf–goat–cabbage puzzle? The same letter contained a more complicated river-crossing puzzle, which may have been invented by the Venerable Bede fifty or so years earlier. It came to prominence in Claude-Gaspar Bachet's seventeenth-century compilation *Pleasant and Delectable Problems*, where it was posed as a problem about jealous husbands who did not trust their wives in the company of other men.

It goes like this. Three jealous husbands with their wives must cross a river, and find a boat with no boatman. The boat can carry only two of them at once. How can they all cross the river so that no wife is left in the company of other men without her husband being present?

Both men and women may row. All husbands are jealous in the extreme: they do not trust their unaccompanied wives to be with another man, *even if the other man's wife is also present.*

Answer on page 273

● ●

Wherefore Art Thou Borromeo?

Three rings can be linked together in such a way that if any one of them is ignored, the remaining two would pull apart. That is, no two of the rings are linked, only the full set of three. This arrangement is generally known as the *Borromean rings*, after the Borromeo family in Renaissance Italy, who used it as a family emblem. However, the arrangement is much older, and can be found in seventh-century Viking relics. Even in Renaissance Italy, it goes back to the Sforza family; Francesco Sforza permitted the Borromeos to use the rings in their coat of arms as a way of thanking them for their support during the defence of Milan.

Emblem of the Borromeo family and its use (bottom, left of centre) in their coat of arms.

On Isola Bella, one of three islands in Lake Maggiore owned by the Borromeo family, there is a seventeenth-century baroque palazzo built by Vitaliano Borromeo. Here the three-ring emblem can be found in numerous locations, indoors and out. A careful observer (such as a topologist) will discover that the rings depicted there are linked in several topologically distinct ways, only one of which has the key feature that no two rings are linked but all three are:

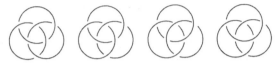

Four variations on the Borromean rings from the family palazzo.

The first version in the picture is the canonical one; it is inset into a floor and also appears in the garden. The second appears on the entrance tickets and on some of the flowerpots. A family crest at the top of the main staircase has the third pattern. Black and white seashells on the floor of a grotto under the palazzo form the fourth pattern. For further information, visit www.liv.ac.uk/~spmr02/rings/

Look at the four versions and explain why they are topologically different.

Can you find an analogous arrangement of four rings, such that if any ring is removed the remaining three can be pulled apart, but the full set of four rings cannot be disentangled?

Answer on page 274

• •

Percentage Play

Alphonse bought two bicycles. He sold one to Bettany for £300, making a loss of 25%, and one to Gemma, also for £300, making a profit of 25%. Overall, did he break even? If not, did he make a profit or a loss, and by how much?

Answer on page 275

• •

Kinds of People

There are 10 kinds of people in the world: those who understand binary numerals, and those who don't.

● ●

The Sausage Conjecture

This is one of my favourite unsolved mathematical problems, and it is absolutely weird, believe me.

As a warm-up, suppose you are packing a lot of identical circles together in the plane, and 'shrink-wrapping' them by surrounding the lot with the shortest curve you can. With 7 circles, you could try a long 'sausage':

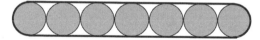

Sausage shape and wrapping.

However, suppose that you want to make the total *area* inside the curve – circles and the spaces between them – as small as possible. If each circle has radius 1, then the area of the sausage is 27.141. But there is a better arrangement of the circles, a hexagon with a central circle, and now the area is 25.533, which is smaller:

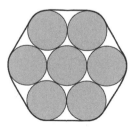

Hexagonal shape and wrapping.

Curiously, if you use identical spheres instead of circles, and shrink-wrap them with the surface of smallest possible area, then for 7 spheres the long sausage shape leads to a smaller total volume than the hexagonal arrangement. This sausage pattern gives the smallest volume inside the wrapping for any number of spheres up to 56. But with 57 spheres or more, the minimal arrangements are more rotund.

Less intuitive still is what happens in spaces of four or more dimensions. The arrangement of 4-dimensional spheres whose wrapping gives the smallest 4-dimensional 'volume' is a sausage for any number of spheres up to 50,000. It's *not* a sausage for 100,000 spheres, though. So the packing of smallest volume uses very long thin strings of spheres until you get an awful lot of them. Nobody knows the precise number at which 4-dimensional sausages cease to be the best.

The really fascinating change *probably* comes at five dimensions. You might imagine that in five dimensions sausages are best for, say, up to 50 billion spheres, but then something more rotund gives a smaller 5-dimensional volume; and for six dimensions the same sort of thing holds up to 29 squillion spheres, and so on. But in 1975 Laszlo Fejes Tóth formulated the *sausage conjecture*, which states that for five or more dimensions, the arrangement of spheres that occupies the smallest volume when shrink-wrapped is *always* a sausage – however large the number of spheres may be.

In 1998 Ulrich Betke, Martin Henk and Jörg Wills proved that Tóth was right for any number of dimensions greater than or equal to 42. To date, that's the best we know.

● ●

Tom Fool's Knot

This trick lets you tie a decorative knot while everybody watches. When you challenge them to do the same, they fail. No matter how many times you demonstrate the method, they seem unable to copy it successfully.

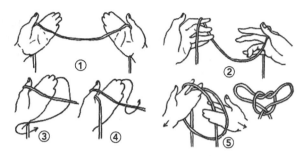

Stages in tying the Tom Fool's knot.

Take a length of soft cord about two metres long and hold it across your palms as in the first diagram, with your hands about half a metre apart. Let the two long ends hang down to counterbalance the weight of the length between the palms. Now bring your hands slowly together, all the while twiddling the fingers of the right hand. The finger twiddles have nothing at all to do with the method of tying the knot, but they distract spectators from the important moves, all of which happen with the left hand. Make the movements of your right hand seem as purposeful as you can.

With the left hand, first slide your thumb under the cord and pick it up, as in the second diagram. Then rapidly withdraw your fingers and replace them behind the hanging end, as shown by the arrow in the second diagram, to reach the position of the third figure. Without stopping, flip your fingers under the horizontal length of cord, as shown by the arrow in the third diagram, and withdraw your thumb. You should now have reached the position shown in the fourth diagram. Finally, use the tips of the fingers of each hand to grasp the end of cord hanging from the other hand, as in the fifth diagram. Holding on to the cord, pull the hands apart, and the lovely symmetrical knot of the final diagram appears.

Practice the method until you can perform it as a single, rhythmic movement. The knot unties if you just pull on the ends of the cord, so you can tie it over and over again. The trick becomes more mysterious every time you do it.

● ●

New Merology

Let him that hath understanding count the number of the beast; for it is the number of a man; and his number is Six hundred threescore and six. Revelation of St John 13:18

Or maybe not. The Oxyrhynchus Papyri – ancient documents found at Oxyrhynchus in Upper Egypt – include a fragment of the Book of Revelation from the third or fourth century which contains the earliest known version of some sections. The number that this papyrus assigns to the Beast is 616, not 666. So much for barcodes being symbols of evil.* No matter, for this puzzle is not about the Beast. It is about an idea that its inventor, Lee Sallows, calls 'new merology'. Let me make it clear that his proposal is not serious, except as a mathematical problem.[†]

The traditional method for assigning numbers to names, known as *gematria*, sets $A = 1$, $B = 2$ up to $Z = 26$. Then you add up all the numbers corresponding to the letters in the name. But there are lots of different systems of this kind, and lots of alphabets. Sallows suggested a more rational method based on words that denote numbers. For instance, with the numbering just described, the word ONE becomes $15 + 14 + 5 = 34$. However,

---

* The middle of a supermarket barcode bears lines that would represent the number 666, except that they have an entirely different function – they are 'guard bars' that help to correct errors. Each guard bar has the binary pattern 101, which on a barcode represents 6. Whence 666. Except that genuine barcode numbers actually have *seven* binary digits, so that 6 is 1010000, and ... oh well. This led some American fundamentalists to denounce barcodes as the work of the Devil. Since it now seems that the number of the Beast is actually 616, even the numerology is dodgy.

† I really shouldn't need to say this–but given the previous footnote ...

the number corresponding to ONE surely ought to be 1. Worse, *no* English number word denotes its numerological total, a property we will call 'perfect'.

Sallows wondered what happens if you assign a whole number to each letter, so that as many as possible of the number-words ONE, TWO, and so on are perfect. To make the problem interesting, different letters must be given different values. So you get a whole pile of equations like

$$O + N + E = 1$$
$$T + W + O = 2$$
$$T + H + R + E + E = 3$$

in algebraic unknowns O, N, E, T, W, H, R, ... And you must solve them in integers, all distinct.

The equation $O + N + E = 1$ tells us that some of the numbers have to be negative. Suppose, for example, that $E = 1$ and $N = 2$. Then the equation for ONE tells us that $O = -2$, and similar equations with other number-words imply that $I = 4$, $T = 7$ and $W = -3$. To make THREE perfect we must assign values to H and R. If $H = 3$, then R has to be -9. FOUR involves two more new letters, F and U. If $F = 5$, then $U = 10$. Now, $F + I + V + E = 5$ leads to $V = -5$. Since SIX contains two new letters, we try SEVEN first, which tells us that $S = 8$. Then we can fill in X from SIX, getting $X = -6$. The equation for EIGHT leads to $G = -7$. Now all the number names from ONE to TEN are perfect.

The only extra letter in ELEVEN and TWELVE is L. Remarkably, $L = 11$ makes them *both* perfect. But
$T + H + I + R + T + E + E + N = 7 + 3 + 4 + (-9) + 7 + 1 + 1 + 2$, which is 16, so we get stuck at this point.

In fact, we always get stuck at this point: if THIRTEEN is perfect, then

THREE + TEN = THIRTEEN

and we can remove common letters from both sides. This leads to $E = I$, violating the rule that different letters get different values.

However, we can go the other way and try to make ZERO perfect as well as ONE to TWELVE. Using the choices above, $Z + E + R + O = 0$ leads to $Z = 10$, but that's the same as U.

Can you find a different assignment of positive or negative whole-number values to letters, so that all words from ZERO to TWELVE are perfect?

Answer on page 275

Numerical Spell

Lee Sallows also applied new merology to magic, inventing the following trick. Select any number on the board shown below. Spell it out, letter by letter. Add together the corresponding numbers (subtracting those on black squares, adding those on white squares). The result will always be plus or minus the number you chose. For instance, TWENTY-TWO leads to

$$20 - 25 - 4 - 2 + 20 + 11 + 20 - 25 + 7 = 22$$

Board for Lee Sallows's magic trick.

Spelling Mistakes

'Thare are five mistukes im this centence.'
True or false?

Answer on page 275

• •

Expanding Universe

The starship *Indefensible* starts from the centre of a spherical universe of radius 1,000 light years, and travels radially at a speed of one light year per year – the speed of light. How long will it take to reach the edge of the universe? Clearly, 1,000 years. Except that I forgot to tell you that this universe is expanding. Every year, the universe expands its radius *instantly* by precisely 1,000 light years. Now, how long will it take to reach the edge? (Assume that the first such expansion happens exactly one year after the *Indefensible* starts its voyage, and successive expansions occur at intervals of exactly one year.)

It might seem that the *Indefensible* never gets to the edge, because that is receding faster than the ship can move. But at the instant that the universe expands, the ship is carried along with the space in which it sits, so its distance from the centre expands proportionately. To make these conditions clear, let's look at what happens for the first few years.

In the first year the ship travels 1 light year, and there are 999 light years left to traverse. Then the universe instantly expands to a radius of 2,000 light years, and the ship moves with it. So it is then 2 light years from the centre, and has 1,998 left to travel.

In the next year it travels a further light year, to a distance of 3 light years, leaving 1,997. But then the universe expands to a radius of 3,000 light-years, multiplying its radius by 1.5, so the ship ends up 4.5 light years from the centre, and the remaining distance increases to 2,995.5 light years.

Does the ship ever get to the edge? If so, how long does it take?

Hint: it will be useful to know that the nth harmonic number

$$H_n = 1 + \frac{1}{2} + \frac{1}{3} + \frac{1}{4} + \cdots + \frac{1}{n}$$

is approximately equal to

$$\log n + \gamma$$

where γ is Euler's constant, which is roughly $0.577\,215\,664\,9$.

Answer on page 276

What is the Golden Number?

The ancient Greek geometers discovered a useful idea which they called 'division in extreme and mean ratio'. By this they meant a line AB being cut at a point P, so that the ratios AP : AB and PB : AP are the same. Euclid used this construction in his work on regular pentagons, and I'll shortly explain why. But first, since nowadays we have the luxury of replacing ratios by numbers, let's turn the geometric recipe into algebra. Take PB to be of length 1, and let AP $= x$, so that AB $= 1 + x$. Then the required condition is

$$\frac{1 + x}{x} = \frac{x}{1}$$

so that $x^2 - x - 1 = 0$. The solutions of this quadratic equation are

$$\phi = \frac{1 + \sqrt{5}}{2} = 1.618\,034\ldots$$

and

$$1 - \phi = \frac{1 - \sqrt{5}}{2} = -0.618\,034\ldots$$

Here the symbol ϕ is the Greek letter phi. The number ϕ, known as the *golden number*, has the pleasant property that its reciprocal is

$$\frac{1}{\phi} = \frac{-1 + \sqrt{5}}{2} = 0.618\,034\ldots = \phi - 1$$

The golden number, in its geometric form as 'division in extreme and mean ratio', was the starting point for the Greek geometry of regular pentagons and anything associated with these, such as the dodecahedron and the icosahedron. The connection is this: if you draw a pentagon with sides equal to 1, then the long diagonals have length φ:

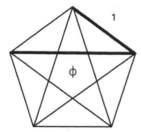

How φ appears in a
regular pentagon.

The golden ratio is often associated with aesthetics; in particular, the 'most beautiful' rectangle is said to be one whose sides are in the ratio φ : 1. The actual evidence for such statements is weak. Moreover, various methods of presenting numerical data exaggerate the role of the golden ratio, so that it is possible to 'deduce' the presence of the golden ratio in data that bear no relation to it. Similarly, claims that famous ancient buildings such as the Great Pyramid of Khufu or the Parthenon were designed using the golden ratio are probably unfounded. As with all numerology, you can find whatever you are looking for if you try hard enough. (Thus 'Parthenon' has 8 letters, 'Khufu' has 5, and $8/5 = 1.6$ – very close to φ.*)

Another common fallacy is to suppose that the golden ratio occurs in the spiral shell of a nautilus. This beautiful shell is – to great accuracy – a type of spiral called a logarithmic spiral. Here each successive turn bears a fixed ratio to the previous one. There is a spiral of this kind for which this ratio equals the golden ratio. But the ratio observed in the nautilus is *not* the golden ratio.

---

* Well, actually 'Parthenon' has 9 letters, but for a moment I had you there. And 1.8 is a lot closer to φ than many alleged instances of this number.

The nautilus shell is a logarithmic spiral but its growth rate is not the golden number.

The term 'golden number' is relatively modern. According to the historian Roger Herz-Fischler, it was first used by Martin Ohm in his book 1835 *Die Reine Elementar-Mathematik* ('Pure Elementary Mathematics') as *Goldene Schnitt* ('golden section'). It does not go back to the ancient Greeks.

The golden number is closely connected with the famous Fibonacci numbers, which come next.

● ●

What are the Fibonacci Numbers?

Many people met the Fibonacci numbers for the first time in Dan Brown's bestseller *The Da Vinci Code*. These numbers have a long and glorious mathematical history, which has very little overlap with anything mentioned in the book.

It all began in 1202 when Leonardo of Pisa published the *Liber Abbaci*, or 'Book of Calculation', an arithmetic text which concentrated mainly on financial computations and promoted the use of Hindu-Arabic numerals – the forerunner of today's familiar system, which uses just ten digits, 0 to 9, to represent all possible numbers.

One of the exercises in the book seems to have been Leonardo's own invention. It goes like this: 'A man put a pair of rabbits in a place surrounded on all sides by a wall. How many pairs of rabbits are produced from that pair in a year, if it is supposed that every month each pair produces a new pair, which from the second month onwards becomes productive?'

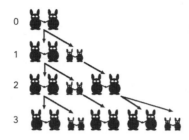

Family tree of of
Fibonacci's rabbits.

Say that a pair is *mature* if it can produce offspring, and *immature* if not.

At the start, month 0, we have 1 mature pair.

At month 1 this pair produces an immature pair, so we have 1 mature pair and 1 immature pair – 2 altogether.

At month 2 the mature pair produces another immature pair; the immature pair matures but produces nothing. So now we have 2 mature pairs and 1 immature pair – 3 in total.

At month 3 the 2 mature pairs produce 2 more immature pairs; the immature pair matures but produces nothing. So now we have 3 mature pairs and 2 immature pairs – 5 in total.

At month 4 the 3 mature pairs produce 3 more immature pairs; the 2 immature pairs mature but produce nothing. So now we have 5 mature pairs and 3 immature pairs – 8 in total.

Continuing step by step, we obtain the sequence

$$1, \ 2, \ 3, \ 5, \ 8, \ 13, \ 21, \ 34, \ 55, \ 89, \ 144, \ 233, \ 377$$

for months 0, 1, 2, 3, ..., 12. Here each term after the second is the sum of the previous two. So the answer to Leonardo's question is 377.

Some time later, probably in the eighteenth century, Leonardo was given the nickname Fibonacci – 'son of Bonaccio'. This name was more catchy than Leonardo Pisano Bigollo, which is what he used, so nowadays he is generally known as Leonardo Fibonacci, and his sequence of numbers is known as the *Fibonacci sequence*. The usual modern convention is to put the numbers 0, 1 in front, giving

$$0, \ 1, \ 1, \ 2, \ 3, \ 5, \ 8, \ 13, \ 21, \ 34, \ 55, \ 89, \ 144, \ 233, \ 377$$

although sometimes the initial 0 is omitted. The symbol for the nth Fibonacci number is F_n, starting from $F_0 = 0$.

The Fibonacci numbers as such are pretty useless as a model of the growth of real rabbit populations, although more general processes of a similar kind, called Leslie models, are used to understand the dynamics of animal and human populations. Nevertheless, the Fibonacci numbers are important in several areas of mathematics, and they also turn up in the natural world – though less widely than is often suggested. Extensive claims have been made for their occurrence in the arts, especially architecture and painting, but here the evidence is mostly inconclusive, except when Fibonacci numbers are used deliberately – for instance, in the architect Le Corbusier's 'modulor' system.

The Fibonacci numbers have strong connections with the golden number, which you'll recall is

$$\phi = \frac{1 + \sqrt{5}}{2} = 1.618\,034 \ldots$$

Ratios of successive Fibonacci numbers, such as 8/5, 13/8, 21/13, and so on become ever closer to ϕ as the numbers get bigger. Or, as mathematicians would say, $> F_{n+1}/F_n$ tends to ϕ as n tends to infinity. For instance, $377/233 = 1.618\,025 \ldots$. In fact, for integers of a given size, these Fibonacci fractions provide the best possible approximations to the golden number. There is even a formula for the nth Fibonacci number in terms of ϕ:

$$F_n = \frac{\phi^n - (1 - \phi)^n}{\sqrt{5}}$$

and this implies that F_n is the integer closest to $\phi^n/\sqrt{5}$.

If you make squares whose sides are the Fibonacci numbers they fit together very tidily, and you can draw quarter-circles in them to create an elegant *Fibonacci spiral*. Because F_n is close to ϕ^n, this spiral is very close to a logarithmic spiral, which grows in size by ϕ every quarter-turn. Contrary to many claims, this spiral is *not* the same shape as the Nautilus shell's spiral. Look at the picture on page 98 – the Nautilus is more tightly wound.

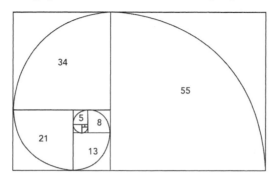

Fibonacci spiral.

However, there is a genuine – and striking – occurrence of Fibonacci numbers in living creatures, namely plants. The flowers of surprisingly many species have a Fibonacci number of petals. Lilies have 3 petals, buttercups have 5, delphiniums have 8, marigolds have 13, asters have 21, and most daisies have 34, 55 or 89. Sunflowers often have 55, 89 or 144.

Other numbers of petals do occur, but much less frequently. Mostly these are twice a Fibonacci number, or a power of 2. Sometimes numbers are from the related *Lucas sequence*:

$$1, \ 3, \ 4, \ 7, \ 11, \ 18, \ 29, \ 47, \ 76, \ 123, \ \ldots$$

where again each number after the second is the sum of the previous two, but the start of the sequence is different.

There seem to be genuine biological reasons for these numbers to occur. The strongest evidence can be seen in the heads of daisies and sunflowers, when the seeds have formed. Here the seeds arrange themselves in spiral patterns:

The head of
a daisy.

In the daisy illustrated, the eye sees one family of spirals that twist clockwise, and a second family of spirals that twist anticlockwise. There are 21 clockwise spirals and 34 anti-clockwise spirals – successive Fibonacci numbers. Similar numerical patterns, also involving successive Fibonacci numbers, occur in pine cones and pineapples.

The precise reasons for Fibonacci numerology in plant life are still open to debate, though a great deal is understood. As the tip of the plant shoot grows, long before the flowers appear, regions of the shoot form tiny bumps, called primordia, from which the seeds and other key parts of the flower eventually grow. Successive primordia form at angles of 137.5° – or 222.5° if we subtract this from 360°, measuring it the other way round. This is a fraction $\phi - 1$ of the full circle of 360°. This occurrence of the golden ratio can be predicted mathematically if we assume that the primordia are packed as efficiently as possible. In turn, efficient packing is a consequence of elastic properties of the growing shoot – the forces that affect the primordia. The genetics of the plant is also involved. Of course, many real plants do not quite follow the ideal mathematical pattern. Nevertheless, the mathematics and geometry associated with the Fibonacci sequence provide significant insights into these numerical features of plants.

The Plastic Number

The plastic number is a little-known relative of the famous golden number. We've just seen how the Fibonacci numbers create a spiralling system of squares, related to the golden number. There is a similar spiral diagram for the plastic number, but composed of equilateral triangles. In the diagram below, the initial triangle is marked in black and successive triangles spiral in a clockwise direction: the spiral shown is again roughly logarithmic. In order to make the shapes fit, the first three triangles all have side 1. The next two have side 2, and then the numbers go 4, 5, 7, 9, 12, 16, 21, and so on.

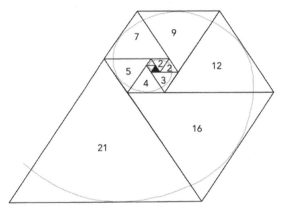

Padovan spiral.

Again there is a simple rule for finding these numbers, analogous to that for Fibonacci numbers: each number in the sequence is the sum of the previous number *but one*, together with the one before that. For example,

$$12 = 7 + 5, 16 = 9 + 7, 21 = 12 + 9$$

This pattern follows from the way the triangles fit together. If P_n is the nth Padovan number (starting from $P_0 = P_1 = P_2 = 1$), then

$$P_n = P_{n-2} + P_{n-3}$$

The first twenty numbers in the sequence are:

1, 1, 1, 2, 2, 3, 4, 5, 7, 9, 12, 16,

21, 28, 37, 49, 65, 86, 114, 151

I call this sequence the *Padovan numbers* because the architect Richard Padovan told me about them, although he denies any responsibility. Curiously, 'Pádova' is the Italian form of 'Padua', and Fibonacci was from Pisa, roughly a hundred miles away. I am tempted to rename the Fibonacci numbers 'Pisan numbers' to reflect the Italian geography, but as you can see I managed to resist.

The *plastic number*, which I denote by p, is roughly 1.324718. It is related to the Padovan numbers in the same way that the golden number is related to the Fibonacci numbers. That is, ratios of successive Padovan numbers, such as 49/37 or 151/114, give good approximations to the plastic number. The pattern of the sequence of Padovan numbers leads to the equation $p^3 - p - 1 = 0$, and p is the unique real solution of this cubic equation. The Padovan sequence increases much more slowly than the Fibonacci sequence because p is smaller than ϕ. There are many interesting patterns in the Padovan sequence. For example, the diagram shows that $21 = 16 + 5$, because triangles adjacent along a suitable edge have to fit together; similarly, $16 = 12 + 4, 12 = 9 + 3$, and so on. Therefore

$$P_n = P_{n-1} + P_{n-5}$$

which is an alternative rule for deriving further terms of the sequence. This equation implies that $p^5 - p^4 - 1 = 0$, and it is not immediately obvious that p, defined as a solution of a cubic equation, must also satisfy this *quintic* (fifth-degree) equation.

● ●

Family Occasion

'It was a wonderful party,' said Lucilla to her friend Harriet.

'Who was there?'

'Well – there was one grandfather, one grandmother, two fathers, two mothers, four children, three grandchildren, one brother, two sisters, two sons, two daughters, one father-in-law, one mother-in-law and one daughter-in-law.'

'Wow! Twenty-three people!'

'No, it was less than that. A lot less.'

What is the *smallest* size of party that is consistent with Lucilla's description?

Answer on page 277

• •

Don't Let Go!

Topology is a branch of mathematics in which two shapes are 'the same' if one can be continuously deformed into the other. So you can bend, stretch, and shrink, but not cut. This ancient topological chestnut still has many attractions – in particular, not everyone has seen it before. What you have to do is pick up a length of rope, with the left hand holding one end and the right hand holding the other, and tie a knot in the rope *without letting go of the ends*.

Answer on page 277

• •

Theorem: All Numbers are Interesting

Proof: For a contradiction, suppose not. Then there is a smallest uninteresting number. But being the *smallest* one singles it out among all other numbers, making it special, hence interesting – contradiction.

• •

Theorem: All Numbers are Boring

Proof: For a contradiction, suppose not. Then there is a smallest non-boring number.

  And your point is—?

• •

The Most Likely Digit

If you look at a list of numerical data, and count how often a given digit turns up as the *first* digit in each entry, which digit is most likely? The obvious guess is that every digit has the same chance of occurring as any other. But it turns out that for most kinds of data, this is wrong.

  Here's a typical data set – the areas of 18 islands in the Bahamas. I've given the figures in square miles and in square kilometres, for reasons I'll shortly explain.

Island	Area (sq mi)	Area (sq km)
Abaco	649	1681
Acklins	192	497
Berry Islands	2300	5957
Bimini Islands	9	23
Cat Island	150	388
Crooked and Long Cay	93	241
Eleuthera	187	484
Exuma	112	290
Grand Bahama	530	1373
Harbour Island	3	8
Inagua	599	1551
Long island	230	596
Mayaguana	110	285
New Providence	80	207
Ragged Island	14	36
Rum Cay	30	78
San Salvador	63	163
Spanish Wells	10	26

For the square mile data, the number of times that a given first digit (shown in brackets) occurs goes like this:

(1)7 (2) 2 (3) 2 (4) 0 (5) 2 (6) 2 (7) 0 (8) 1 (9) 2

and 1 wins hands down. In square kilometres, the corresponding numbers are

(1) 4 (2) 6 (3) 2 (4) 2 (5) 2 (6) 0 (7) 1 (8) 1 (9) 0

and now 2 wins, but only just.

In 1938 the physicist Frank Benford observed that for long enough lists of data, the numbers encountered by physicists and engineers are most likely to start with the digit 1 and least likely to start with 9. The frequency with which a given initial digit occurs – that is, the probability that the first digit takes a given value – *decreases* as the digits increase from 1 to 9. Benford discovered empirically that the probability of encountering n as the first decimal digit is

$$\log_{10}(n+1) - \log_{10}(n)$$

where the subscript 10 means that the logarithms are to base ten. (The value $n = 0$ is excluded because the initial digit is by definition the first *non-zero* digit.) Benford called this formula the law of anomalous numbers, but nowadays it's usually known as *Benford's law*.

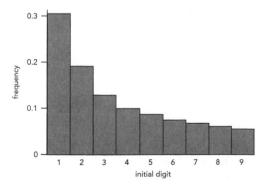

Theoretical frequencies according to Benford's law.

For the Bahama Island data, the frequencies look like this:

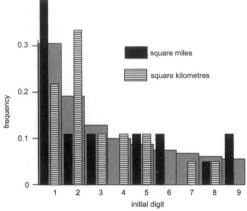

Observed frequencies for the areas of the Bahama Islands, compared with Benford's theoretical ideal.

There are some differences between theory and reality, but the data sets here are fairly small, so we would expect that. Even with only 18 numbers, there is a strong prevalence of 1's and 2's – which, according to Benford's law, should between them occur just over half the time.

Benford's formula is far from obvious, but a little thought shows that the nine frequencies are unlikely to be identical. Think of a street of houses, numbered from 1 upwards. The probability of a given digit coming first varies considerably with the number of houses on the street. If there are nine houses, each digit occurs equally often. But, if there are 19, then the initial digit is 1 for houses 1 and 10–19, a frequency of 11/19, or more than 50%. As the length of the street increases, the frequency with which a given first digit occurs wanders up and down in a complicated but computable manner. The nine frequencies are the same *only* when the number of houses is 9, 99, 999, and so on.

Benford's formula is distinguished by a beautiful property: it is *scale-invariant*. If you measure the areas of Bahamian islands in square miles or square kilometres, if you multiply house numbers by 7 or 93, then – provided you have a big enough sample – the same law applies. In fact, Benford's Law is the *only* scale-invariant frequency law. It is unclear why nature prefers scale-invariant frequencies, but it seems reasonable that the natural world should not be affected by the units in which humans choose to measure it.

Tax collectors use Benford's law to detect fake figures in tax forms, because people who invent fictitious numbers tend to use the same initial digits equally often. Probably because they think that's what should happen for genuine figures!

● ●

Why Call It a Witch?

Maria Agnesi was born in 1718 and died in 1799. She was the daughter of a wealthy silk merchant, Pietro Agnesi (often wrongly said to have been a professor of mathematics at Bologna), and the eldest of his 21 children. Maria was precocious, and published an essay advocating higher education for women when she was nine years old. The essay was actually written by one of her tutors, but she translated it into Latin and delivered it from memory to an academic gathering in the garden of the family home. Her father also arranged for her to debate philosophy in the presence of prominent scholars and public figures. She disliked making a public spectacle of herself and asked her father for permission to become a nun. When he refused, she extracted an agreement that she could attend church whenever she wished, wear simple clothing, and be spared from all public events and entertainments.

Maria Gaetana Agnesi.

From that time on, she focused on religion and mathematics. She wrote a book on differential calculus, printed privately around 1740. In 1748 she published her most famous work, *Instituzioni Analitiche ad Uso Della Gioventù Italiana* ('Analytical Institutions for the Use of the Youth of Italy'). In 1750 Pope Benedict XIV invited her to become professor of mathematics at the University of Bologna, and she was officially confirmed in the role, but she never actually attended the university because this would not have been in keeping with her humble lifestyle. As a result, some sources say she was a professor and others say she wasn't. Was she, or wasn't she? Yes.

There is a famous curve, called the 'witch of Agnesi', which has the equation

$$y = a^3/(a^2 + x^2)$$

where *a* is a constant. The curve looks remarkably *unlike* a witch – it isn't even pointy:

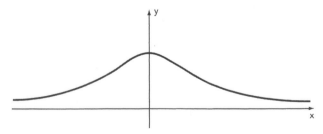

Witch of Agnesi.

So how did this strange name get attached to the curve? Fermat was the first to discuss this curve, in about 1700. Maria Agnesi wrote about the curve in her book *Instituzione Analitiche*. The word 'witch' was a mistake in translation. In 1718 Guido Grandi named the curve 'versoria', a Latin term for a rope that turns a sail, because that's what it looked like. In Italian this term became 'versiera', which is what Agnesi called it. But John Colson, who translated various mathematics books into English, mistook 'la versiera' for 'l'aversiera', meaning 'the witch'.

It could have been worse. Another meaning is 'she-devil'.

. . . .

Möbius and His Band

There are some pieces of mathematical folklore that you really should be reminded about, even though they're 'well known' – just in case. An excellent example is the Möbius band.

Augustus Möbius was a German mathematician, born 1790, died 1868. He worked in several areas of mathematics, including geometry, complex analysis and number theory. He is famous for his curious surface, the *Möbius band*. You can make a Möbius band by taking a strip of paper, say 2 cm wide and 20 cm long, bending it round until the ends meet, then twisting one end through 180°, and finally gluing the ends together. For comparison, make a cylinder in the same way, omitting the twist.

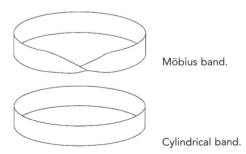

Möbius band.

Cylindrical band.

The Möbius band is famous for one surprising feature: it has only one side. If an ant crawls around on a cylindrical band, it can cover only half the surface – one side of the band. But if an ant crawls around on the Möbius band, it can cover the entire surface. The Möbius band has only one side.

You can check these statements by painting the band. You can paint the cylinder so that one side is red and the other is blue, and the two sides are completely distinct, even though they are separated only by the thickness of the paper. But if you start to paint the Möbius band red, and keep going until you run out of band to paint, the *whole thing* ends up red.

In retrospect, this is not such a surprise, because the 180° twist connects each side of the original paper strip to the other. If you don't twist before gluing, the two sides stay separate. But until Möbius (and a few others) thought this one up, mathematicians didn't appreciate that there are two distinct kinds of surface: those with two sides, and those with one. This turned out to be important in topology. And it showed how careful you have to be about making 'obvious' assumptions.

There are lots of Möbius band recreations. Here are three.

- If you cut the cylindrical band along the middle with scissors, it falls apart into two cylindrical bands. What happens if you try this with a Möbius band?
- Repeat, but this time make the cut about one-third of the way across the width of the band. Now what happens to the cylinder, and to the band?
- Make a band like a Möbius band but with a 360° twist. How many sides does it have? What happens if you cut it along the middle?

The Möbius band is also known as a Möbius strip, but this can lead to misunderstandings, as in a Limerick written by science fiction author Cyril Kornbluth:

A burleycue dancer, a pip
Named Virginia, could peel in a zip;
    But she read science fiction
    and died of constriction
Attempting a Möbius strip.

A more politically correct Möbius limerick, which gives away one of the answers, is:

A mathematician confided
That a Möbius strip is one-sided.
    You'll get quite a laugh
    if you cut it in half,
For it stays in one piece when divided.

Answers on page 277

Golden Oldie

Why did the chicken cross the Möbius band?
    To get to the other ... um ...

Three More Quickies

(1) If five dogs dig five holes in five days, how long does it take ten dogs to dig ten holes? Assume that they all dig at the same rate all the time and all holes are the same size.

(2) A woman bought a parrot in a pet-shop. The shop assistant, who always told the truth, said, 'I guarantee that this parrot will repeat every word it hears.' A week later, the woman took the parrot back, complaining that it hadn't spoken a single word. 'Did anyone talk to it?' asked the suspicious assistant. 'Oh, yes.' What is the explanation?

(3) The planet Nff-Pff in the Anathema Galaxy is inhabited by precisely two sentient beings, Nff and Pff. Nff lives on a large

continent, in the middle of which is an enormous lake. Pff lives on an island in the middle of the lake. Neither Nff nor Pff can swim, fly or teleport: their only form of transport is to walk on dry land. Yet each morning, one walks to the other's house for breakfast. Explain.

Answers on page 277

● ●

Miles of Tiles

Bathroom walls and kitchen floors provide everyday examples of tiling patterns, using real tiles, plastic or ceramic. The simplest pattern is made from identical square tiles, fitted together like the squares of a chessboard. Over the centuries, mathematicians and artists have discovered many beautiful tilings, and mathematicians have gone a stage further by seeking all possible tilings with particular features.

For instance, exactly three regular polygons tile the entire infinite plane – that is, identical tiles of that shape cover the plane without overlaps or gaps. These polygons are the equilateral triangle, the square and the hexagon:

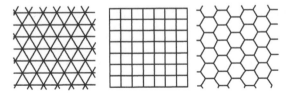

The three regular polygons that tile the plane.

We can be confident that no other *regular* polygon tiles the plane, by thinking about the angles at which the edges of the tiles meet. If several tiles meet at a given point, the angles involved must add to 360°. So the angle at the corner of a tile is 360° divided by a whole number, say $360/m$. As m gets larger, this angle gets smaller. In contrast, as the number of sides of a regular polygon increases, the angle at each corner gets bigger. The effect of this is to 'sandwich' m within very narrow limits, and this in turn restricts the possible polygons.

The details go like this. When $m = 1, 2, 3, 4, 5, 6, 7$, and so on, $360/m$ takes the values 360, 180, 120, 90, 72, 60, $51\frac{3}{7}$, and so on. The angle at the corner of a regular n-gon, for $n = 3, 4, 5, 6, 7$, and so on, is 60, 90, 108, 120, $128\frac{4}{7}$, and so on. The only places where these lists coincide are when $m = 3, 4$, and 6; here $n = 6, 4$ and 3.

Actually, this proof as stated has a subtle flaw. *What have I forgotten to say?*

The most striking omission from my list is the regular pentagon, which does not tile the plane. If you try to fit regular pentagonal tiles together, they don't fit. When three of them meet at a common point, the total angle is $3 \times 108° = 324°$, less than 360°. But if you try to make four of them meet, the total angle is $4 \times 108° = 432°$, which is too big.

Irregular pentagons can tile the plane, and so can innumerable other shapes. In fact, 14 distinct types of convex pentagon are known to tile the plane. It is probable, but not yet proved, that there are no others. You can find all 14 patterns at www.mathpuzzle.com/tilepent.html mathworld.wolfram.com/PentagonTiling.html

The mathematics of tilings has important applications in crystallography, where it governs how the atoms in a crystal can be arranged, and what symmetries can occur. In particular, crystallographers know that the possible rotational symmetries of a regular lattice of atoms is tightly constrained. There are 2-fold, 3-fold, 4-fold and 6-fold symmetries – meaning that the arrangement of atoms looks identical if the whole thing is rotated through $\frac{1}{2}$, $\frac{1}{3}$, $\frac{1}{4}$ or $\frac{1}{6}$ of a full turn (360°). However, 5-fold symmetries are impossible – just as the regular pentagon cannot tile the plane.

There the matter stood until 1972, when Roger Penrose discovered a new type of tiling, using two types of tile, which he called *kites* and *darts*:

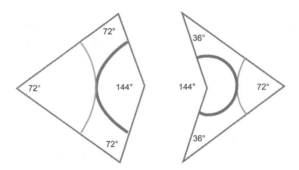

A kite (left) and dart (right). The matching rules require the thick and thin arcs to meet at any join – see the pictures below.

These shapes are derived from the regular pentagon, and the associated tilings are required to obey certain 'matching rules' where several tiles meet, to avoid simple repetitive patterns. Under these conditions, the two shapes can tile the plane, but not by forming a repetitive lattice pattern. Instead, they form a bewildering variety of complicated patterns. Precisely two of these, called the star and sun patterns, have exact fivefold rotational symmetry.

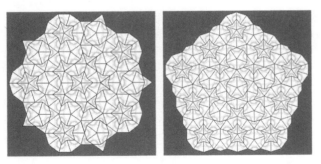

The two fivefold-symmetric Penrose tilings. Left: star pattern; right: sun pattern. Tinted lines illustrate the matching rules. Black lines are edges of tiles.

It then turned out that nature knows this trick. Some chemical compounds can form 'quasicrystals' using Penrose patterns for their atoms. These forms of matter are not regular lattices, but they can occur naturally. So Penrose's discovery changed our ideas about natural arrangements of atoms in crystal-like structures.

The detailed mathematics and crystallography are too complicated to describe here. To find out more, go to: en.wikipedia.org/wiki/Penrose_tiling

Answers on page 278

• •

Chaos Theory

If you want your friends to accept you as a 'rocket scientist', you have to be able to spout about *chaos theory*. You will casually mention the butterfly effect, and then you get to talk about where Pluto (no longer a planet but a mere dwarf planet) will be in 200 million years' time, and how really good dishwashers work.

Chaos theory is the name given by the media to an important new discovery in dynamical systems theory – the mathematics of systems that change over time according to specific rules. The name refers to a surprising and rather counterintuitive type of behaviour known as deterministic chaos. A system is called *deterministic* if its present state completely determines its future behaviour; if not, the system is called *stochastic* or *random*. Deterministic chaos – universally shortened to 'chaos' – is apparently random behaviour in a deterministic dynamical system. At first sight, this seems to be a contradiction in terms, but the issues are quite subtle, and it turns out that some features of deterministic systems can behave randomly.

Let me explain why.

You may remember the bit in Douglas Adams's *The Hitch Hiker's Guide to the Galaxy* that parodies the concept of determinism. No, really, you do – remember the supercomputer Deep Thought? When asked for the answer to the Great Question of Life, the Universe and Everything, it ruminates for five million years, and finally delivers the answer as 42. The philosophers then realise that they didn't actually understand the *question*, and an even greater computer is given the task of finding it.

Deep Thought is the literary embodiment of a 'vast intellect' envisaged by one of the great French mathematicians of the eighteenth century, the Marquis de Laplace. He observed that the laws of nature, as expressed mathematically by Isaac Newton and his successors, are deterministic, saying that: 'An intellect which at a certain moment knew all forces that set nature in motion, and all positions of all items of which nature is composed, if this intellect were also vast enough to submit these data to analysis, it would embrace in a single formula the movements of the greatest bodies of the universe and those of the tiniest atom; for such an intellect nothing would be uncertain and the future just like the past would be present before its eyes.'

In effect, Laplace was telling us that any deterministic system is inherently *predictable* – in principle, at least. In practice, however, we have no access to a Vast Intellect of the kind he had in mind, so we can't carry out the calculations that are needed to predict the system's future. Well, maybe for a short period, if we're lucky. For example, modern weather forecasts are fairly accurate for about two days, but a ten-day forecast is often badly wrong. (When it isn't, they've got lucky.)

Chaos raises another objection to Laplace's vision: even if his Vast Intellect existed, it would have to know 'all positions of all items' with *perfect* accuracy. In a chaotic system, any uncertainty about the present state grows very rapidly as time passes. So we quickly lose track of what the system will be doing. Even if this initial uncertainty first shows up in the millionth decimal place of some measurement – with the previous 999,999 decimal places absolutely correct – the predicted future based on one value for that millionth decimal place will be utterly different from a prediction based on some other value.

In a non-chaotic system such uncertainties grow fairly slowly, and very long-term predictions can be made. In a chaotic system, inevitable errors in measuring its state *now* mean that its state a short time ahead may be completely uncertain.

A (slightly artificial) example may help to clarify this effect. Suppose that the state of some system is represented by a real number – an infinite decimal – between 0 and 10. Perhaps its current value is 5.430874, say. To keep the maths simple, suppose that time passes in discrete intervals – 1, 2, 3, and so on. Let's call these intervals 'seconds' for definiteness. Further, suppose that the rule for the future behaviour is this: to find the 'next' state – the state one second into the future – you take the current state, multiply by 10, and ignore any initial digit that would make the result bigger than 10. So the current value 5.430874 becomes 54.30874, and you ignore the initial digit 5 to get the next state, 4.30874. Then, as time ticks on, successive states are:

> 5.430874
> 4.30874
> 3.0874
> 0.874
> 8.74
> 7.4

and so on.

Now suppose that the initial measurement was slightly inaccurate, and should have been 5.430824 – differing in the fifth decimal place. In most practical circumstances, this is a very tiny error. Now the predicted behaviour would be:

> 5.430824
> 4.30824
> 3.0824
> 0.824
> **2**.4

See how that **2** moves one step to the left at each step – making the error ten times as big each time. After a mere 5 seconds, the first prediction of 7.4 has changed to 2.4 – a significant difference.

If we had started with a million-digit number, and changed the final digit, it would have taken a million seconds for the change to affect the predicted *first* digit. But a million seconds is only $11\frac{1}{2}$ days. And most mathematical schemes for predicting the future behaviour of a system work with much smaller intervals of time – thousandths or millionths of seconds.

If the rule for moving one time-step into the future is different, this kind of error may not grow as quickly. For example, if the rule is 'divide the number by 2', then the effect of such a change dies away as we move further and further into the future. So what makes a system chaotic, or not, is the *rule* for forecasting its next state. Some rules exaggerate errors, some filter them out.

The first person to realise that sometimes the error can grow rapidly – that the system may be chaotic, despite being deterministic – was Henri Poincaré, in 1887. He was competing for a major mathematical prize. King Oscar II of Norway and Sweden offered 2,500 crowns to anyone who could calculate whether the solar system is stable. If we wait long enough, will the planets continue to follow roughly their present orbits, or could something dramatic happen – such as two of them colliding, or one being flung away into the depths of interstellar space?

This problem turned out to be far too difficult, but Poincaré managed to make progress on a simpler question – a hypothetical solar system with just three bodies. The mathematics, even in this simplified set-up, was still extraordinarily difficult. But Poincaré was up to the task, and he convinced himself that this 'three-body' system sometimes behaved in an irregular, unpredictable manner. The equations were deterministic, but their solutions were erratic.

He wasn't sure what to do about that, but he knew it must be true. He wrote up his work, and won the prize.

Complicated orbits for three bodies moving under gravity.

And that was what everyone thought until recently. But in 1999 the historian June Barrow-Green discovered a skeleton in Poincaré's closet. The published version of his prizewinning paper was not the one he submitted, not the one that won the prize. The version he submitted – which was printed in a major mathematical journal – claimed that *no* irregular behaviour would occur. Which is the exact opposite of the standard story.

Barrow-Green discovered that shortly after winning the prize, an embarrassed Poincaré realised he had blundered. He withdrew the winning memoir and paid for the entire print run of the journal to be destroyed. Then he put his error right, and the official published version is the corrected one. No one knew that there had been a previous version until Barrow-Green discovered a copy tucked away among the archives of the Mittag-Leffler Institute in Stockholm.

Anyway, Poincaré deserves full credit as the first person to appreciate that deterministic mathematical laws do not always imply predictable, regular behaviour. Another famous advance was made by the meteorologist Edward Lorenz in 1961. He was running a mathematical model of convection currents on his computer. The machines available in those days were very slow and cumbersome compared with what we have now – your mobile phone is a far more powerful computer than the top research machine of the 1960s. Lorenz had to stop his computer in the middle of a long calculation, so he printed out all the numbers it had found. Then he went back several steps, input the numbers at that point, and restarted the calculation. The reason for backtracking was to check that the new calculation agreed with the old one, to eliminate errors when he fed the old figures back in.

It didn't.

At first the new numbers were the same as the old ones, but then they started to differ. What was wrong? Eventually Lorenz discovered that he hadn't typed in any wrong numbers. The difference arose because the computer stored numbers to a few more decimal places than it printed out. So what it stored as 2.371 45, say, was printed out as 2.371. When he typed that number in for the second run, the computer began calculating using 2.371 00, not 2.371 45. The difference grew – chaotically – and eventually became obvious.

When Lorenz published his results, he wrote: 'One meteor-ologist remarked that if the theory were correct, one flap of a

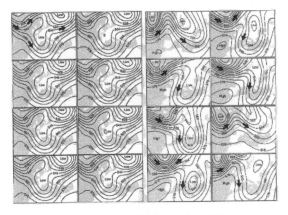

(Left) Initial conditions for eight weather forecasts, apparently identical but with tiny differences. (Right) The predicted weather a week later – the initial differences have grown enormously. Italian weather is more predictable than British. [Courtesy of the European Medium Range Weather Forecasting Centre, Reading.]

seagull's wings could change the course of weather for ever.' The objection was intended as a put-down, but we now know that this is exactly what happens. Weather forecasters routinely make a whole 'ensemble' of predictions, with slightly different initial conditions, and then take a majority vote on the future, so to speak.

Before you rush out with a shotgun, I must add that there are billions of seagulls, and we don't get to run the weather twice. What we end up with is a random selection from the range of possible weathers that might have happened instead.

Lorenz quickly replaced the seagull by a butterfly, because that sounded better. In 1972 he gave a lecture with the title 'Does the flap of a butterfly's wings in Brazil set off a tornado in Texas?' The title was invented by Philip Merilees when Lorenz failed to provide one. Thanks to this lecture, the mathematical point concerned became known as the butterfly effect. It is a characteristic feature of chaotic systems, and it is why they are unpredictable, despite being deterministic. The slightest change to the current state of the system can grow so rapidly that it changes the future behaviour. Beyond some relatively small 'prediction horizon', the future must remain mysterious. It may be predetermined, but we can't find out what has been predetermined, except by waiting to see what happens. Even a big increase in computer speed makes little difference to this horizon, because the errors grow so fast.

For weather, the prediction horizon is about two days ahead. For the solar system as a whole, it is far longer. We can predict that in 200 million years' time, Pluto will still be in much the same orbit as it is today; however, we have no idea on which side of the Sun it will be by then. So some features are predictable, others are not.

Although chaos is unpredictable, it is not random. This is the whole point. There are hidden 'patterns', but you have to know how to find them. If you plot the solutions of Lorenz's model in three dimensions, they form a beautiful, complicated shape called a strange attractor. If you plotted random data that way, you'd just get a fuzzy mess.

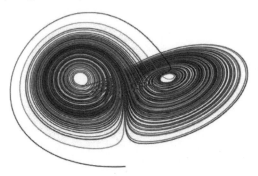

The Lorenz attractor, a geometric representation of Lorenz's calculations.

Chaos may seem a useless phenomenon, on the grounds that it prevents practical predictions. Even is this objection were correct, chaos would still exist. The real world is not obliged to behave in ways that are convenient for humans. As it happens, there are ways to make use of chaos. For a time, a Japanese company marketed a chaotic dishwasher, with two rotary arms spraying water on its contents. The resulting irregular spray cleaned the dishes better than the regular spray from a single rotating arm would have done.

And, of course, a dishwasher based on chaos theory was obviously very scientific and advanced. The marketing people must have loved it.

Après-le-Ski

The little-known Alpine village of Après-le-Ski is situated in a deep mountain valley with vertical cliffs on both sides. The cliffs are 600 metres high on one side and 400 metres high on the other. A cable car runs from the foot of each cliff to the top of the other cliff, and the cables are perfectly straight. At what height above the ground do the two cables cross?

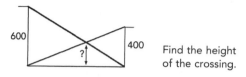

Find the height of the crossing.

Answer on page 278

● ●

Pick's Theorem

Here is a *lattice polygon*: a polygon whose vertices lie on the points of a square lattice. Assuming that the points are spaced at intervals of one unit, what is the area of the polygon?

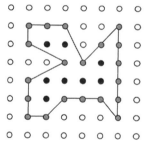

A lattice polygon.

There's a wonderfully simple way to find such areas, however complicated the polygon may be – by using *Pick's theorem*. It was proved by Georg Pick in 1899. For any lattice polygon, the area A can be calculated from the number of boundary points B (grey) and interior points I (black) by the formula

$$A = \tfrac{1}{2}B + I - 1$$

Here $B = 20$ and $I = 8$, so the area is $\tfrac{1}{2} \times 20 + 8 - 1 = 17$ square units.

What is the area of the lattice polygon in the second diagram?

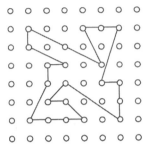

Find the area.

Answer on page 279

Mathematical Prizes

There is no Nobel Prize in mathematics, but there are several equally prestigious prizes and a vast range of smaller ones, among them:

Fields Medal
The Fields Medal was instituted by the Canadian mathematician John Charles Fields and was first awarded in 1936. Every four years the International Mathematical Union selects for the award up to four of the world's leading research mathematicians, who must be under 40 years old. The prize consists of a gold medal and a small sum of money – currently around $13,500 – but is considered equivalent to a Nobel Prize in prestige.

Abel Prize
In 2001 the Norwegian government commemorated the 200th anniversary of the birth of Niels Henrik Abel – one of the all-time greats of mathematics – with a new prize. Each year, one or more mathematicians share a prize in the region of $1,000,000, which is comparable to the sum that Nobel Prize winners receive. The King of Norway presents the award at a special ceremony.

Shaw Prize

Sir Run Run Shaw, a prominent figure in Hong Kong's media and a long-standing philanthropist, established an annual prize for three areas of science: astronomy, life sciences and medicine, and mathematics. The total value awarded each year is $1,000,000, and there is also a medal. The first Shaw Prize was awarded in 2002.

Clay Millennium Prizes

The Clay Mathematics Institute in Cambridge, Massachusetts, founded by Boston businessman Landon T. Clay and Lavinia D. Clay, offers seven prizes, each of $1,000,000, for the definitive solution of seven major open problems. These 'Millennium Prize Problems' were selected to represent some of the biggest challenges facing mathematicians. For the record, they are:

- The Birch and Swinnerton-Dyer Conjecture in algebraic number theory.
- The Hodge Conjecture in algebraic geometry.
- The existence of solutions, valid for all time, to the Navier–Stokes equations of fluid dynamics.
- The $P = NP$? problem in computer science.
- The Poincaré Conjecture in topology.
- The Riemann Hypothesis in complex analysis and the theory of prime numbers.
- The mass gap hypothesis and associated issues for the Yang–Mills equations in quantum field theory.

None of the prizes has yet been awarded, but the Poincaré Conjecture has now been proved. The main breakthrough was made by Grigori Perelman, and many details were clarified by other mathematicians. For details of the seven problems, see www.claymath.org/millennium/

King Faisal International Prize

Between 1977 and 1982 the King Faisal Foundation instituted prizes for service to Islam, Islamic studies, Arabic literature, medicine and science. The science prize is open to, and has been won by, mathematicians. The winner receives a certificate, a gold medal and SR 750 000 ($200,000).

Wolf Prize

Since 1978 this prize has been awarded by the Wolf Foundation, set up by Ricardo Wolf and his wife Francisca Subirana Wolf. It covers five areas of science: agriculture, chemistry, mathematics, medicine and physics. The prize consists of a diploma and $100,000.

Beal Prize

In 1993 Andrew Beal, a Texan with a passion for number theory, was led to conjecture that if $a^p + b^q = c^r$, where a, b, c, p, q and r are positive integers, and p, q and r are all greater than 2, then a, b and c must have a common factor. In 1997 he offered a prize, later increased to $100,000, for a proof or disproof.

• •

Why No Nobel for Maths?

Why didn't Alfred Nobel set up a mathematics prize? There's a persistent story that Nobel's wife had an affair with the Swedish mathematician Gosta Mittag-Leffler, so Nobel hated mathematicians. But there's a problem with this theory, because Nobel never married. Some versions of the story replace the hypothetical wife with a fiancée or a mistress. Nobel may have had a mistress – a Viennese lady called Sophie Hess – but there's no evidence that she had anything to do with Mittag-Leffler.

An alternative theory holds that Mittag-Leffler, who became quite wealthy himself, did something to annoy Nobel. Since Mittag-Leffler was the leading Swedish mathematician of the time, Nobel realised that he was very likely to win a prize for mathematics, and decided not to set one up. However, in 1985 Lars Gårding and Lars Hörmander noted that Nobel left Sweden in 1865, to live in Paris, and seldom returned – and in 1865 Mittag-Leffler was a young student. So there was little opportunity for them to interact, which casts doubt on both theories.

It's true that late in Nobel's life, Mittag-Leffler was chosen to negotiate with him about leaving to the Stockholm Högskola (which later became the University) a significant amount of money in his will, and this attempt eventually failed – but presumably Mittag-Leffler wouldn't have been chosen if he'd already offended Nobel. In any case, Mittag-Leffler wasn't likely to win a mathematical Nobel if one existed – there were plenty of more prominent mathematicians around. So it seems more likely that it simply never occurred to Nobel to award a prize for mathematics, or that he considered the idea and rejected it, or that he didn't want to spend even more cash.

Despite this, several mathematicians and mathematical physicists have won the prize for work in other areas – physics, chemistry, physiology/medicine, even literature. They have also won the 'Nobel' in economics – the Prize in Economic Sciences in Memory of Alfred Nobel, established by the Sveriges Riksbank in 1968.

Is There a Perfect Cuboid?

It is easy to find rectangles whose sides and diagonals are whole numbers – this is the hoary old problem of Pythagorean triangles, and it has been known since antiquity how to find all of them (page 58). Using the classical recipe, it is not too hard to find a cuboid – a box with rectangular sides – such that its sides, and the diagonals of all its faces, are whole numbers. The first set of values given below achieves this. But what no one has yet been able to find is a *perfect* cuboid – one in which the 'long diagonal' between opposite corners of the cuboid is also a whole number.

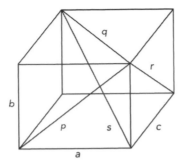

Make all lengths integers.

With the notation in the diagram, and bearing Pythagoras in mind, we have to find a, b and c so that all four of the numbers $a^2 + b^2, a^2 + c^2, b^2 + c^2$ and $a^2 + b^2 + c^2$ are perfect squares – equal, respectively, to p^2, q^2, r^2 and s^2. The existence of such numbers has neither been proved nor disproved, but some 'near misses' have been found:

> $a = 240, \quad b = 117, \quad c = 44, \quad p = 267, \quad q = 244, \quad r = 125,$
> but s is not an integer
> $a = 672, \quad b = 153, \quad c = 104, \quad q = 680, \quad r = 185, \quad s = 697,$
> but p is not an integer
> $a = 18,720, \quad b = 211,773,121, \quad c = 7,800, \quad p = 23,711,$
> $q = 20,280, \quad r = 16,511, \quad s = 24,961,$ but b is not an integer

If there is a perfect cuboid, it involves big numbers: it has been proved that the smallest edge is at least $2^{32} = 4,294,967,296$.

● ●

Paradox Lost

In mathematical logic, a *paradox* is a self-contradictory statement – the best known is 'This sentence is a lie.' Another is Bertrand Russell's 'barber paradox'. In a village there is a barber who shaves everyone who does not shave themselves. So who shaves the barber? Neither 'the barber' nor 'someone else' is logically acceptable. If it is the barber, then he shaves himself – but we are told that he doesn't. But if it's someone else, then the barber does not shave himself ... but we are told that he shaves all such people, so he does shave himself.

In the real world, there are plenty of get-outs (are we talking about shaving beards here, or legs, or what? Is the barber a woman? Can such a barber actually exist anyway?) But in mathematics, a more carefully stated version of Russell's paradox ruined the life's work of Gottlob Frege, who attempted to base the whole of mathematics on set theory – the study of collections of objects, and how these can be combined to form other collections.

Here's another famous (alleged) paradox:

Protagoras was a Greek lawyer who lived and taught in the fifth century BC. He had a student, and it was agreed that the student would pay for his teaching after he had won his first case. But the student didn't get any clients, and eventually Protagoras threatened to sue him. Protagoras reckoned that he would win either way: if the court upheld his case, the student would be required to pay up, but if Protagoras lost, then by their agreement the student would have to pay anyway. The student argued exactly the other way round: if Protagoras won, then by their agreement the student did *not* have to pay, but if Protagoras lost, the court would have ruled that the student did not have to pay.

Is this a genuine logical paradox or not?

Answer on page 279

• •

When Will My MP3 Player Repeat?

You have 1,000 songs on your MP3 player. If it plays songs 'at random', how long would you expect to wait before the same song is repeated?

It all depends on what 'at random' means. The leading MP3 player on the market 'shuffles' songs just like someone shuffling a pack of cards. Once the list has been shuffled, all songs are played in order. If you don't reshuffle, it will take 1,001 songs to get a repeat. However, it is also possible to pick a song at random, and keep repeating this procedure without eliminating that song. If so, the same song might – just – come up twice in a row. I'll assume that all songs appear with the same probability, though some MP3 players bias the choice in favour of songs you play a lot.

You've probably met the same problem with birthdays replacing songs. If you ask people their birthday, one at a time, then on average how many do you have to ask to get a repeat? The answer is 23, remarkably small. There is a second, super-ficially similar problem: how many people should there be at a party so that the probability that at least two share a birthday is bigger than $\frac{1}{2}$? Again the answer is 23. In both calculations we ignore leap years and assume that any particular birthday occurs with probability 1/365. This isn't quite accurate, but it simplifies the sums. We also assume that all individuals have statistically independent birthdays, which would not be the case if, say, the party included twins.

I'll solve the second birthday problem, because the sums are easier to understand. The trick is to imagine the people entering the room one at a time, and to work out, at each stage, the probability that all birthdays so far are *different*. Subtract the result from 1 and you get the probability that at least two are equal. So we want to continue allowing people to enter until the probability that all birthdays are different drops *below* $\frac{1}{2}$.

When the first person enters, the probability that their birthday is different from that of anyone else present is 1, because nobody else is present. I'll write that as the fraction

$$\frac{365}{365}$$

because it tells us that out of the 365 possible birthdays, all 365 have the required outcome.

When the second person enters, their birthday has to be

different, so there are now only 364 choices out of 365. So the probability we want is now

$$\frac{365}{365} \times \frac{364}{365}$$

When the third person enters, they have only 363 choices, and the probability of no duplication so far is

$$\frac{365}{365} \times \frac{364}{365} \times \frac{363}{365}$$

The pattern should now be clear. After k people have entered, the probability that all k birthdays are distinct is

$$\frac{365}{365} \times \frac{364}{365} \times \frac{363}{365} \times \cdots \times \frac{365 - k + 1}{365}$$

and we want the first k for which this is less than $\frac{1}{2}$. Each fraction, other than the first, is smaller than 1, so the probability decreases as k increases. Direct calculation shows that when $k = 22$ the fraction equals 0.524 305, and when k is 23 it equals 0.492 703. So the required number of people is 23.

This number seems surprisingly small, which may be because we confuse the question with a different one: how many people do you have to ask for the probability that one of them has the same birthday as *you* to be bigger than $\frac{1}{2}$? The answer to that is much bigger – in fact, it's 253.

The same calculation with 1,000 songs on an MP3 player shows that if each song is chosen at random then you have to play a mere 38 songs to make the probability of a repeat bigger than $\frac{1}{2}$. The *average* number of songs you have to play to get a repeat is 39 – slightly more.

These sums are all very well, but they don't provide much insight. What if you had a million songs? It's a big sum – a computer can do it, though. But is there a simpler answer? We can't expect an exact formula, but we ought to be able to find a good approximation. Let's say we have n songs. Then it turns out that on average we have to play approximately

$$\sqrt{(\tfrac{1}{2}\pi)}\sqrt{n} = 1.2533\sqrt{n}$$

songs to get a repeat (of *some* song already played, not necessarily the first one). To make the probability of a repeat greater than $\tfrac{1}{2}$, we have to play approximately

$$\sqrt{(\log 4)}\sqrt{n}$$

songs, which is

$$1.1774\sqrt{n}$$

This is about 6% smaller.

Both numbers are proportional to the square root of n, which grows much more slowly than n. This is why we get quite small answers when n is large. If you did have a million songs on your MP3 player, then on average you would have to play only 1,253 of them to get a repeat (the square root of a million is 1,000). And to make the probability of a repeat greater than $\tfrac{1}{2}$, you would have to play approximately 1,177 of them. The exact answer, according to my computer, is 1,178.

• •

Six Pens

Farmer Hogswill has run into another mathematico-agricultural problem. He had carefully assembled 13 identical fence panels to create 6 identical pens for his rare-breed Alexander-horned pigs. But during the night some antisocial person stole one of his panels. So now he needs to use 12 fence panels to create 6 identical pens. How can he achieve this? All 12 panels must be used.

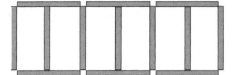

13 panels
making 6 pens.

Answer on page 280

. . . .

Patented Primes

Because of their importance in encryption algorithms, prime numbers have commercial significance. In 1994 Roger Schlafly obtained US Patent 5,373,560 on two primes. The patent states them as hexadecimal (base-16) numbers, but I've converted them into decimal. They are:

> 7,994,412,097,716,110,548,127,211,733,331,600,522,933,
> 776,757,046,707,649,963,673,962,686,200,838,432,950,239,
> 103,981,070,728,369,599,816,314,646,482,720,706,826,018,
> 360,181,196,843,154,224,748,382,211,019

and

> 103,864,912,054,654,272,074,839,999,186,936,834,171,066,
> 194,620,139,675,036,534,769,616,693,904,589,884,931,513,
> 925,858,861,749,077,079,643,532,169,815,633,834,450,952,
> 832,125,258,174,795,234,553,238,258,030,222,937,772,878,
> 346,831,083,983,624,739,712,536,721,932,666,180,751,292,
> 001,388,772,039,413,446,493,758,317,344,413,531,957,900,
> 028,443,184,983,069,698,882,035,800,332,668,237,985,846,
> 170,997,572,388,089

He did this to publicise deficiencies in the US patent system.

Legally, you can't use these numbers without Schlafly's permission. Hmmm ...

. . . .

The Poincaré Conjecture

Towards the end of the nineteenth century, mathematicians succeeded in finding all possible 'topological types' of surfaces. Two surfaces have the same topological type if one of them can be continuously deformed into the other. Imagine that the surface is made from flexible dough. You can stretch it, squeeze it, twist it – but you can't tear it, or squash different bits together.

To keep the story simple, I'll assume that the surface has no boundary, that it's orientable (two-sided, unlike the Möbius band) and that it's of finite extent. The nineteenth-century mathematicians proved that every such surface is topologically equivalent to a sphere, a torus, a torus with two holes, a torus with three holes, and so on.

Sphere.   Torus.   Two-holed torus.

'Surface' here really does refer only to the *surface*. A topologist's sphere is like a balloon – an infinitely thin sheet of rubber. A torus is shaped like an inner tube for a tyre (for those of you who know what an inner tube is). So the 'dough' I just mentioned is really a very thin sheet, not a solid lump. Topologists call a solid sphere a 'ball'.

To achieve their classification of all surfaces, the topologists had to characterise them 'intrinsically', without reference to any surrounding space. Think of an ant living on the surface, ignorant of any surrounding space. How can it work out which surface it inhabits? By 1900 it was understood that a good way to answer such questions is to think about closed loops in the surface, and how these loops can be deformed. For example, on a sphere (by which I mean just the surface, not the solid interior) any closed loop can be continuously deformed to a point – 'shrunk'. For example, the circle running round the equator can be gradually moved towards the south pole, becoming smaller and smaller until it coincides with the pole itself:

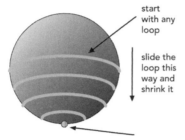

start with any loop

slide the loop this way and shrink it

How to shrink a loop on a sphere continuously to a point.

In contrast, every surface that is not equivalent to a sphere contains loops, which cannot be deformed to points. Such loops 'pass through a hole', and the hole prevents them from being shrunk. So the sphere can be characterised as the *only* surface on which any closed loop can be shrunk to a point.

Observe, however, that the 'hole' that we see in a picture is not actually part of the surface. By definition, it's a place where the surface *isn't*. If we think intrinsically, we can't talk sensibly about holes if we try to visualise them in the usual manner. The ant who lives on the surface and knows no other universe can't see that his torus has a dirty great hole in it – any more than we can look along a fourth dimension. So although I'm using the word 'hole' to explain why the loop can't be shrunk, a topological proof runs along different lines.

a loop like this one can't be shrunk

On all other surfaces, loops can get stuck.

In 1904 Henri Poincaré was trying to take the next step, and understand 'manifolds'– three-dimensional analogues of surfaces – and for a time he assumed that the characterisation of a sphere in terms of shrinking loops is also true in three dimensions, where there is a natural analogue of the sphere called the 3-sphere. A 3-sphere is *not* just a solid ball, but it can be visualised – if that's the word – by taking a solid ball and pretending that its entire surface is actually just a single point.

Imagine doing the same with a circular disc. The rim closes up like the top of a bag as you draw a string tight round its edge, and the result is topologically a sphere. Now go up a dimension ...

take a disc...   ...scrunge the edge together...   ...and you get a sphere   Turning a disc into a sphere.

At first, Poincaré thought that this characterisation of the 3-sphere should be obvious, or at least easily proved, but later he realised that one plausible version of this statement is actually wrong, while another closely related formulation seemed difficult to prove but might well be true. He posed a deceptively simple question: if a three-dimensional manifold (without boundary, of finite extent, and so on) had the property that any loop in it can be shrunk to a point, must that manifold be topologically equivalent to the 3-sphere?

Subsequent attempts to answer this question failed dismally, although after a huge effort by the world's topologists, the answer has proved to be 'yes' for all versions in every dimension *higher* than 3. The belief that the same answer applies in three dimensions became known as the *Poincaré Conjecture*, famous as one of the eight Millennium Prize Problems (page 127).

In 2002 a Russian-born mathematician, Grigori Perelman, caused a sensation by placing several papers on arXiv.org, an informal website for current research in physics and mathematics. His papers were ostensibly about various properties of the 'Ricci flow', but it became clear that if the work was correct, it implied that the Poincaré Conjecture is also correct. The idea of using the Ricci flow dates to 1982, when Richard Hamilton introduced a new technique based on mathematical ideas used by Albert Einstein in general relativity. According to Einstein, spacetime can be considered as curved, and the curvature describes the force of gravity. Curvature is measured by something called the 'curvature tensor', and this has a simpler relative known as the 'Ricci tensor' after its inventor, Gregorio Ricci-Curbastro.

According to general relativity, gravitational fields can change the geometry of the universe as time passes, and these changes are governed by the Einstein equations, which say that the stress tensor is proportional to the curvature. In effect, the gravitational bending of the universe tries to smooth itself out as time passes, and the Einstein equations quantify that idea.

The same game can be played using the Ricci version of curvature, and it leads to the same kind of behaviour: a surface that obeys the equations for the Ricci flow will naturally tend to simplify its own geometry by redistributing its curvature more evenly. Hamilton showed that the familiar two-dimensional version of the Poincaré Conjecture, characterising the sphere, can be proved using the Ricci flow. Basically, a surface in which all loops shrink simplifies itself so much as it follows the Ricci flow that it ends up being a perfect sphere. Hamilton suggested generalising this approach to three dimensions, but he hit some difficult obstacles.

The main complication in three dimensions is that 'singularities' can develop, where the manifold pinches together and the flow breaks down. Perelman's new idea was to cut the surface apart near such a singularity, cap off the resulting holes, and then allow the flow to continue. If the manifold manages to simplify itself completely after only finitely many singularities have arisen, then not only is the Poincaré Conjecture true, but a more far-reaching result, the Thurston Geometrisation Conjecture, is also true. And that tells us about *all possible* three-dimensional manifolds.

Now the story takes a curious turn. It is generally accepted that Perelman's work is correct, although his arXiv papers leave a lot of gaps that have to be filled in correctly, and that has turned out to be quite difficult. Perelman had his own reasons for not wanting the prize – indeed, any reward save the solution itself – and decided not to expand his papers into something suitable for publication, although he was generally willing to explain how to fill in various details if anyone asked him. Experts in the area were forced to develop their own versions of his ideas.

Perelman was also awarded a Fields Medal at the Madrid International Congress of Mathematicians in 2006, the top prize in mathematics. He turned that down, too.

●●

Hippopotamian Logic

I won't eat my hat.
If hippos don't eat acorns, then oak trees will grow in Africa.
If oak trees don't grow in Africa, then squirrels hibernate in winter.
If hippos eat acorns and squirrels hibernate in winter, then I'll eat my hat.
Therefore – *what?*

Answer on page 280

●●

Langton's Ant

Langton's ant was invented by Christopher Langton, and it shows how amazingly complex simple ideas can be. It leads to one of the most baffling unsolved problems in the whole of mathematics, and all from astonishingly simple ingredients.

The ant lives on an infinite square grid of black and white cells, and it can face in one of the four compass directions: north, south, east or west. At each tick of a clock it moves one cell forward, and then follows three simple rules:

- If it lands on a black cell it makes a 90° turn to the left.
- If it lands on a white cell it makes a 90° turn to the right.
- The cell that it has just vacated then changes colour, from white to black, or vice versa.

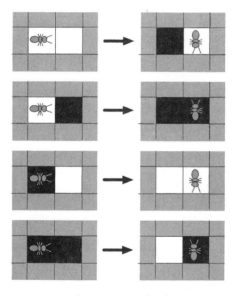

Effect of the ant moving. Grey cells can be any colour and do not change on this move.

As a warm-up, the ant starts by facing east on a completely white grid. Its first move takes it to a white square, while the square it started from turns black. Because it is on a white square, the ant's next move is a right turn, so now it faces south. That takes it to a new white square, and the square it has just vacated turns black. After a few more moves the ant starts to revisit earlier squares that have turned black, so it then turns to the left instead. As time passes, the ant's motion gets quite complicated, and so does the ever-changing pattern of black and white squares that trails behind it.

Jim Propp discovered that the first few hundred moves occasionally produce a nice, symmetrical pattern. Then things get rather chaotic for about ten thousand moves. After that, the ant gets trapped in a cycle in which the same sequence of 104 moves is repeated indefinitely, each cycle moving it two squares diagonally. It continues like this for ever, systematically building a broad diagonal 'highway'.

Langton's ant builds a highway.

This 'order out of chaos' behaviour is already puzzling, but computer experiments suggest something more surprising. If you scatter any finite number of black squares on the grid, before the ant sets off, it *still* ends up building a highway. It may take longer to do so, and its initial movements may be very different, but ultimately that's what will happen. As an example, the second diagram shows a pattern that forms when the ant starts inside a solid rectangle. Before building its highway, the ant builds a 'castle' with straight walls and complicated crenellations. It keeps destroying and rebuilding these structures in a curiously purposeful way, until it gets distracted and wanders off – building a highway.

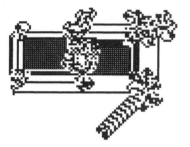

Pattern created by Langton's ant when it starts inside a black rectangle. The highway is at the lower right. Small white dots mark squares of the original rectangle that have never been visited.

The problem that is baffling mathematicians is to prove that the ant *always* ends up building a highway, for every initial configuration of finitely many black squares. Or disprove that, if it's wrong. We do know that the ant can never get trapped inside any bounded region of the grid – it always escapes if you wait long enough. But we don't know that it escapes along a highway.

• •

Pig on a Rope

Farmer Hogswill owns a field, which is a perfect equilateral triangle, each side 100 metres long. His prize pig Pigasus is tied to one corner, so that the portion of the field that Pigasus can reach is exactly half the total area. How long is the rope?

You may – indeed, must – assume that the pig has zero size (which admittedly is pretty silly) and that the rope is infinitely thin and any necessary knots can be ignored.

Pigs may safely graze . . . over half the area of the field.

Answer on page 280

• •

The Surprise Examination

This paradox is so famous that I nearly left it out. It raises some intriguing issues.

Teacher tells the class that there will be a test one day next week (Monday to Friday), and that it will be a surprise. This seems reasonable: the teacher can choose any day out of five, and there is no way that the students can know which day it will be. But the students don't see things that way at all. They reason that the test can't be on Friday – because if it was, then as soon as Thursday passed without a test, they'd know it had to be Friday, so no surprise. And once they've ruled out Friday, they apply the same reasoning to the remaining four days of the week, so the test can't be on Thursday, either. In which case it can't be on Wednesday, so it can't be on Tuesday, so it can't be on Monday. Apparently, no surprise test is possible.

That's all very well, but if the teacher decides to set the test on Wednesday, there seems to be no way that the students could actually *know* the day ahead of time! Is this a genuine paradox or not?

Answer on page 281

Answer on page 281

Antigravity Cone

In defiance of the Law of Gravity, this double cone *rolls uphill.* Here's how to make it.

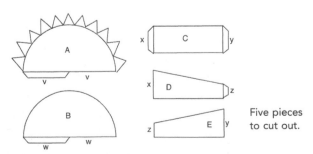

Five pieces to cut out.

Copy the five shapes on to a thin sheet of card, two or three times the size shown here, and cut them out. On piece A, glue flap v to edge v to make a cone. On piece B, glue flap w to edge w to make a second cone. Then glue the two cones base to base using the triangular flaps on A.

Glue flap x of C to edge x of D, and flap y of C to edge y of E. Finally, glue flap z of D to edge z of E to make a triangular 'fence'.

Place the double cone at the lower end of this triangle, and let go. It will appear to roll uphill.

How can this happen?

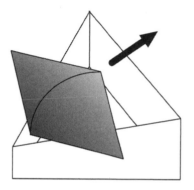

A rolling cone gathers no gravity.

Answer on page 282

Mathematical Jokes 2

An engineer, a physicist, and a mathematician are staying in a hotel. The engineer wakes up and smells smoke. He goes into the hallway, sees a fire, fills the wastepaper basket from his room with water, and pours it on the fire, putting it out.

Later, the physicist wakes up and smells smoke. He goes into the hallway and sees a (second) fire. He pulls a fire hose off the wall. Having calculated the temperature of the exothermic reaction, the velocity of the flame front, the water pressure in the hose, and so on, he uses the hose to put out the fire with the minimum expenditure of energy.

Later, the mathematician wakes up and smells smoke. He goes into the hallway and sees a (third) fire. He notices the fire hose on the wall, and thinks for a moment ... Then he says, 'OK, a solution exists!' – and goes back to bed.

. . . .

Why Gauss Became a Mathematician

Carl Friedrich Gauss.

Carl Friedrich Gauss was born in Brunswick in 1777 and died in Göttingen in 1855. His parents were uneducated manual workers, but he became one of the greatest mathematicians ever; many consider him the best. He was precocious – he is said to have pointed out a mistake in his father's financial calculations when he was three. At the age of nineteen he had to decide whether to study mathematics or languages, and the decision was made for him when he discovered how to construct a regular 17-sided polygon using the traditional Euclidean tools of an unmarked ruler and a compass.

This may not sound like much, but it was totally unprece-dented, and the discovery led to a new branch of number theory. Euclid's *Elements* contains constructions for regular polygons (all sides equal length, all angles equal) with 3, 4, 5, 6 and 15 sides, and the ancient Greeks knew that the number of sides could be doubled as often as you wish. Up to 100, the number of sides in a constructible (regular) polygon – as far as the Greeks knew – must be

2, 3, 4, 5, 6, 8, 10, 12, 15, 16, 20, 24, 30, 32, 40, 48, 60, 64, 80, 96

For more than two thousand years, everyone assumed that no other polygons were constructible. In particular, Euclid does not tell us how to construct 7-gons or 9-gons, and the reason is that he had no idea how this might be done. Gauss's discovery was a bombshell, adding 17, 34 and 68 to the list. Even more amazingly, his methods prove that other numbers, such as 7, 9, 11 and 13, are impossible. (The polygons do exist, but you can't construct them by Euclidean methods.)

Gauss's construction depends on two simple facts about the number 17: it is prime, and it is one greater than a power of 2. The whole problem pretty much reduces to finding which prime numbers correspond to constructible polygons, and powers of 2 come into the story because every Euclidean construction boils down to taking a series of square roots – which in particular implies that the lengths of any lines that feature in the construction must satisfy algebraic equations whose degree is a power of two. The key equation for the 17-gon is

$$x^{16} + x^{15} + x^{14} + x^{13} + x^{12} + x^{11} + x^{10} + x^9 + x^8 + x^7 + x^6$$
$$+ x^5 + x^4 + x^3 + x^2 + x + 1 = 0$$

where x is a *complex* number. The 16 solutions, together with the number 1, form the vertices of a regular 17-gon in the complex plane. Since 16 is a power of 2, Gauss realised that he was in with a chance. He did some clever calculations, and proved that the

17-gon can be constructed provided you can construct a line whose length is

$$\frac{1}{16}\left[-1+\sqrt{17}+\sqrt{34-2\sqrt{17}}+\right.$$
$$\left.\sqrt{68+12\sqrt{17}-16\sqrt{34+2\sqrt{17}}-2(1-\sqrt{17})(\sqrt{34-2\sqrt{17}})}\right]$$

Since you can always construct square roots, this effectively solves the problem, and Gauss didn't bother to describe the precise steps needed – the formula itself does that. Later, other mathematicians wrote down explicit constructions. Ulrich von Huguenin published the first in 1803, and H.W. Richmond found a simpler one in 1893.

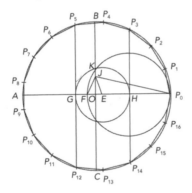

Richmond's method for constructing a regular 17-gon. Take two perpendicular radii, AOP_0 and BOC, of a circle. Make $\frac{OJ}{4OB}=1$ and angle $\frac{OJE}{4OJP_0}=1$. Find F Such that angle EJF is 45°. Draw a circle with FP_0 as diameter, meeting OB at K. Draw the circle with centre E through K, cutting AP_0 in G and H. Draw HP_3 and GP_5 perpendicular to AP_0. Then P_0, P_3 and P_5 are respectively the 0th, 3rd and 5th vertices of a regular 17-gon, and the other vertices are now easily constructed.

Gauss's method proves that a regular n-gon can be constructed whenever n is a prime of the form $2^k + 1$. Primes like this are called *Fermat primes*, because Fermat investigated them. In particular he noticed that k must itself be a power of 2 if $2^k + 1$ is going to be prime. The values $k = 1, 2, 4, 8$ and 16 yield the Fermat primes 3, 5, 17, 257 and 65,537. However, $2^{32} + 1 = 4{,}294{,}967{,}297 = 641 \times 6{,}700{,}417$ is not prime. Gauss was aware that the regular n-gon is constructible if and only if n is a power of 2, or a power of 2 multiplied by *distinct* Fermat primes. But he didn't give a complete proof – probably because to him it was obvious.

His results prove that it is impossible to construct regular 7-, 11- or 13-gons by Euclidean methods, because these are prime but not of Fermat type. The analogous equation for the 7-gon, for instance, is $x^6 + x^5 + x^4 + x^3 + x^2 + x + 1 = 0$, and that has degree 6, which is not a power of 2. The 9-gon is not constructible because 9 is not a product of distinct Fermat primes – it is 3×3, and 3 is a Fermat prime, but the same prime occurs twice here.

The Fermat primes just listed are the only *known* ones. If there is another, it must be absolutely gigantic: in the current state of knowledge the first candidate is $2^{33{,}554{,}432} + 1$, where $33{,}554{,}432 = 2^{25}$. Although we're still not sure exactly which regular polygons are constructible, the only obstacle is the possible existence of very large Fermat primes. A useful website for Fermat primes is mathworld.wolfram.com/FermatNumber.html

In 1832 Friedrich Julius Richelot published a construction for the regular 257-gon. Johann Gustav Hermes of Lingen University devoted ten years to the 65,537-gon, and his unpublished work can be found at the University of Göttingen, but it probably contains errors.

With more general construction techniques, other numbers are possible. If you use a gadget for trisecting angles, then the 9-gon is easy. The 7-gon turns out to be possible too, but that's nowhere near as obvious.

• •

What Shape is a Crescent Moon?

The Moon is low in the sky shortly after sunset or before dawn; the bright part of its surface forms a beautiful crescent. The two curves that form the boundary of the crescent resemble arcs of circles, and are often drawn that way. Assuming the Moon to be a perfect sphere, and the Sun's rays to be parallel, *are* they arcs of circles?

A crescent formed by two arcs of circles. Is the crescent Moon like this?

Answer on page 283

Famous Mathematicians

All the people listed below – except one – either started a degree (or joint degree) in mathematics, or studied under famous mathematicians, or were professional mathematicians in their other life. What are they famous for? Which person does not belong on the list?

Pierre Boulez
Sergey Brin
Lewis Carroll
J.M. Coetzee
Alberto Fujimori
Art Garfunkel
Philip Glass
Teri Hatcher
Edmund Husserl

Michael Jordan
Theodore Kaczynski
John Maynard Keynes
Carole King
Emanuel Lasker
J.P. Morgan
Larry Niven
Alexander Solzhenitsyn
Bram Stoker

Leon Trotsky
Eamon de Valera
Carol Vorderman
Answers on page 284

Virginia Wade
Ludwig Wittgenstein
Sir Christopher Wren

• •

What is a Mersenne Prime?

A *Mersenne number* is a number of the form $2^n - 1$. That is, it is one less than a power of 2. A *Mersenne prime* is a Mersenne number that happens also to be prime. It is straightforward to prove that in this case the exponent n must itself be prime. For the first few primes, $n = 2, 3, 5$ and 7, the corresponding Mersenne numbers 3, 7, 31 and 127 are all prime.

Interest in Mersenne numbers goes back a long way, and initially it was thought that they are prime whenever n is prime. However, in 1536 Hudalricus Regius proved that this assumption is false, pointing out that $2^{11} - 1 = 2,047 = 23 \times 89$. In 1603 Pietro Cataldi noted that $2^{17} - 1$ and $2^{19} - 1$ are prime, which is correct, and claimed that $n = 23, 29, 31$ and 37 also lead to primes. Fermat proved that he was wrong for 23 and 37, and Euler demolished his claim for 29. But Euler later proved that $2^{31} - 1$ is prime.

In his 1644 book *Cogitata Physico-Mathematica*, the French monk Marin Mersenne stated that $2^n - 1$ is prime when n is 2, 3, 5, 7, 13, 17, 19, 31, 67, 127 and 257 – and for no other values in that range. Using the methods then available, he could not have tested most of these numbers, so his claims were mainly guesswork, but his name became associated with the problem.

In 1876 Édouard Lucas developed a cunning way to test Mersenne numbers to see if they are prime, and showed that Mersenne was right for $n = 127$. By 1947 all cases in Mersenne's range had been checked, and it turned out that he had mistakenly included 67 and 257. He had also omitted 61, 89 and

107. Lucas improved his test, and in the 1930s Derrick Lehmer found further improvements. The Lucas–Lehmer test uses the sequence of numbers

$$4, \ 14, \ 194, \ 37634, \ \ldots$$

in which each number is the square of the previous one, decreased by 2. It can be proved that the nth Mersenne number is prime if and only if it divides the $(n-1)$th term of this sequence. This test can prove that a Mersenne number is composite without finding any of its prime factors, and it can prove the number is prime without testing for any prime factors. There's a trick to keep all numbers involved in the test smaller than the Mersenne number concerned.

Looking for new, larger Mersenne primes is an amusing way to try out new, fast computers, and over the years prime-hunters have extended the list. It now includes 44 primes:

n	Year	Discoverer
2	—	known from antiquity
3	—	known from antiquity
5	—	known from antiquity
7	—	known from antiquity
13	1456	anonymous
17	1588	Pietro Cataldi
19	1588	Pietro Cataldi
31	1772	Leonhard Euler
61	1883	Ivan Pervushin
89	1911	R.E. Powers[*]
107	1914	R.E. Powers
127	1876	Édouard Lucas
521	1952	Raphael Robinson
607	1952	Raphael Robinson
1,279	1952	Raphael Robinson
2,203	1952	Raphael Robinson
2,281	1952	Raphael Robinson
3,217	1957	Hans Riesel

[*] Powers is a rather obscure, possibly amateur, mathematician. I haven't been able to locate his first name.

4,253	1961	Alexander Hurwitz
4,423	1961	Alexander Hurwitz
9,689	1963	Donald Gillies
9,941	1963	Donald Gillies
11,213	1963	Donald Gillies
19,937	1971	Bryant Tuckerman
21,701	1978	Landon Noll and Laura Nickel
23,209	1979	Landon Noll
44,497	1979	Harry Nelson and David Slowinski
86,243	1982	David Slowinski
110,503	1988	Walter Colquitt and Luther Welsh
132,049	1983	David Slowinski
216,091	1985	David Slowinski
756,839	1992	David Slowinski *et al.*
859,433	1994	David Slowinski and Paul Gage
1,257,787	1996	David Slowinski and Paul Gage
1,398,269	1996	Joel Armengaud *et al.*
2,976,221	1997	Gordon Spence *et al.*
3,021,377	1998	Roland Clarkson *et al.*
6,972,593	1999	Nayan Hajratwala *et al.*
13,466,917	2001	Michael Cameron *et al.*
20,996,011	2003	Michael Shafer *et al.*
24,036,583	2004	Josh Findley *et al.*
25,964,951	2005	Martin Nowak *et al.*
30,402,457	2005	Curtis Cooper *et al.*
32,582,657	2006	Curtis Cooper *et al.*
37,156,667	2008	Hans-Michael Elvenich
43,112,609	2008	Edson Smith

Up to and including the 39th Mersenne prime ($n = 13,466,917$) the list is complete, but there may be undiscovered Mersenne primes in the gaps between the known ones after that. The 46th known Mersenne prime, $2^{43,112,609} - 1$, has 12,978,189 decimal digits and is currently (November 2008) the largest known prime. Mersenne primes generally hold this record, thanks to the Lucas–Lehmer test; however, we know from Euclid that there is no largest prime. For up-to-date information, go to the Mersenne Primes website primes.utm.edu/mersenne/; you can also join the Great Internet Mersenne Prime Search (GIMPS) at www.mersenne.org/

The Goldbach Conjecture

In 2000, as a publicity stunt for Apostolos Doxiadis's novel *Uncle Petros and Goldbach's Conjecture*, the publisher Faber & Faber offered a million-dollar prize for a proof of the conjecture, provided it was submitted before April 2002. The prize was never claimed, which mathematicians did not find surprising, because the problem has resisted all efforts for more than 250 years.

It began in 1742, when Christian Goldbach wrote to Leonhard Euler, suggesting that every even integer is the sum of two primes. (Apparently René Descartes had come across the same idea a little earlier, but no one had noticed.) At that time the number 1 was considered to be prime, so $2 = 1 + 1$ was acceptable, but nowadays we reformulate the the *Goldbach Conjecture* thus: every even integer greater than 2 is the sum of two primes – often in several different ways. For example,

$$4 = 2 + 2$$
$$6 = 3 + 3$$
$$8 = 5 + 3$$
$$10 = 7 + 3 = 5 + 5$$
$$12 = 7 + 5$$
$$14 = 11 + 3 = 7 + 7$$

Euler replied that he was sure Goldbach must be right, but he couldn't find a proof – and that remains true today. We do know that every even integer is the sum of at most six primes – proved by Olivier Ramaré in 1995. In 1973 Chen Jing-Run proved that every sufficiently large even integer is the sum of a prime and a semiprime (either a prime or a product of two primes).

In 1998 Jean-Marc Deshouillers, Yannick Saouter and Herman te Riele verified Goldbach's Conjecture for all even numbers up to 10^{14}. By 2007, Oliveira e Silva had improved that to 10^{18}, and his computations continue. If the Riemann Hypothesis (page 215) is true, then the Odd Goldbach Conjecture – that every odd integer greater than 5 is the sum of three primes – is a consequence of the 1998 result.

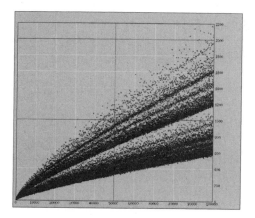

Graph showing in how many ways (vertical axis) a given even number (horizontal axis) can be expressed as a sum of two primes. The lowest points in the graph move upwards as we go from left to right, indicating that there are many ways to achieve this. However, for all we know an occasional point might fall on the horizontal axis. Just one such point would disprove the Goldbach Conjecture.

In 1923 Godfrey Hardy and John Littlewood obtained a heuristic formula – one that they could not prove rigorously, but looked plausible – for the number of different ways to write a given even integer as a sum of two primes. This formula, which agrees with numerical evidence, indicates that when the number gets large, there are many ways to write it as a sum of two primes. Therefore we may expect the smallest of the two primes to be relatively tiny. In 2001 Jörg Richstein observed that for numbers up to 10^{14}, the smaller prime is at most 5,569, and this occurs for

$$389,965,026,819,938 = 5,569 + 389,965,814,369$$

• •

Turtles All the Way Down

Infinity is a slippery idea. People talk fairly casually of 'eternity' – an infinite period of time. According to the Big Bang theory, the universe came into being about 13 billion years ago. Not only was there no universe before then – there was no 'before' before then.* Some people worry about that, and most of them seem much happier with the idea that the universe 'has always existed'. That is, its past has already been infinitely long.

This alternative seems to solve the difficult question of the origin of the universe, by denying that it ever had an origin. If something has always been here, it's silly to ask why it's here now. Isn't it?

Probably. But that still doesn't explain *why it's always been here*.

This can be a difficult point to grasp. To bring it into perspective, let me compare it with a rather different proposal. There is an amusing (and very likely true) tale that a famous scientist – Stephen Hawking is often mentioned because he told the story in *A Brief History of Time* – was giving a lecture about the universe, and a lady in the audience pointed out that the Earth floats in space because it rests on the back of four elephants, which in turn rest on the back of a turtle.

'Ah, but what supports the turtle?' the scientist asked.

'Don't be silly,' she said. *'It's turtles all the way down!'*

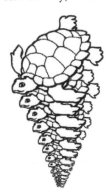

Turtles all the way down.

---

* Some cosmologists now think that there could have been something before the Big Bang after all – our universe may be part of a 'multiverse' in which individual universes could come into existence or fade away again. The theory is nice, but it's difficult to find any way to test it.

All very amusing, and we don't buy that explanation. A self-supporting pile of turtles is ludicrous, and not just because it's turtles. Each turtle being supported by a previous one just doesn't look like an explanation of how *the whole pile* stays up.

Very well. But now replace the Earth by the present state of the universe, and replace each turtle by the previous state of the universe. Oh, and change 'support' to 'cause'. Why does the universe exist? Because a previous one did. Why did that one exist? Because a previous one did. Did it all start a finite time in the past? No, it's *universes all the way back.**

So a universe that has always existed is at least as puzzling as one that has not.

● ●

Hilbert's Hotel

Among the paradoxes concerning the infinite are a series of bizarre events at Hilbert's Hotel. David Hilbert was one of the world's leading mathematicians around 1900. He worked in the logical foundations of mathematics and took a particular interest in infinity. Anyway, Hilbert's Hotel has infinitely many rooms, numbered 1, 2, 3, 4, and so on – every positive integer.

One bank holiday weekend, the hotel was completely full. A traveller without a reservation arrived at reception wanting a room. In any finite hotel, no matter how big, the traveller would be out of luck – but not in Hilbert's Hotel.

'No problem, sir,' said the manager. 'I'll ask the person in Room 1 to move to Room 2, the person in Room 2 to move to Room 3, the person in Room 3 to move to Room 4, and so on. The person in Room n will move to Room $n + 1$. Then Room 1 will be free, so I'll put you there.'

---

* Part of the appeal of the multiverse approach is that it revives the 'it's always been here' point of view. Our universe hasn't, but the surrounding multiverse has. It's multiverses all the way back ...

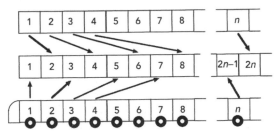

All move up one, and Room 1 is free.

This trick works in an infinite hotel. In a finite hotel it goes wrong, because the person in the room with the biggest number has nowhere to go. But in Hilbert's Hotel there is *no* biggest room number. Problem sorted.

Ten minutes later, an Infinity Tours coach arrived, with infinitely many passengers sitting in seats 1, 2, 3, 4, and so on.

'Well, I can't fit you in by asking every other guest to move up some number of places,' said the manager. 'Even if they all moved up a million places, that would only free up a million rooms.' He thought for a moment. 'Nevertheless, I can still fit you in. I'll ask the person in Room 1 to move to Room 2, the person in Room 2 to move to Room 4, the person in Room 3 to move to Room 6, and so on. The person in Room n will move to Room $2n$. That frees up all the odd-numbered rooms, so now I can put the person in Seat 1 of your bus into Room 1, the person in Seat 2 into Room 3, the person in Seat 3 into Room 5, and so on. The person in Seat n will move to Room $2n - 1$.'

How to accommodate an infinite bus-load.

However, the manager's troubles were still not over. Ten minutes later, he was horrified to see infinitely many Transfinity Travel buses arriving in his (infinite) car park.

He rushed out to meet them. 'We're full – but I can *still* fit you all in!'

'How?' asked the driver of Bus 1.

'I'll reduce you to a problem I've already solved,' said the manager. 'I want you to move everyone into Bus 1.'

'But Bus 1 is full! And there are infinitely many other buses!'

'No problem. Line up all your buses side by side, and renumber all the seats using a diagonal order.'

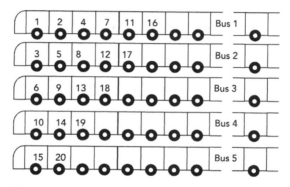

The Manager's 'diagonal' order – the numbers 2–3, 4–5–6, 7–8–9–10, and so on slant to the left.

'What does that achieve?' asked the driver.

'Nothing – yet. But notice: each passenger, in each of your infinitely many buses, is assigned a new number. Every number occurs exactly once.'

'And your point is—?'

'Move each passenger to the seat in Bus 1 that corresponds to their new number.'

The driver did so. Then everyone was sitting in Bus 1, and all the other buses were empty – so they drove away.

'Now I've got a full hotel and just one extra bus-load,' said the manager. 'And I already know how to deal with that.'

Continuum Coaches

You won't be surprised to hear that the Hilbert's Hotel eventually ran into an accommodation problem that the manager could *not* solve. This time the hotel was completely empty – not that this ever seemed to make much difference. Then one of Cantor's Continuum Coaches stopped at the front door.

Georg Cantor was the first to sort out the mathematics of infinite sets. And he discovered something remarkable about the 'continuum' – the real number system. A *real number* is one that can be written as a decimal, which can either stop after finitely many digits, like 1.44, or go on for ever, like π. Here's what Cantor found.

The seats of the Continuum Coach were numbered using real numbers, not positive integers.

'Well,' the manager thought, 'one infinity is just like any other, right?' So he assigned passengers to rooms, and eventually the Hotel was full and the lobby was empty. The manager sighed with relief. 'Everyone has a room,' he said to himself.

Then a forlorn figure came in through the revolving doors.

'Good evening,' said the manager.

'My name is Mr Diagonal. Geddit? Missed-a-diagonal. You've missed me out, mate.'

'Well, I can always bump everyone up one room—'

'No, mate, you said "Everyone has a room" – I heard you. But I don't.'

'Nonsense! You've gone to your room, then nipped out the back and come in the front. I know your kind!'

'No, mate – I can *prove* I'm not in any of your rooms. Who's in Room 1?'

'I can't reveal personal information about guests.'

'What's the first decimal place of their coach seat?'

'I suppose I can reveal that. It's a 2.'

'My first digit is 3. So I'm not the person in Room 1, mate. Agreed?'

'Agreed.'

'What's the *second* decimal place of the coach seat of the person in Room 2?'

'It's a 7.'

'My second digit is 5. So I'm not the person in Room 2.'

'That makes sense.'

'Yeah, mate, and it goes on doing that. What's the *third* decimal place of the coach seat of the person in Room 3?'

'It's a 4.'

'My third digit is 8. So I'm not the person in Room 3.'

'Hmm. I think I see where this is headed.'

'Too right, mate. My nth digit is different from the nth digit of the person in Room n, *for every n*. So I'm not in Room n. Like I said, you left me out.'

'And like *I* said, I can always bump everyone up one place and fit you in.'

'No use, mate. There's infinitely many more just like me out there, sitting in your car park waiting for a room. However you assign passengers to rooms, there's going to be someone on the coach whose nth digit is different from the nth digit of the person in Room n, for every n. Hordes of them, in fact. You'll always miss people out.'

Now, you understand that Cantor didn't quite write his proof in those terms, but that was the basic idea. He proved that the infinite set of real numbers can't be matched, one for one, with the infinite set of whole numbers. Some infinities are bigger than others.

• •

A Puzzling Dissection

'Why are you hacking that chessboard to bits?' asked Innumeratus.

'I want to show you something about areas,' said Mathophila. 'What's the area of the chessboard if each square has area one square unit?'

Innumeratus thought about this, and because he was better

at maths than his name might suggest, he quickly said, 'It's 8 times 8, which is 64 square units.'

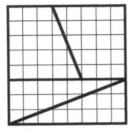

How Mathophila cut up her chessboard ...

'Excellent!' said Mathophila. 'Now, I'm going to rearrange the four pieces to make a rectangle.'

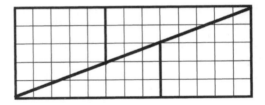

... and how she rearranged the pieces.

'OK,' said Innumeratus.

'What's the area of the rectangle?'

'Er – it must be 64 square units as well! It's made from the same pieces.'

'Right ... but what size is the rectangle?'

'Let me see – 13 by 5.'

'And what is 13 times 5?'

'65,' replied Innumeratus. He paused. 'So its area must be 65 square units. That's strange. The area can't change when the pieces are reassembled in a different way ...'

So what's happened?

Answer on page 286

A *Really* Puzzling Dissection

'The area can't change when the pieces are reassembled in a different way.'

Hmmm.

In 1924 two Polish mathematicians, Stefan Banach and Alfred Tarski, proved that it is possible to dissect a sphere into finitely many pieces, which can then be rearranged to make two spheres – each the same size as the original. No overlaps, no missing bits – the pieces fit together perfectly. This result has become known as the *Banach–Tarski paradox*, although it's a perfectly valid theorem and the only element of paradox is that it seems to be obviously false.

It *can* be done – but not with pieces like these.

Hang on, though. Surely, if you cut a sphere into several pieces, the total volume of the pieces must be the same as that of the sphere. So however you reassemble the pieces, the total volume can't change. But two identical spheres have twice the volume of a single sphere (of the same size). You don't have to be a genius to see that it can't be done! In fact, if it *could* be done, then you could start with a gold sphere, cut it up, fit the pieces back together, and end up with twice as much gold. Then repeat ... But you can't get something for nothing.

Hang on, though. Let's not be so hasty.

The argument about gold is inconclusive, because mathematical concepts do not always model the real world exactly. In mathematics, volumes can be subdivided into indefinitely small pieces. In the real world, you hit problems at the atomic scale. This could spoil things if we try to use gold.

In contrast, the argument about volumes looks watertight. But there's a tiny loophole in the logic: the tacit assumption that the separate pieces *have* well-defined volumes. 'Volume' is such a familiar concept that we tend to forget just how tricky it can be.

None of this means that Banach and Tarski were right; it just explains why they are not *obviously* wrong. Unlike the nice polygonal pieces in Mathophila's dissection of a chessboard, the Banach–Tarski 'pieces' are more like disconnected clouds of infinitely small dust specks than solid lumps. They are so complicated, in fact, that their volumes cannot be defined – not if we want them to obey the usual rule 'when you combine several pieces, their volumes add'. And if that rule fails, the argument about volumes comes to bits. The single sphere, and two copies of it, have well-defined volumes. But the intermediate stages, when they are cut into pieces, aren't like that.

What *are* they like? Well ... not like that.

Banach and Tarski realised that this loophole might actually make their paradoxical dissection possible. They proved that:

- You can split a single sphere A into finitely many very complicated, possibly disconnected, parts.
- You can do the same to two spheres B and C, the same size as A.
- You can accomplish all of that in such a way that the parts of B and C together correspond exactly to the parts of A.
- You can arrange for corresponding parts to be perfect copies of one another.

The proof of the Banach–Tarski paradox is complicated and technical, and it requires a set-theoretic assumption known as the axiom of choice. This particular assumption worries some mathematicians. However, the fact that it leads to the Banach–Tarski paradox is not what worries them, and is no reason to reject it. Why not? Because the Banach–Tarski paradox isn't really very paradoxical. With the right intuition, we would expect such paradoxical dissections to be possible.

Let me try to give you that intuition. It all hinges on the weird behaviour of what are called infinite sets. Although a sphere has finite size, it contains infinitely many *points*. That leaves room for the weirdness of infinity to show up in the geometry of the sphere.

A useful analogy involves the English alphabet, the 26 letters A, B, C, ..., Z. These letters can be combined to make *words*, and we list permissible words in a *dictionary*. Suppose we allow all possible sequences of letters, as long or as short as we like. So AAAAVDQX is a word, and so is GNU, and so is ZZZ...Z with ten million Z's. We can't print such a dictionary, but to mathematicians it is a well-defined set which contains infinitely many words.

Now, we can dissect this dictionary into 26 pieces. The first piece contains all words starting with A, the second contains all words starting with B, and so on, with the 26th piece containing all words starting with Z. These pieces do not overlap, and every word occurs in exactly one piece.

Each piece, however, has exactly the same structure as the original dictionary. The second piece, for example, contains the words BAAAAVDQX, BGNU and BZZZ...Z. The third contains CAAAAVDQX, CGNU and CZZZ...Z. You can convert each piece into the entire dictionary by lopping off the first letter from every word.

In other words: we can cut the dictionary apart, and reassemble the pieces to make 26 exact copies of the dictionary.

Banach and Tarski found a way to do the same kind of thing with the infinite set of all points in a solid sphere. Their alphabet consisted of two different rotations of the sphere; their words were sequences of these rotations. By playing a more complicated version of the dictionary game with the rotations, you can create an analogous dissection of the sphere. Since there are now two 'letters' in the alphabet, we convert the original sphere into two identical copies.

Careful readers will observe that I've cheated slightly in the interests of simplicity. When I chop off the initial letter B from the second piece, for instance, I not only get the entire original dictionary: I also get the 'empty' word that arises when the initial B is deleted from the word B. So really my dissection turns the dictionary into 26 copies of itself, plus 26 extra words A, B, C, ..., Z of length 1. To keep everything neat and tidy, we have to absorb the extra 26 words into the pieces. A similar problem occurs in Banach and Tarski's construction – but this is a very fine point. If we ignore it, we still double the sphere – we just have a few extra points *left over*. Which is just as suprising.

After Banach and Tarski proved their theorem, mathematicians began to wonder how few pieces you could get away with. In 1947 Abraham Robinson proved that it can be done with five pieces, but no fewer. If you are willing to ignore a single point at the centre of the sphere, this number reduces to four.

The Banach–Tarski paradox isn't really about dissecting spheres. It is about the impossibility of defining a sensible concept of 'volume' for really complicated shapes.

● ●

Nothing Up My Sleeve . . .

How can you remove a loop of string from your arm without taking your hand out of your jacket pocket?

More precisely: take a two-metre length of string and tie its ends together to form a closed loop. Put on your jacket, button it up, and put your arm through a loop and into the side pocket. Now you have to remove the loop without taking your hand out of your pocket – and without sliding the loop into the pocket to sneak it out over the ends of your fingers.

Remove the string without removing your hand from your pocket.

Answer on page 286

● ●

Nothing Down My Leg . . .

Once your audience has learned to solve the previous problem, ask someone to try the same thing, still wearing the jacket, but with his hand in his *trouser* pocket.

Answer on page 287

● ●

Two Perpendiculars

Euclidean geometry is renowned for its logical consistency: no two theorems contradict each other. Actually, there are errors in Euclid. Here's a case in point.

One of Euclid's theorems proves that if we have a line, and a point not on the line, then there is *exactly one* 'perpendicular' from the point to the line. That is, there is a line through the point that meets the original line at right angles – and there is only one such line. (If there were two, they would be parallel, so they couldn't both pass through the same point.)

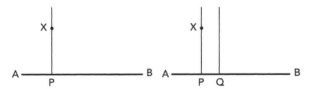

Given AB and X, we can find P such that PX is perpendicular to AB. There can't be another point Q like that, because the line through Q is parallel to PX so it can't pass through X.

A second Euclidean theorem proves that if you take a circle and join the two ends of a diameter to a point on the circumference, you get a right angle.

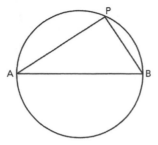

If AB is a diameter of the circle, angle APB is a right angle.

Let's put these two theorems together and see what happens.

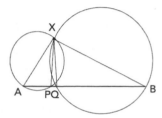

How to find two perpendiculars.

Given the line AB and the point X, draw circles with diameters AX and BX. Let the line AB meet the first circle at P and the second circle at Q. Then the angle APX is a right angle, since AX is a diameter of the first circle. Similarly, the angle BQX is a right angle. So there are *two* perpendiculars XP and XQ from X to AB.

Which of Euclid's two theorems is wrong?

Answer on page 288

• •

Can You Hear the Shape of a Drum?

The backcloth depicted a striking scene: the Rhine valley by moonlight. In the pit, the orchestra was rehearsing Wagner's *Götterdämmerung*. The story had reached the tragic death of Siegfried, and the conductor, Otto Fenderbender, raised his baton for the beginning of the 'Funeral March'. First, just tympani, an intricate repeated rhythm in a low C sharp . . .

'No, no, no!' screamed Fenderbender, hurling his baton to the floor. 'Not like that, you incompetent pigs!'

The leading tympanist, somewhat unwisely, protested. 'But Herr Fenderbender, the rhythm was absolutely pre—'

'Rhythm, schmythm!' said Fenderbender.

'The tempo was exactly as the score indic—'

'I am *not* complaining about the *tempo*!' screamed the conductor.

'The pitch was a perfect C shar—'

'Pitch? *Pitch?* Of *course* the pitch was perfect! I heard that for myself when the orchestra was tuning up! I have an inherent sense of pitch!'

'Then what—'

'The *shape*, you fool! The shape!'

The lead tympanist looked perplexed. It was hard to describe. Otto tried to express what he had heard. 'One of the drums sounded too ... well, too *square*,' he said. 'The other tympani had their usual ... *rounded* sound, but one of them— well, it had *corners*.'

'Come now, Herr Fenderbender – surely you're not claiming you can *hear* the shape of a drum?'

'I heard what I heard,' Otto said doggedly. 'One of the drums is too square.'

And, what do you know? He was right. It's the Bessel functions, you see.

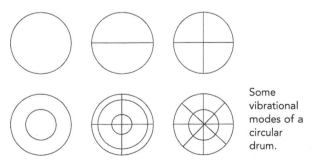

Some vibrational modes of a circular drum.

Let me explain. When a drum beats, it produces several

different notes at once, each note corresponding to a different *mode* of vibration. Each has its own frequency, or equivalently, pitch. Euler calculated the vibrational spectrum of a circular drum – the list of frequencies of these basic modes – using mathematical gadgets called Bessel functions. For a square drum you get sines and cosines instead. In both cases there are characteristic patterns of *nodal lines*, where the drum remains stationary. At any given instant, the drum is displaced upwards on one side of a nodal line, and downwards on the other. As the drum vibrates, each region between the nodal lines oscillates up and down. Fast oscillations create a high-pitched sound, slow oscillations make the pitch lower.

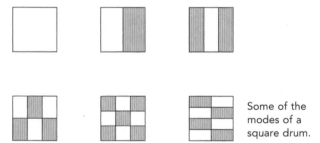

Some of the modes of a square drum.

The mathematics of vibrations proves that the shape of a drum determines its list of frequencies – basically, what it can sound like. But can we go the other way, and deduce the shape from the sound? In 1966 Mark Kac made that question precise: given the spectrum, is it possible to find the shape of the drum?

Kac's question is much more important than its quirky formulation might suggest. When an earthquake hits, the entire Earth rings like a bell, and seismologists deduce a great deal about the internal structure of our planet from the 'sound' that it produces and the way those sounds echo around as they bounce off different layers of rock. Kac's question is the simplest one we can ask about such techniques. 'Personally, I believe that one cannot "hear" the shape,' Kac wrote. 'But I may well be wrong and I am not prepared to bet large sums either way.'

The first significant evidence that Kac was right showed up in a higher-dimensional analogue of the problem. John Milnor wrote a one-page paper proving that two distinct 16-dimensional tori (generalised doughnuts, basically) have the same spectrum. The first results for ordinary 2-dimensional drums were in a more positive direction: various features of the shape *can* be deduced from the spectrum. Kac himself proved that the spectrum of a drum determines its area and perimeter. A curious consequence is that you can hear whether or not a drum is circular, because a circle has the smallest perimeter for a given area. If you know the area A and the perimeter p, and it so happens that $p^2 = 4\pi A$ – as it is for a circle – then the drum *is* a circle, and vice versa. So when Fenderbender said that tympani should have a nice 'rounded' sound, he knew what he was talking about.

In 1989 Carolyn Gordon, David Webb and Scott Wolpert answered Kac's question by constructing two distinct mathematical drums that produce an identical range of sounds. Since then, simpler examples have been found. So now we know that there are limits to what information can be deduced from a shape's vibrational spectrum.

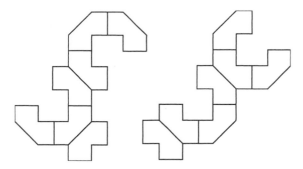

The first example of two sound-alike drums with different shapes.

What is e, and Why?

The number e, which is approximately 2.7182, is the 'base of natural logarithms', a term that refers to its historical origins. One way to see how it arises is to see how a sum of money grows when compound interest is applied at increasingly fine intervals. Suppose that you deposit £1 in the Bank of Logarithmania—

No, no, no. This is the twenty-first century. People don't deposit savings in banks, they borrow.

OK, suppose you borrow £1 on your Logarithmania credit card. (More likely it would be £4,675.23, but £1 is easier to think about.) Once the 0% balance transfer deal has lapsed – about a week after you sign up for the card – the bank applies an interest rate of 100%, paid annually. Then after one year you will owe them

£1.00 borrowed + £1.00 interest = £2.00 total

If instead you paid 50% interest every six months, *compounded* (so that interest becomes payable on previous interest) then after one year you would owe

£1.00 invested + £0.50 interest + £0.75 interest = £2.25 total

This is $(1 + \frac{1}{2})^2$, and the pattern continues like that. So, for example, if you paid interest of 10% at intervals of one-tenth of a year, you would end up owing

$$\left(1 + \frac{1}{10}\right)^{10} = 2.5937$$

pounds. The Bank likes the way these sums are going, so it decides to apply the interest rate ever more frequently. If you paid interest of 1% at intervals of one-hundredth of a year, you would end up owing

$$\left(1 + \frac{1}{100}\right)^{100} = 2.7048$$

pounds. If you paid interest of 0.1% at intervals of one-thousandth of a year, you would end up owing

$$\left(1 + \frac{1}{1,000}\right)^{1,000} = 2.7169$$

pounds. And so on.

As the intervals become ever finer, the amount you owe does not increase without limit. It just seems that way. The amount owed gets closer and closer to 2.7182 pounds – and this number is given the symbol e. It's one of those weird numbers which, like π, turn up naturally in mathematics but can't be expressed exactly as a fraction, so it gets a special symbol. It is especially important in calculus, and it is widely used in scientific applications.

• •

May Husband and Ay ...

In a single move, a chess queen can travel any number of squares in a straight line – horizontally, vertically or diagonally. (Unless another piece stops her, but we ignore that in this puzzle.)

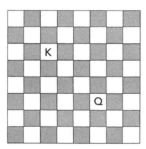

Move the queen from Q to K, visiting each square exactly once and in as few moves as possible.

She starts on square Q and wishes to visit the king on square K. Along the way, she wants to visit all of her other subjects, who live on the other 62 squares. Just passing through, you appreciate – she doesn't stop on every square, but she does has to stop now and again. How can she visit all the squares and finish at the king's square, without passing through any square twice – in the smallest number of moves?

Answer on page 288

•••

Many Knees, Many Seats

A polyhedron is a solid with finitely many flat (that is, planar) *faces*. Faces meet along lines called *edges*; edges meet at points called *vertices*. The climax of Euclid's *Elements* is a proof that there are precisely five *regular polyhedrons*, meaning that every face is a regular polygon (equal sides, equal angles), all faces are identical, and each vertex is surrounded by exactly the same arrangement of faces. The five regular polyhedrons (also called regular solids) are:

- the tetrahedron, with 4 triangular faces, 4 vertices and 6 edges
- the cube or hexahedron, with 6 square faces, 8 vertices and 12 edges
- the octahedron, with 8 triangular faces, 6 vertices and 12 edges
- the dodecahedron, with 12 pentagonal faces, 20 vertices and 30 edges
- the icosahedron, with 20 triangular faces, 12 vertices and 30 edges.

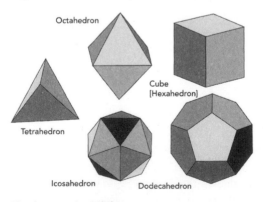

The five regular solids.

The names start with the Greek word for the number of faces, and 'hedron' means 'face'. Originally it meant 'seat', which isn't quite the same thing. While we're discussing linguistics, the '-gon' in 'polygon' originally meant 'knee' and later acquired the technical meaning of 'angle'. So a polygon has many knees, and a polyhedron has many seats.

The regular solids arise in nature – in particular, they all occur in tiny organisms known as radiolarians. The first three also occur in crystals; the dodecahedron and icosahedron don't, although irregular dodecahedra are sometimes found.

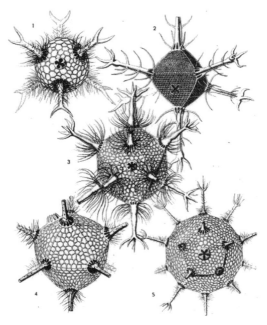

Radiolarians shaped like the regular solids.

It's quite easy to make models of polyhedrons out of card, by cutting out a connected set of faces – called the *net* of the solid – folding along edges, and gluing or taping appropriate pairs of edges together. It helps to add flaps to one edge of each such pair, as shown.

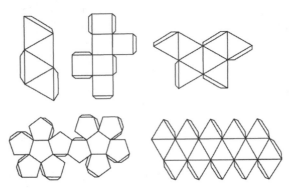

Nets of the regular solids.

Here's a bit of arcane lore: if the edges are of unit length, then the volumes of these solids (in cubic units) are:

- Tetrahedron: $\frac{\sqrt{2}}{12} \sim 0.117\,851$
- Cube: 1
- Octahedron: $\frac{\sqrt{2}}{3} \sim 0.471\,405$
- Dodecahedron: $\frac{\sqrt{5}}{2}\phi^4 \sim 7.663\,12$
- Icosahedron: $\frac{\sqrt{5}}{6}\phi^2 \sim 2.181\,69$

Here ϕ is the golden number (page 96), which turns up whenever you have pentagons around – just as π turns up whenever you have spheres or circles. And $\sim$ means 'approximately equals'.

Analogues of the regular polyhedrons can be defined in spaces of 4 or more dimensions, and are called *polytopes*. There are six regular polytopes in 4 dimensions, but only three regular polytopes in 5 dimensions or more.

● ●

Euler's Formula

The regular solids have a curious pattern which turns out to be far more general. If F is the number of faces, E the number of edges and V the number of vertices, then

$$F - E + V = 2$$

for all five solids. In fact, the same formula holds for any polyhedron that has no 'holes' in it – one that is topologically equivalent to a sphere. This relation is called *Euler's formula*, and its generalisations to higher dimensions are important in topology.

The formula also applies to a map in the plane, provided we consider the infinite region outside the map to be an extra face – or ignore this 'face' and replace the formula by

$$F - E + V = 1$$

which amounts to the same thing but is easier to think about. I'll call this expression *Euler's formula for maps*.

The diagram shows, using a typical example, why this formula is true. The value of $F - E + V$ is written underneath each step in the process. The method of proof is to simplify the map, one step at a time. If we choose a face adjacent to the outside of the map, and remove that face and an adjacent outside edge, then both F and E decrease by 1. This leaves $F - E$ unchanged. Since we haven't altered V, it also leaves $F - E + V$ unchanged. We can keep erasing a face and a corresponding edge until all faces have been removed. We are left with a network of edges and vertices, and this always forms a 'tree' – there are no closed loops of edges. In the example in the diagram, this stage is reached at the sixth step, when $F - E + V = 0 - 7 + 8$.

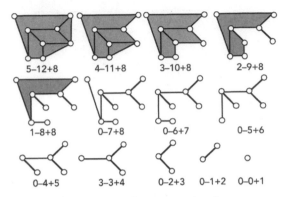

5–12+8	4–11+8	3–10+8	2–9+8
1–8+8	0–7+8	0–6+7	0–5+6
0–4+5	3–3+4	0–2+3	0–1+2 0–0+1

Proof of Euler's formula for maps in the plane.

Now we simplify the tree by snipping off one 'branch' – an edge on the end of the tree plus the vertex on the outside end of that edge – at a time. Now F remains at 0, while E and V both decrease by 1 with each snip. Again, $F - E + V$ remains unchanged. Eventually, a single vertex remains. Now $F = 0$, $E = 0$ and $V = 1$. So at the end of the process, $F - E + V = 1$. Since the process does not change this quantity, it must also have been equal to 1 when we started.

The proof explains why the signs alternate – plus, minus, plus – as we go from faces to edges to vertices. A similar trick works for higher-dimensional topology, for much the same reason.

There is a hidden topological assumption in the proof: the map is drawn in the plane. Equivalently, when considering polyhedrons, they must be 'drawable' on the surface of a sphere. If the polyhedron or map lives on a surface that is topologically distinct from a sphere, such as a torus, then the method of proof can be adapted but the final result is slightly different. For example, the formula for polyhedrons becomes

$$F - E + V = 0$$

when the polyhedron is topologically equivalent to a torus. As an example, this 'picture frame' polyhedron has $F = 16$, $E = 32$ and $V = 16$:

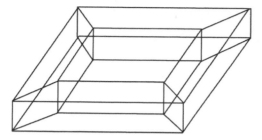

Picture frame polyhedron.

On a surface with g holes, the formula becomes

$$F - E + V = 2 - 2g$$

so we can calculate the number of holes by drawing a polyhedron on the surface. In this manner, an ant that inhabited the surface and could not perceive it 'from outside' would still be able to work out the topology of the surface. Today's cosmologists are trying to work out the topological shape of our own universe – which *we* can't observe 'from outside' – by using more elaborate topological ideas of a similar kind.

• •

What Day is It?

Yesterday, Dad got confused about which day of the week it was. 'Whenever we go on holiday, I forget,' he said.

'Friday,' said Darren.

'Saturday,' his twin sister Delia contradicted.

'What day is it tomorrow, then?' asked Mum, trying to sort out the dispute without too much stress.

'Monday,' said Delia.

'Tuesday,' said Darren.

'Oh, for Heaven's sake! What day was it yesterday, then?'

'Wednesday,' said Darren.

'Thursday,' said Delia.

'Grrrrrrrr!' said Mum, doing her famous Marge Simpson

impression. 'Each of you has given one correct answer and two wrong ones.'

What day is it today?

Answer on page 288

• •

Strictly Logical

Only an elephant or a whale gives birth to a creature that weighs more than 100 kilograms.

The President weighs 150 kilograms.

Therefore ...

(I learned this one from the writer and publisher Stefan Themerson.)

• •

Logical or Not?

If pigs had wings, they'd fly.

Pigs don't fly if the weather is bad.

If pigs had wings, the sensible person would carry an umbrella.

Therefore:

If the weather is bad, the sensible person would carry an umbrella.

Is the deduction logically valid?

Answer on page 288

• •

A Question of Breeding

Farmer Hogswill went to the village fete, where he met five of his friends: Percy Catt, Dougal Dogge, Benjamin Hamster, Porky Pigge and Zoe Zebra. By a remarkable coincidence – which was a constant source of amusement – each of them was an expert breeder of one type of animal: cat, dog, hamster, pig and zebra. Between them, they bred all five types. None bred an animal that sounded like their surname.

'Congratulations, Percy!' said Hogswill. 'I hear you've just won third prize in the pig-breeding competition!'

'That's right,' said Zoe.

'And Benjamin got second for dogs!'

'No,' said Benjamin. 'You knows fine well I never touches no dogs. Nor zebras, neither.'

Hogswill turned to the person whose surname sounded like the animals that Zoe bred. 'And did you win anything?'

'Yes, a gold medal for my prize hamster.'

Assuming that all statements except the alleged second for dogs are true, who breeds what?

Answer on page 289

• •

Fair Shares

In 1944, as the Russian army fought to reclaim Poland from the Germans, the mathematician Hugo Steinhaus, trapped in the city of Lvov, sought distraction in a puzzle. As you do.

The puzzle was this. Several people want to share a cake (by all means replace that by a pizza if you wish). And they want the procedure to be fair, in the sense that no one will feel that they have got less than their fair share.

Steinhaus knew that for two people there is a simple method: one person cuts the cake into two pieces, and the other chooses which one they want. The second person can't complain, because they made the choice. The first person also can't complain – if they do, it was their fault for cutting the cake wrongly.

How can *three* people divide a cake fairly?

Answer on page 289

• •

The Sixth Deadly Sin

It's envy, and the problem is to avoid it.

Stefan Banach and Bronislaw Knaster extended Steinhaus's method of fair cake division to any number of people, and simplified it for three people. Their work pretty much summed up the whole area until a subtle flaw emerged: the procedure may be fair, but it takes no account of envy. A method is *envy-free* if no one thinks that anyone else has got a bigger share than they have. Every envy-free method is fair, but a fair method need not be envy-free. And neither Steinhaus's method, nor that of Banach and Knaster, is envy-free.

For example, Belinda may think that Arthur's division is fair. Then Steinhaus's method stops after step 3, and both Arthur and Belinda consider all three pieces to be of size 1/3. Charlie must think that his own piece is at least 1/3, so the allocation is proportional. But if Charlie sees Arthur's piece as 1/6 and Belinda's as 1/2, then he will envy Belinda, because Belinda got first crack at a piece that Charlie *thinks* is bigger than his.

Can you find an envy-free method for dividing a cake among three people?

Answer on page 290

• •

Weird Arithmetic

'No, Henry, you can't do that,' said the teacher, pointing to Henry's exercise book, where he had written

$$\frac{1}{4} \times \frac{8}{5} = \frac{18}{45}$$

'Sorry, sir,' said Henry. 'What's wrong? I checked it on my calculator and it seemed to work.'

'Well, Henry, the *answer* is right, I guess,' the teacher admitted. 'Though you should probably cancel the 9's to get $\frac{2}{5}$, which is simpler. What's wrong is—'

Explain the mistake to Henry. Then find all such sums, with single non-zero digits in the first two fractions, that are correct.

Answer on page 291

● ●

How Deep is the Well?

In one episode of the television series *Time Team*, the indefatigable archaeologists want to measure the depth of a mediaeval well. They drop something into it and time its fall, which takes an amazingly long six seconds. You hear it clattering its way down for ages. They come dangerously close to calculating the depth using Newton's laws of motion, but cop out at the last moment and use three very long tape measures joined together instead.

The formula they very nearly state is

$$s = \tfrac{1}{2}gt^2$$

where s is the distance travelled under gravity, falling from rest, and g is the acceleration due to gravity. It applies when air resistance can be ignored. This formula was discovered experimentally by Galileo Galilei and later generalised by Isaac Newton to describe motion under the influence of *any* force.

Taking $g = 10\,\text{m s}^{-2}$ (metres per second per second), *how deep is the well?*

You've got three days to do it.

Answer on page 292

● ●

Mcmahon's Squares

This puzzle was invented by the combinatorialist* P.A. McMahon in 1921. He was thinking about a square that has been divided into four triangular regions by diagonals. He wondered how many different ways there are to colour the various regions,

* A combinatorialist is someone who invents this kind of thing.

using three colours. He discovered that if rotations and reflections are regarded as the same colouring, there are exactly 24 possibilities. Find them all.

Now, a 6 × 4 rectangle contains 24 1 × 1 squares. Can you fit the 24 squares together to make such a rectangle, so that adjacent regions have the same colour, and the entire perimeter of the rectangle has the same colour?

Answer on page 292

. . . .

What is the Square Root of Minus One?

The square root of a number is a number whose square is the given one. For instance, the square root of 4 is 2. If we allow negative numbers, then −2 is a second square root of 4 because minus times minus makes plus. Since plus times plus also makes plus, the square of any number – positive or negative – is always positive. So it looks as though negative numbers, in particular −1, can't have square roots.

Despite this, mathematicians (and physicists and engineers and indeed anyone working in any branch of science) have found it useful to provide −1 with a square root. This is not a number in the usual sense, so it is given a new symbol, which is i if you are a mathematician, and j if you are an engineer.

Square roots of negative numbers first showed up in mathematics around 1450, in an algebra problem. In those days the idea was a huge puzzle, because people thought of a number as something real. Even negative numbers caused a great deal of head-scratching, but people quickly got accustomed to them when they realised how useful they could be. Much the same happened with i, but it took a lot longer.

A big issue was how to visualise i geometrically. Everyone had got used to the idea of the number line, like an infinitely long ruler, with positive numbers on the right and negative ones on the left, and fractions and decimals in between:

The 'real' number line.

Collectively, these familiar kinds of number became known as *real* numbers, because they correspond directly to physical quantities. You can *observe* 3 cows or 2.73 kilograms of sugar.

The puzzle was that there seemed to be nowhere on the real number line for the 'new' number i. Eventually, mathematicians realised that *it didn't have to go on the real number line*. In fact, being a new kind of number, it *couldn't* go there. Instead, i had to live on a second line, at right angles to the real number line:

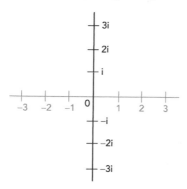

The 'imaginary' number line, placed at right angles to the real one.

And if you added an imaginary number to a real one, the answer had to live in the plane defined by the two lines:

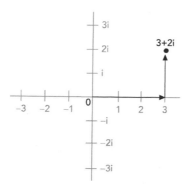

A *complex* number is a real one plus an imaginary one.

Multiplication was more complicated. The main point was that multiplying a number by i rotated it around the origin O through a right angle, anticlockwise. For instance, 3 multiplied by i is 3i, and that's what you get when you rotate the point labelled 3 through 90°.

The new numbers extended the familiar real number line to a larger space, a number plane. Three mathematicians discovered this idea independently: the Norwegian Caspar Wessel, the Frenchman Jean-Robert Argand and the German Carl Friedrich Gauss.

Complex numbers don't turn up in everyday situations, such as checking the supermarket bill or measuring someone for a suit. Their applications are in things like electrical engineering and aircraft design, which lead to technology that we can use without having to know the underlying mathematics.

The engineers and designers need to know it, though.

The Most Beautiful Formula

Occasionally people hold polls for the most beautiful mathematical formula of all time – yes, they really do, I'm not making this up, honest – and nearly always the winner is a famous formula discovered by Euler, which uses complex numbers to link the two famous constants e and π. The formula is

$$e^{i\pi} = -1$$

and it is extremely influential in a branch of maths known as complex analysis.

Why is Euler's Beautiful Formula True?

I often get asked whether there is a simple way to explain why Euler's formula $e^{i\pi} = -1$ is true. It turns out that there is, but some preparation is needed – about two years of undergraduate mathematics.

This is uncomfortably like the joke about the professor who says in a lecture that some fact is obvious, and when challenged goes away for half an hour and returns to say 'yes, it is obvious,' and then continues lecturing without further explanation. It just takes two years instead of half an hour. So I'm going to give you the explanation. Skip this bit if it doesn't make sense – but it does illustrate how higher mathematics sometimes gains new insights by putting different ideas together in unexpected ways. The necessary ingredients are some geometry, some differential equations and a bit of complex analysis.

The main idea is to solve the differential equation

$$\frac{dz}{dt} = iz$$

where z is a complex function of time t, with the initial condition $z(0) = 1$. It is standard in differential equations courses that the solution is

$$z(t) = e^{it}$$

Indeed, you can *define* the exponential function e^w this way.

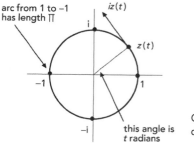

Geometry of the differential equation.

Now let's interpret the equation geometrically. Multiplication by i is the same as rotation through a right angle, so iz is at right angles to z. Therefore the tangent vector $iz(t)$ to the solution at any point $z(t)$ is always at right angles to the 'radius vector' from 0 to $z(t)$ and has length 1. Therefore the solution $z(t)$ always lies in the unit circle, and the point $z(t)$ moves round this circle with angular velocity 1 measured in radians per second. (The *radian measure* of an angle is the length of the arc of the unit circle corresponding to that angle.) The circumference of the unit circle is 2π, so $t = \pi$ is halfway round the circle. But halfway round is visibly the point $z = -1$. Therefore $e^{i\pi} = -1$, which is Euler's formula.

All the ingredients of this proof are well known, but the overall package seems not to get much prominence. Its big advantage is to explain why circles (leading to π) have anything to do with exponentials (defined using e). So given the right background, Euler's formula ceases to be mysterious.

● ●

Your Call May be Monitored for Training Purposes

'The number you have dialled is imaginary. Please rotate your phone 90 degrees and try again.'

● ●

Archimedes, You Old Fraud!

'Give me a place to stand, and I will move the Earth.' So, famously, said Archimedes, dramatising his newly discovered law of the lever. Which in this case takes the form

> Force exerted by Archimedes
> × distance from Archimedes to fulcrum
> *equals*
> Mass of Earth × distance from Earth to fulcrum

The *fulcrum* is the pivot – the black triangle in the picture:

The law of the lever.

Now, I don't think Archimedes was interested in the position of the Earth in space, but he did want the fulcrum to be fixed. (I know he *said* 'a place to stand', but if the fulcrum moves, all bets are off, so presumably that's what he meant.) He also needed a perfectly rigid lever of zero mass, and he probably didn't realise that he also needed uniform gravity, contrary to astronomical fact, to convert mass to weight. No matter. I don't want to get into discussions about inertia or other quibbles. Let's grant him all those things. My question is: when the Earth moves, how *far* does it move? And can Archimedes achieve the same result more easily?

Answer on page 293

• •

Fractals – The Geometry of Nature

Every so often, an entire new area of mathematics arises. One of the best known in recent times is *fractal geometry*, pioneered by Benoît Mandelbrot, who coined the term 'fractal' in 1975. Roughly speaking, it is a mathematical method for coming to grips with apparent irregularities in the natural world, and revealing hidden structure. The subject is best known for its beautiful, complex computer graphics, but it goes far deeper than that.

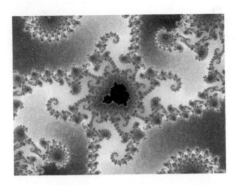

Part of the Mandelbrot set, a famous fractal.

The traditional shapes of Euclidean geometry are triangles, squares, circles, cones, spheres, and the like. These shapes are simple, and in particular they have no fine structure. If you magnify a circle, for instance, any portion of it looks more and more like a featureless straight line. Shapes like this have played a prominent role in science – for instance, the Earth is close to a sphere, and for many purposes that level of detail is good enough.

Many natural shapes are far more complex. Trees are a mass of branches, clouds are fuzzy and convoluted, mountains are jagged, coastlines are wiggly ... To understand these shapes mathematically, and to solve problems about them, we need new ingredients. The supply of problems, by the way, is endless – how do trees dissipate the energy of the wind, how do waves erode a coastline, how does water run off mountains into rivers? These are practical issues, often related to ecology and the environment, not just theoretical problems.

Coastlines are a good example. They are wiggly curves, but you can't use any old wiggly curve. Coastlines have a curious property: they look much the same on any scale of map. If the map shows more detail, extra wiggles can be distinguished. The exact shape changes, but the 'texture' seems pretty much the same. The jargon here is 'statistically self-similar'. All statistical features of a coastline, such as what proportion of bays have a given relative size, are the same no matter what scale of magnification you work on.

Mandelbrot introduced the word *fractal* to describe any shape that has intricate structure no matter how much you magnify it. It doesn't have to be statistically self-similar – but such fractals are easier to understand. And those that are exactly self-similar are even nicer, which is how the subject started.

About a century ago, mathematicians invented a spate of weird shapes, for various esoteric purposes. These shapes were not just statistically self-similar – they were exactly self-similar. When suitably magnified, the result looked identical to the original. The most famous is the *snowflake curve*, invented by Helge von Koch in 1904. It can be assembled from three copies of the curve shown in the right-hand diagram.

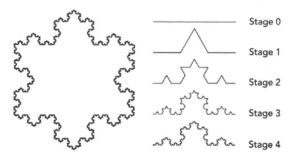

The snowflake curve and successive stages in its construction.

This component curve (though not the whole snowflake) is exactly self-similar. You can see that each stage in the construction is made from four copies of the previous stage, each one-third as big. The four copies are fitted together as in Stage 1. Passing to the infinite limit, we obtain an infinitely intricate curve that is built from four copies of itself, each one-third the size – so it is self-similar.

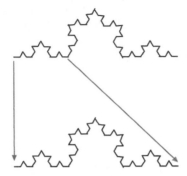

Each quarter of the
curve, blown up to three
times the size, looks like
the original curve.

This shape is too regular to represent a real coastline, but it
has about the right degree of wiggliness, and less regular curves
formed in a similar way do look like genuine coastlines. The
degree of wiggliness can be represented by a number, called the
fractal dimension.

To see how this goes, I'm going to take some simpler *non-*
fractal shapes and see how they fit together at different scales of
magnification. If I break a line into pieces 1/5 the size, say, then I
need 5 of them to reconstruct the line. With a square, I need 25
pieces, which is 5^2. And with cubes I need 125, which is 5^3.

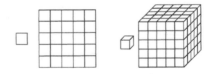

Effect of scaling on 'cubes' in 1, 2 and 3 dimensions.

The power of 5 occurring here is the same as the dimension
of the shape concerned: 1 for a line, 2 for a square, 3 for a cube.
In general, if the dimension is d and we have to fit k pieces of size
$1/n$ together to reassemble the original shape, then $k = n^d$.
Taking logarithms, we find that

$$d = \frac{\log k}{\log n}$$

Now, taking a deep breath, we try this formula out on the snowflake. Here we need $k = 4$ pieces each 1/3 the size, so $n = 3$. Therefore our formula yields

$$d = \frac{\log 4}{\log 3}$$

which is roughly 1.2618. So the 'dimension' of the snowflake curve is not a whole number!

That would be bad if we wanted to think of 'dimension' in the conventional way, as the number of independent directions available. But it's fine if we want a numerical measure of wiggliness, based on self-similarity. A curve with dimension 1.2618 is more wiggly than a curve of dimension 1, such as a straight line; but it is less wiggly than a curve of dimension 1.5, say.

There are dozens of technically distinct ways to define the dimension of a fractal. Most of them work when it is not self-similar. The one used by mathematicians is called the *Hausdorff–Besicovitch dimension*. It's a pig to define and a pig to calculate, but it has pleasant properties. Physicists generally use a simpler version, called the *box dimension*. This is easy to calculate, but lacks nearly all the pleasant properties of the Hausdorff–Besicovitch dimension. Despite that, the two dimensions are often the same. So the term *fractal dimension* is used to mean either of them.

Fractals need not be curves: they can be highly intricate surfaces or solids, or higher-dimensional shapes. The fractal dimension then measures how *rough* the fractal is, and how effectively it fills space. The fractal dimension turns up in most applications of fractals, both in the theoretical calculations and in experimental tests. For example, the fractal dimension of real coastlines is generally close to 1.25 – surprisingly close to that of the snowflake curve.

Fractals have come a long way, and they are now routinely used as mathematical models throughout the sciences. They are also the basis of an effective method for compressing computer files of video images. But their most interesting role is as 'the' geometry of many natural forms. A striking example is a kind of cauliflower called Romanesco broccoli. You can find it in most supermarkets. Each tiny floret has much the same form as the whole cauliflower, and everything is arranged in a series of ever-smaller Fibonacci spirals. This example is the tip of an iceberg – the fractal structure of plants. While much remains to be sorted out, it is already clear that the fractal structure arises from the way plants grow, which in turn is regulated by their genetics. So the geometry here is more than just a visual pun.

Romanesco broccoli – you can't get much more self-similar than that!

The applications of fractals are extensive, ranging from the fine structure of minerals to the form of the entire universe. Fractal shapes have been used to make antennas for mobile phones – such shapes are more efficient. Fractal image compression techniques cram huge quantities of data on to CDs and DVDs. There are even medical applications: for example, fractal geometry can be used to detect cancerous cells, whose surfaces are wrinkly and have a higher fractal dimension than normal cells.

About ten years ago a team of biologists (Geoffrey West, James Brown and Brain Enquist) discovered that fractal geometry can explain a long-standing puzzle about patterns in living creatures. The patterns concerned are statistical 'scaling laws'. For example, the metabolic rates of many animals seem to be proportional to the $\frac{3}{4}$th power of their masses, and the time it takes for the embryo to develop is proportional to the $-\frac{1}{4}$th power of the mass of the adult. The main enigma here is that fraction $\frac{1}{4}$. A power law with the value $\frac{1}{3}$ could be explained in terms of volume, which is proportional to the cube of the creature's length. But $\frac{1}{4}$, and related fractions such as $\frac{3}{4}$ or $-\frac{1}{4}$, are harder to explain.

The team's idea was an elegant one: a basic constraint on how organisms can grow is the transport of fluids, such as blood, around the body. Nature solves this problem by building a branching network of veins and arteries. Such a network obeys three basic rules: it should reach all regions of the body, it should transport fluids using as little energy as possible, and its smallest tubes should all be much the same size (because the tube can't be smaller than a single blood cell, or the blood can't flow). What shapes satisfy these conditions? Space-filling fractals – with the fine structure cut off at the limiting size, that of the single cell. This approach – which takes into account some important physical and biological details, such as the flexibility of the tubes and the occurrence of pulses in the blood as the heart beats – predicts that elusive $\frac{1}{4}$th power.

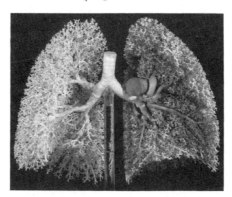

Fractal branching of blood vessels in the lungs.

The Missing Symbol

Place a standard mathematical symbol between 4 and 5 to get a number greater than 4 and less than 5.

Answer on page 294

• •

Where There's a Wall, There's a Way

In the county of Hexshire, fields are separated by walls built from the local stones – which for some reason are all made from identical hexagonal lumps joined together. Perhaps they origi-nated as basalt columns like the ones in the Giant's Causeway. Anyway, Farmer Hogswill has seven stones, each formed from four hexagons. In fact, they are precisely the seven possible combinations of four hexagons:

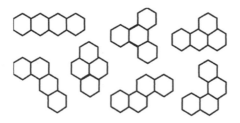

Seven stones to make a wall.

He has to make a wall shaped like this:

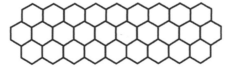

The required wall.

How can he do it? (He can rotate the stones and turn them over to obtain their mirror images if necessary.)

Answer on page 294

• •

Constants to 50 Places

π 3.141 592 653 589 793 238 462 643 383 279 502 884 197 169 399 375 11

e 2.718 281 828 459 045 235 360 287 471 352 662 497 757 247 093 699 96

$\sqrt{2}$ 1.414 213 562 373 095 048 801 688 724 209 698 078 569 671 875 376 95

$\sqrt{3}$ 1.732 050 807 568 877 293 527 446 341 505 872 366 942 805 253 810 38

$\log 2$ 0.693 147 180 559 945 309 417 232 121 458 176 568 075 500 134 360 26

ϕ 1.618 033 988 749 894 848 204 586 834 365 638 117 720 309 179 805 76

γ 0.577 215 664 901 532 860 606 512 090 082 402 431 042 159 335 939 94

δ 4.669 201 609 102 990 671 853 203 820 466 201 617 258 185 577 475 76

Here ϕ is the golden number (page 96), γ is Euler's constant (page 96), and δ is the *Feigenbaum constant*, which is important in chaos theory (page 117). See
en.wikipedia.org/wiki/Logistic_map
mathworld.wolfram.com/FeigenbaumConstant.html

● ●

Richard's Paradox

In 1905 Jules Richard, a French logician, invented a very curious paradox. In the English language, some sentences define positive integers and others do not. For example 'The year of the Declaration of Independence' defines 1776, whereas 'The historical significance of the Declaration of Independence' does not define a number. So what about this sentence: 'The smallest number that cannot be defined by a sentence in the English language containing fewer than 20 words.' Observe that whatever this number may be, we have just defined it using a sentence in the English language containing only 19 words. Oops.

A plausible way out is to say that the proposed sentence does not actually define a specific number. However, it ought to. The English language contains a finite number of words, so the number of sentences with fewer than 20 words is itself finite. Of course, many of these sentences make no sense, and many of those that do make sense don't define a positive integer – but that just means that we have fewer sentences to consider. Between them, they define a finite set of positive integers, and it is a standard theorem of mathematics that in such circumstances there is a unique smallest positive integer that is not in the set. So on the face of it, the sentence does define a specific positive integer.

But logically, it can't.

Possible ambiguities of definition such as 'A number which when multiplied by zero gives zero' don't let us off the logical hook. If a sentence is ambiguous, then we rule it out, because an ambiguous sentence doesn't *define* anything. Is the troublesome sentence ambiguous, then? Uniqueness is not the issue: there can't be *two* distinct smallest-numbers-not-definable-(etc.), because one must be smaller than the other.

One possible escape route involves how we decide which sentences do or do not define a positive integer. For instance, if we go through them in some kind of order, excluding bad ones in turn, then the sentences that survive depend on the order in which they are considered. Suppose that two consecutive sentences are:

(1) The number in the next valid sentence plus one.

(2) The number in the previous valid sentence plus two.

These sentences cannot both be valid – they would then contradict each other. But once we have excluded one of them, the other one *is* valid, because it now refers to a different sentence altogether.

Forbidding this type of sentence puts us on a slippery slope, with more and more sentences being excluded for various reasons. All of which strongly suggests that the alleged sentence does not, in fact, define a specific number – even though it seems to.

Connecting Utilities

Three houses have to be connected to three utility companies – water, gas and electricity. Each house must be connected to *all three* utilities. Can you do this without the connections crossing?

(Work 'in the plane' – there is no third dimension in which pipes can be passed over or under cables. And you are not allowed to route cables or pipes through a house or a utility company building.)

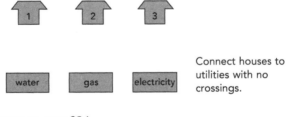

Connect houses to utilities with no crossings.

Answer on page 294

Are Hard Problems Easy?
or
How to Win a Million Dollars by Proving the Obvious

Naturally, it's not *that* obvious. TANSTAAFL, as science fiction author Robert A. Heinlein used to say – There Ain't No Such Thing As A Free Lunch. But we can all dream.

I'm referring here to one of the seven Millennium Prize

Problems (page 127), whose solution will leave some lucky person a million dollars better off. Technically, it is known as 'P=NP?' which is a pretty silly name. But what it's about is of vital importance: inherent limits to the efficiency of computers.

Computers solve problems by running programs, which are lists of instructions. A program that always stops with the right answer (assuming that the computer is doing what its designers think it should) is called an 'algorithm'. The name honours the Arabic mathematician Abu Ja'far Muhammad ibn Musa al-Khwarizmi, who lived around AD 800 in present-day Iraq. His book *Hisab al-jabr w'al-muqabala* gave us the word 'algebra', and it consists of a series of procedures – algorithms – for solving algebraic equations of various kinds.

An algorithm is a method for solving a specific type of problem, but it is useless in practice unless it delivers the answer reasonably quickly. The theoretical issue here is not how fast the computer is, but how many calculations the algorithm has to perform. Even for a specific problem – to find the shortest route that visits a number of cities in turn, say – the number of calculations depends on how complicated the question is. If there are more cities to visit, the computer will have to do more work to find an answer.

For these reasons, a good way to measure the efficiency of an algorithm is to work out how many computational steps it takes to solve a problem of a given size. There is a natural division into 'easy' calculations, where the size of the calculation is some fixed power of the input data, and 'hard' ones, where the growth rate is much faster, often exponential. Multiplying two n-digit numbers together, for example, can be done in about n^2 steps using good old-fashioned long multiplication, so this calculation is 'easy'. Finding the prime factors of an n-digit number, on the other hand, takes about 3^n steps if you try every possible divisor up to the square root of n, which is the most obvious approach, so this calculation is 'hard'. The algorithms concerned are said to run in *polynomial time* (class P) and *non-polynomial time* (not-P), respectively.

Working out how quickly a given algorithm runs is relatively straightforward. The hard bit is to decide whether some other algorithm might be faster. The hardest of all is to show that what you've got is the fastest algorithm that will work, and basically we don't know how to do that. So problems that we think are hard might turn out to be easy if we found a better method for solving them, and this is where the million dollars comes in. It will go to whoever manages to prove that some specific problem is unavoidably hard – that no polynomial-time algorithm exists to solve it. Or, just possibly, to whoever proves that There Ain't No Such Thing As A Hard Problem – though that doesn't seem likely, the universe being what it is.

Before you rush out to get started, though, there are a couple of things you should bear in mind. The first is that there is a 'trivial' type of problem that is automatically hard, simply because the size of the *output* is gigantic. 'List all ways to rearrange the first n numbers' is a good example. However fast the algorithm might be, it takes at least $n!$ steps to print out the answer. So this kind of problem has to be removed from consideration, and this is done using the concept of a *nondeterministic polynomial time*, or NP, problem. (Note that NP is different from not-P.) These are the problems where you can verify a proposed *answer* in polynomial time – that is, easily.

My favourite example of an NP problem is solving a jigsaw puzzle. It may be very hard to find a solution, but if someone shows you an allegedly completed puzzle you can tell instantly whether they've done it right. A more mathematical example is finding a factor of a number: it is much easier to divide out and see whether some number works than it is to find that number in the first place.

The P=NP? problem asks whether every NP problem is P. That is, if you can check a proposed answer easily, can you *find* it easily? Experience suggests very strongly that the answer should be 'no' – the hard part is to find the answer. But, amazingly, no one knows how to prove that, or even whether it's correct. And that's why you can pocket a million bucks for proving that P is different from NP, or indeed for proving that, on the contrary, the two are equal.

As a final twist, it turns out that all likely candidates to show that P $\neq$ NP are in some sense equivalent. A problem is called *NP-complete* if a polynomial-time algorithm to solve that particular problem automatically leads to a polynomial-time algorithm to solve *any* NP problem. Almost any reasonable candidate for proving that P $\neq$ NP is known to be NP-complete. The nasty consequence of this fact is that no particular candidate is likely to be more approachable than any of the others – they all live or die together. In short: we know *why* P=NP? must be a very hard problem, but that doesn't help us to solve it.

I suspect that there are far easier ways to make a million.

• •

Don't Get the Goat

There used to be an American game show, hosted by Monty Hall, in which the guest had to choose one of three doors. Behind one was an expensive prize – a sports car, say. Behind the other two were booby prizes – goats.

After the contestant had chosen, Hall would open one of the *other* doors to reveal a goat. (With two doors to choose from, he could always do this – he knew where the car was.) He would then offer the contestant the chance to change their mind and choose the other unopened door.

Hardly anyone took this opportunity – perhaps with good reason, as I'll eventually explain. But for the moment let's take the problem at face value, and assume that the car has equal probability (one in three) of being behind any given door. We'll assume also that everyone knows ahead of time that Hall *always* offers the contestant a chance to change their mind, after revealing a goat. Should they change?

The argument against goes like this: the two remaining doors are equally likely to conceal a car or a goat. Since the odds are fifty–fifty, there's no reason to change.

Or is there?

Answer on page 296

● ●

All Triangles are Isosceles

This puzzle requires some knowledge of Euclidean geometry, which nowadays isn't taught ... Ho hum. It's still accessible if you're prepared to take a few facts on trust.

An *isosceles* triangle has two sides equal. (The third could also be equal: this makes the triangle *equilateral*, but it still counts as isosceles too.) Since it is easy to draw triangles with all three sides different, the title of this section is clearly *false*. Nevertheless, here is a geometric proof that it is true.

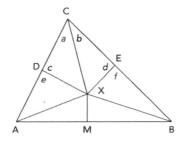

This triangle is isosceles – except that it clearly isn't.

(1) Take any triangle ABC.

(2) Draw a line CX that cuts the top angle in half, so that angles *a* and *b* are equal. Draw a line MX at right angles to the bottom edge at its midpoint, so that AM = MB. This meets the previous line, CX, somewhere inside the triangle at the point X.

(3) Draw lines from X to the other two corners A and B. Draw XD and XE to make angles *c*, *d*, *e* and *f* all right angles.

(4) Triangles CXD and CXE are *congruent* – that is, they have the same shape and size (though one is the other flipped over). The

reason is that angles *a* and *b* are equal, angles *c* and *d* are equal, and the side CX is common to both triangles.

(5)   Therefore lines CD and CE are equal.

(6)   So are lines XD and XE.

(7)   Since M is the midpoint of AB and MX is at right angles to AB, the lines XA and XB are equal.

(8)   But now triangles XDA and XEB are congruent. The reason is that XD = XE, XA = XB and angle *e* equals angle *f*.

(9)   Therefore DA = EB.

(10)   Combining steps 5 and 9: CA = CD + DA = CE + EB = CB. So lines CA and CB are equal, and triangle ABC is isosceles.

What's wrong here? (Hint: it's not the use of congruent triangles.)

Answer on page 298

• •

Square Year

It was midnight on 31 December 2001, and Alfie and Betty – both of whom were aged less than sixty – were talking about the calendar.

'At some time in the past, the year was the square of my father's age,' said Betty proudly. 'He died at the age of a hundred!'

'And at some time in the future, the year will be the square of *my* age,' Alfie replied. 'I don't know whether I'll reach a hundred, though.'

In which years were Betty's father and Alfie born?

Answer on page 298

• •

Gödel's Theorems

In 1931 the mathematical logician Kurt Gödel proved two important theorems, of great originality, which placed unavoidable limits on the power of formal reasoning in mathematics. Gödel was responding to a research programme initiated by David Hilbert, who was convinced that the whole of mathematics could be placed on an axiomatic basis. Which is to say it should be possible to state a list of basic assumptions, or 'axioms', and deduce the rest of mathematics from the axioms. Additionally, Hilbert expected to be able to prove two key properties:

- The system is *logically consistent* – it is not possible to deduce two statements that contradict each other.
- The system is *complete* – every statement has either a proof or a disproof.

The kind of axiomatic 'system' that Hilbert had in mind was more basic than, say, arithmetic – something like the theory of sets introduced by Georg Cantor in 1879 and developed over the next few years. Starting from sets, there are ways to define whole numbers, the usual operations of arithmetic, negative and rational numbers, real numbers, complex numbers, and so on. So placing set theory on an axiomatic basis would automatically do the same for the rest of mathematics. And proving that the axiomatic system for set theory is consistent and complete would also do the same for the rest of mathematics. Since set theory is conceptually simpler than arithmetic, this seemed a sensible way to proceed. In fact, there was even a candidate axiomatisation of set theory, developed by Betrand Russell and Alfred North Whitehead in their three-volume epic *Principia Mathematica*. There were various alternatives, too.

Hilbert pushed a substantial part of his programme through successfully, but there were still some gaps when Gödel arrived on the scene. Gödel's 1931 paper 'On Formally Undecidable Propositions in *Principia Mathematica* and Related Systems I' left

Hilbert's programme in ruins, by proving that no such approach could ever succeed.

Gödel went to great lengths to place his proofs in a rigorous logical context, and to avoid several subtle logical traps. In fact most of his paper is devoted to setting up these background ideas, which are very technical – 'recursively enumerable sets'. The climax to the paper can be stated informally, as two dramatic theorems:

- In a formal system that is rich enough to include arithmetic, there exist *undecidable* statements – statements that can neither be proved nor disproved within that system.
- If a formal system that is rich enough to include arithmetic is logically consistent, then it is impossible to prove its consistency within that system.

The first theorem does not just indicate that finding a proof or disproof of the appropriate statement is difficult. It established that no proof exists, *and* no disproof exists. It means that the logical distinction between 'true' and 'false' is *not* identical to that between 'provable' and 'disprovable'. In conventional logic – including that used in *Principia Mathematica* – every statement is either true or false, and cannot be both. Since the negation not-P of any true statement P is false, and the negation of a false statement is true, conventional logic obeys the 'law of the excluded middle': given any statement P, then exactly one of P and not-P is true, and the other is false. Either 2+2 is equal to 4, or 2+2 is not equal to 4. It has to be one or the other, and it can't be both.

Now, if P has a proof, then P must be true – this is how mathematicians establish the truth (in a mathematical sense) of their theorems. If P has a disproof, then not-P must be true, so P must be false. But Gödel proved that for some statements P, neither P nor not-P has a proof. So a statement can be provable, disprovable – or *neither*. If it's neither, it is said to be 'undecidable'. So now there *is* a third possibility, and the 'middle' is no longer excluded.

Before Gödel, mathematicians had happily assumed that anything true was provable, and anything false was disprovable. *Finding* the proof or disproof might be very hard, but there was no reason to doubt that one or the other must exist. So mathematicians considered 'provable' to be the same as 'true', and 'disprovable' to be the same as 'false'. And they felt happier with practical concepts of proof and disproof than with deep and tricky philosophical concepts like truth and falsity, so mostly they settled for proofs and disproofs. And so it was disturbing to discover that these left a gap, a kind of logical no-man's-land. And in ordinary arithmetic, too!

Gödel set up his undecidable statement by finding a formal version of the logical paradox 'this statement is false', or more accurately of 'this statement has no proof'. However, in mathematical logic a statement is not permitted to refer to itself – in fact, 'this statement' is not something that has a meaning within the formal system concerned. Gödel found a cunning way to achieve much the same result without breaking the rules, by associating a numerical *code* with each formal statement. Then a proof of any statement corresponded to some sequence of transformations of the corresponding code number. So the formal system could model arithmetic – but arithmetic could also model the formal system.

Within this set-up, and assuming the formal system to be logically consistent, the statement P whose interpretation was basically 'this statement has no proof' must be undecidable. If P has a proof, then P is true, so by its defining property P has no proof – a contradiction. But the system is assumed to be consistent, so that can't happen. On the other hand, if P has no proof, then P is true. Therefore not-P has no proof. So neither P nor not-P has a proof.

From here it is a short step to the second theorem – if the formal system is consistent, then there can't be a proof that it is. I've always thought this to be rather plausible. Think of arithmetic as a used car salesman. Hilbert wanted to ask the salesman 'are you honest?' and get an answer that guaranteed that he was. Gödel basically argued that if you ask him this question and he says 'Yes, I am,' that is no guarantee of honesty. Would *you* believe that someone is telling the truth because they tell you they are? A court of law certainly would not.

Because of the technical complications, Gödel proved his theorems within one specific formal system for arithmetic, the one in *Principia Mathematica*. So a possible consequence might have been that this system is inadequate, and something better is needed. But Gödel pointed out in the introduction to his paper that a similar line of reasoning would apply to any alternative formal system for arithmetic. Changing the axioms wouldn't help. His successors filled in the necessary details, and Hilbert's programme was a dead duck.

Several important mathematical problems are now known to be undecidable. The most famous is probably the halting problem for Turing machines – which in effect asks for a method to determine in advance whether a computer program will eventually stop with an answer, or go on for ever. Alan Turing proved that some programs are undecidable – there is no way to prove that they stop, and no way to prove that they don't.

• •

If π isn't a Fraction, How Can You Calculate It?

The school value 22/7 for π is not exact. It's not even terribly good. But it *is* good for something so simple. Since we know that π is not an exact fraction, it's not obvious how it can be calculated to very high accuracy. Mathematicians achieve this using a variety of cunning formulas for π, all of which are exact, and all of which involve some process that goes on for ever. By stopping before we *get* to 'for ever', a good approximation to π can be found.

In fact, mathematics presents us with an embarrassment of riches, because one of the perennial fascinations of π is its tendency to appear in a huge variety of beautiful formulas. Typically they are infinite series, infinite products or infinite fractions (indicated by the dots ...) – which should not be a surprise since there is no simple finite expression for π, unless you cheat with integral calculus. Here are a few of the high points.

The first formula was one of the earliest expressions for π, discovered by François Viète in 1593. It is related to polygons with $2n$ sides:

$$\frac{2}{\pi} = \sqrt{\frac{1}{2}} \times \sqrt{\frac{1}{2}+\frac{1}{2}\sqrt{\frac{1}{2}}} \times \sqrt{\frac{1}{2}+\frac{1}{2}\sqrt{\frac{1}{2}+\frac{1}{2}\sqrt{\frac{1}{2}}}} \times \cdots$$

The next was found by John Wallis in 1655:

$$\frac{\pi}{2} = \frac{2}{1} \times \frac{2}{3} \times \frac{4}{3} \times \frac{4}{5} \times \frac{6}{5} \times \frac{6}{7} \times \frac{8}{7} \times \frac{8}{9} \times \cdots$$

Around 1675, James Gregory and Gottfried Leibniz both discovered

$$\frac{\pi}{4} = 1 - \frac{1}{3} + \frac{1}{5} - \frac{1}{7} + \frac{1}{9} - \frac{1}{11} + \frac{1}{13} - \cdots$$

This converges too slowly to be of any help in calculating π; that is, a good approximation requires oodles of terms. But closely related series were used to find several hundred digits of π in the eighteenth and nineteenth centuries. In the seventeenth century, Lord Brouncker discovered an infinite 'continued fraction':

$$\pi = \cfrac{4}{1 + \cfrac{1^2}{2 + \cfrac{3^2}{2 + \cfrac{5^2}{2 + \cfrac{7^2}{2 + \cdots}}}}}$$

and Euler discovered a pile of formulas like these:

$$\pi^2 = 1 + \frac{1}{2^2} + \frac{1}{3^2} + \frac{1}{4^2} + \frac{1}{5^2} + \frac{1}{6^2} + \cdots$$

$$\frac{\pi^3}{32} = 1 - \frac{1}{3^3} + \frac{1}{3^3} - \frac{1}{7^3} + \frac{1}{9^3} - \frac{1}{11^3} + \cdots$$

$$\frac{\pi^4}{90} = 1 + \frac{1}{2^4} + \frac{1}{3^4} + \frac{1}{4^4} + \frac{1}{5^4} + \frac{1}{6^4} + \cdots$$

(By the way, there seems to be no such formula for

$$1 + \frac{1}{2^3} + \frac{1}{3^3} + \frac{1}{4^3} + \frac{1}{5^3} + \frac{1}{6^3} + \cdots$$

which is very mysterious and not fully understood. In particular, this sum is not any simple rational number times π^3. We do know that the sum of the series is irrational.)

For the other formulas, we'll need the 'sigma notation' for sums. The idea is that: we can write the series for $\pi^2/6$ in the more compact form

$$\frac{\pi^2}{6} = \sum_{n=1}^{\infty} \frac{1}{n^2}$$

Let me unpack this. The fancy Σ symbol is Greek capital sigma, for 'sum', and it tells you to add together all the numbers to its right, namely $1/n^2$. The '$n = 1$' below the Σ says that we start adding from $n = 1$, and by convention n runs through the positive integers. The symbol ∞ over the Σ which means 'infinity', tells us to keep adding these numbers for ever. So this is the same series for $\pi^2/6$ that we saw earlier, but written as an instruction 'Add the terms $1/n^2$, for $n = 1, 2, 3$, and so on, going on for ever.'

Around 1985, Jonathan and Peter Borwein discovered the series

$$\frac{1}{\pi} = \frac{2\sqrt{2}}{9,801} \sum_{n=0}^{\infty} \frac{(4n)!}{(n!)^4} \times \frac{1,103 + 26,390n}{(4 \times 99)^{4n}}$$

which converges extremely rapidly. In 1997 David Bailey, Peter Borwein and Simon Plouffe found an unprecedented formula,

$$\pi = \sum_{n=0}^{\infty} \left(\frac{4}{8n+1} - \frac{2}{8n+4} - \frac{1}{8n+5} - \frac{1}{8n+6} \right) \left(\frac{1}{16} \right)^n$$

Why is this so special? It allows us to calculate a *specific* digit of π without calculating the preceding digits. The only snag is that these are not decimal digits: they are *hexadecimal* (base 16), from which we can also work out a given digit in base 8 (octal), 4 (quaternary) or 2 (binary). In 1998 Fabrice Ballard used this formula to show that the 100 billionth hexadecimal digit of π

is 9. Within two years, the record had risen to 250 trillion hexadecimal digits (one quadrillion binary digits).

The current record for decimal digits of π is held by Yasumasa Kanada and coworkers, who computed the first 1.2411 trillion digits in 2002.

• •

Infinite Wealth

During the early development of probability theory, a lot of effort was expended – mainly by various members of the Bernoulli family, which had four generations of able mathematicians – on a strange puzzle, the *St Petersburg paradox*.

You play against the bank, tossing a coin until it first lands heads. The longer you keep tossing tails, the more the bank will pay out. In fact, if you toss heads on the first try, the bank pays you £2. If you first toss heads on the second try, the bank pays you £4. If you first toss heads on the third try, the bank pays you £8. In general, if you first toss heads on the nth try, the bank pays you $£2^n$.

The question is: how much should you be willing to pay to take part in the game?

To answer this, you should calculate your 'expected' winnings, in the long run, and the rules of probability tell you how. The probability of a head on the first toss is $\frac{1}{2}$, and you then win £1, so the expected gain on the first toss is $\frac{1}{2} \times 2 = 1$. The probability of a head first arising on the second toss is $\frac{1}{4}$, and you then win £4, so the expected gain on the second toss is $\frac{1}{4} \times 4 = 1$. Continuing in this way, the expected gain on the nth toss is $\frac{1}{2}n \times 2n = 1$. In total, your expected winnings amount to

$$1 + 1 + 1 + 1 + \ldots$$

going on forever, which is infinite. Therefore you should pay the bank an infinite amount to play the game.

What—if anything—is wrong here?

Answer on page 299

• •

Let Fate Decide

Two university mathematics students are trying to decide how to spend their evening.

'We'll toss a coin,' says the first. 'If it's heads, we'll go to the pub for a beer.'

'Great!' says the second. 'If it's tails, we'll go to the movies.'

'Exactly. And if it lands on its edge, we'll study.'

Comment: Twice in my life I have witnessed a coin land on its edge. Once was when I was seventeen, playing a game with some friends, and the coin landed in a groove in the table. The second was in 1997, when I gave the Royal Institution Christmas Lectures on BBC Television. We made a large coin from polystyrene, and a young lady from the audience tossed it in a frying-pan like a pancake. The first time she did so, the coin landed stably on its edge.

Admittedly, it was a rather thick coin.

• •

How Many—

Different sets of bridge hands are there?

$$53,644,737,765,488,792,839,237,440,000$$

if you distinguish hands according to who (N, S, E, W) holds them. If not, divide by 8 (the N–S and E–W pairing has to be maintained) to get

$$6,705,592,220,686,099,104,904,680,000$$

Protons are there in the universe according to Sir Arthur Stanley Eddington?

$$136 \times 2^{256} = 15,747,724,136,275,002,577,605,653,961,181,$$
$$555,468,044,717,914,527,116,709,366,231,425,$$
$$076,185,631,031,296$$

Ways are there to rearrange the first 100 numbers?

93,326,215,443,944,152,681,699,238,856,266,700,490,715,
968,264,381,621,468,592,963,895,217,599,993,229,915,608,
941,463,976,156,518,286,253,697,920,827,223,758,251,185,
210,916,864,000,000,000,000,000,000,000,000

unless you argue that 'rearrange' excludes the usual ordering $1, 2, 3, \ldots, 100$. If so, the number is

93,326,215,443,944,152,681,699,238,856,266,700,490,715,
968,264,381,621,468,592,963,895,217,599,993,229,915,608,
941,463,976,156,518,286,253,697,920,827,223,758,251,185,
210,916,863,999,999,999,999,999,999,999,999

Zeros are there in a googol?

100

Googol is a name invented in 1920 by Milton Sirotta (aged 9), nephew of American mathematician Edward Kasner, who popularised the term in his book *Mathematics and the Imagination*. It is equal to 10^{100}, which is 1 followed by one hundred zeros:

10,000,000,000,000,000,000,000,000,000,000,000,000,
000,000,000,000,000,000,000,000,000,000,000,000,000,000,
000,000,000,000,000,000

Zeros are there in a googolplex?

10^{100}

Googolplex is another invented name, equal to $10^{10^{100}}$, which is 1 followed by 10^{100} zeros. The universe is too small to write it down in full, and the lifetime of the universe is too short anyway. Unless our universe is part of a much larger multiverse, and even then it's hard to see why anyone would bother.

What Shape is a Rainbow?

Why?

We all remember being told what causes rainbows. Sunlight bounces around inside raindrops, which split the white light into its component colours. Whenever you look directly at a rainbow, the Sun will be behind you, and the rain will be falling in front of you. And to knock it on the head, the teacher showed us how a glass prism splits a ray of white light into all the colours of the rainbow.

A neat piece of misdirection, worthy of a conjurer. That explains the colours. But what about the *shape*?

If it's just a matter of light reflecting back from raindrops, why don't we see the colours wherever the rain is coming down? And if that were happening, wouldn't the colours fuzz out back to white, or maybe a muddy grey? Why is the rainbow a series of coloured arcs? And what shape are the arcs?

Answers on page 300

Alien Abduction

Two aliens from the planet Porqupyne want to abduct two Earthlings, but are blissfully unaware that the objects of their attention are actually pigs. In their formal way, the aliens play a game.

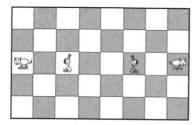

First catch your pig ...

On the first move, each alien moves one square horizontally or vertically (*not* diagonally). Each alien can move in any of the four directions, independently of what the other one does. On the next move, the pigs can do likewise. The aliens get to abduct any pig upon whose square they land. To their surprise, the pigs always seem to get away. *What are the aliens doing wrong?*

Answers on page 302

The Riemann Hypothesis

If there is one single problem that mathematicians would dearly love to solve, it is the Riemann Hypothesis. Entire areas of mathematics would open up if some bright spark could prove this wonderful theorem. And entire areas of mathematics would close down if some bright spark could disprove it. Right now, those areas are in limbo. We can get a glimpse of the Promised Land, but for all we know it might be a mirage.

Oh, there's also a million-dollar prize on offer from the Clay Mathematics Institute.

The story goes back to the time of Gauss, around 1800, and the discovery that although the prime numbers seem rather randomly distributed doing the number line, they have clear *statistical* regularities. Various mathematicians noticed that the number of primes up to some number x, denoted by $\pi(x)$ just to confuse everyone who thought that $\pi = 3.141\,59$, is approximately

$$\pi(x) = \frac{x}{\log x}$$

Gauss found what seemed to be a slightly better approximation, the *logarithmic integral*

$$\mathrm{Li}(x) = \int_2^x \frac{\mathrm{d}x}{\log x}$$

Now, it's one thing to notice this 'prime number theorem', but what counts is proving it, and that turned out to be hard. The most powerful approach is to turn the question into something quite different, in this case complex analysis. The connection between primes and complex functions is not at all obvious, but the key idea was spotted by Euler.

Every positive integer is a product of primes in a unique way. We can formulate this basic property analytically. A first attempt would be to notice that

$$(1 + 2 + 2^2 + 2^3 + \ldots) \times (1 + 3 + 3^2 + 3^3 + \ldots)$$
$$\times (1 + 5 + 5^2 + 5^3 + \ldots) \times \ldots$$

with each bracketed series going on for ever, and taking the product over all primes, is equal to

$$1 + 2 + 3 + 4 + 5 + 6 + 7 + 8 + \ldots$$

summed over all integers. For example, to find out where a number like 360 comes from, we write it as a product of primes

$$360 = 2^3 \times 3^2 \times 5$$

and then pick out from the formula the corresponding terms, here shown in bold:

$$(1 + 2 + 2^2 + \mathbf{2^3} + \ldots) \times (1 + 3 + \mathbf{3^2} + 3^3 + \ldots)$$
$$\times (1 + \mathbf{5} + 5^2 + 5^3 + \ldots) \times \ldots$$

When you 'expand' the brackets, each possible product of prime powers occurs exactly once.

Unfortunately this makes no sense because the series diverge to infinity, and so does the product. However, if we replace each number n by a suitable power n^{-s}, and make s large enough, everything converges. (The minus sign ensures that *large* values of s lead to convergence, which is more convenient.) So we get the formula

$$(1 + 2^{-s} + 2^{-2s} + 2^{-3s} + \ldots) \times (1 + 3^{-s} + 3^{-2s} + 3^{-3s} + \ldots)$$
$$\times (1 + 5^{-s} + 5^{-2s} + 5^{-3s} + \ldots) \times \ldots$$
$$= 1 + 2^{-s} + 3^{-s} + 4^{-s} + 5^{-s} + 6^{-s} + 7^{-s} + 8^{-s} + \ldots$$

(where I've written 1 instead of 1^{-s} because these are equal anyway.) This formula makes perfectly good sense provided s is real and greater than 1. It's true because

$$60^{-s} = 2^{-3s} \times 3^{-2s} \times 5^{-s}$$

and similarly for any positive integer.

In fact, the formula makes perfectly good sense if $s = a+ib$ is complex and its real part a is greater than 1. The final series in the formula is called the *Riemann zeta function* of s, denoted by $\zeta(s)$. Here ζ is the Greek letter zeta.

In 1859 Georg Riemann wrote a brief, astonishingly inventive paper showing that the analytic properties of the zeta function reveal deep statistical features of primes, including Gauss's prime number theorem. In fact, he could do more: he could make the error in the approximation of $\pi(x)$ much smaller, by adding further terms to Gauss's expression. Infinitely many such terms, themselves forming a convergent series, would make the error disappear altogether. Riemann could write down an *exact* expression for $\pi(x)$ as an analytic series.

For the record, here's his formula:

$$\pi(x) + \pi(x^{1/2}) + \pi(x^{1/3}) + \ldots$$
$$= \mathrm{Li}(x) + \int_x^\infty \left[(t^2 - 1)t \log t\right]^{-1} t - \log 2 - \sum_\rho \mathrm{Li}(x^\rho)$$

where ρ runs through the non-trivial zeros of the zeta function. Strictly speaking, this formula is not quite correct when the left-hand side has discontinuities, but that can be fixed up. You can get an even more complicated formula for $\pi(x)$ itself by applying the formula again with x replaced by $x^{1/2}, x^{1/3}$, and so on.

All very pretty, but there was one tiny snag. In order to prove that his series is correct, Riemann needed to establish an apparently straightforward property of the zeta function. Unfortunately, he couldn't find a proof.

All complex analysts learn at their mother's knee (the mother here being Augustin-Louis Cauchy, who along with Gauss first understood the point) that the best way to understand any complex function is to work out where its *zeros* lie. That is: which complex numbers s make $\zeta(s) = 0$? Well, it becomes the best way after some nifty footwork; in the region where the series for $\zeta(s)$ converges, there *aren't* any zeros. However, there is another formula which agrees with the series whenever it converges, but also makes sense when it doesn't. This formula lets us extend the definition of $\zeta(s)$ so that it makes sense for *all* complex numbers s. And this 'analytic continuation' of the zeta function does have zeros. Infinitely many of them.

Some of the zeros are obvious – once you see the formula involved in the continuation process. These 'trivial zeros' are the negative even integers -2, -4, -6, and so on. The other zeros come in pairs $a + ib$ and $a - ib$, and all such zeros that Riemann could find had $a = \frac{1}{2}$. The first three pairs, for instance, are

$$\tfrac{1}{2} \pm 14.13i, \quad \tfrac{1}{2} \pm 21.02i, \quad \tfrac{1}{2} \pm 25.01i$$

Evidence like this led Riemann to conjecture ('hypothesise') that *all* non-trivial zeros of the zeta function must lie on the so-called critical line $\frac{1}{2} + ib$.

If he could prove this statement – the famous *Riemann Hypothesis* – then he could prove that Gauss's approximate formula for $\pi(x)$ is correct. He could improve it to an *exact* – though complicated – formula. Great vistas of number theory would be wide open for development.

But he couldn't, and we still can't.

Eventually, the prime number theorem was proved, independently, by Jacques Hadamard and Charles de la Vallée-Poussin in 1896. They used complex analysis, but managed to find a proof that avoided the Riemann Hypothesis. We now know that the first ten trillion non-trivial zeros of the zeta function lie on the critical line, thanks to Xavier Gourdon and Patrick Demichel in 2004. You might think that ought to settle the matter, but in this area of number theory ten trillion is ridiculously small, and may be misleading.

The Riemann Hypothesis is important for several reasons. If true, it would tell us a lot about the statistical properties of primes. In particular, Helge von Koch proved in 1901 that the Riemann Hypothesis is true if and only if the estimate

$$|\pi(x) - \text{Li}(x)| < C\sqrt{x}\log x$$

for the error in Gauss's formula holds for some constant C. Later, Lowell Schoenfeld proved that we can take $C = 1/8\pi$ for all $x \geqslant 2{,}657$. (Sorry, this area of mathematics does that kind of thing.) The point here is that the error is small compared with x, and it tells us how much the primes fluctuate away from their more typical behaviour.

Riemann's exact formula, of course, would also follow from the Riemann Hypothesis. So would a huge list of other mathematical results – you can find some of them at en.wikipedia.org/wiki/Riemann_hypothesis

However, the main reason why the Riemann Hypothesis is important – apart from 'because it's there' – is that it has a lot of far-reaching analogues and generalisations in algebraic number theory. A few of the analogues have even been proved. There is a feeling that if the Riemann Hypothesis can be proved in its original form, then so can the generalisations. These ideas are too technical to describe, but see mathworld.wolfram.com/RiemannHypothesis.html

I will tell you one deceptively simple statement that is equivalent to the Riemann Hypothesis. Of itself, it looks harmless and unimportant. Not so! Here's how it goes. If n is a whole number, then the sum of its divisors, including n itself, is written as $\sigma(n)$. (Here σ is the lower-case Greek letter 'sigma'.) So

$$\sigma(24) = 1 + 2 + 3 + 4 + 6 + 8 + 12 + 24 = 60$$
$$\sigma(12) = 1 + 2 + 3 + 4 + 6 + 12 = 28$$

and so on. In 2002 Jeffrey Lagarias proved that the Riemann Hypothesis is equivalent to the inequality

$$\sigma(n) \leqslant e^{H_n} \log H_n$$

for every n. Here H_n is the nth harmonic number, equal to

$$1 + \frac{1}{2} + \frac{1}{3} + \frac{1}{4} + \ldots \frac{1}{n}$$

• •

Anti-Atheism

Godfrey Harold Hardy, a Cambridge mathematician who worked mainly in analysis, claimed to believe in God – but unlike most believers, he considered the Deity to be his personal enemy. Hardy had it in for God, and he was convinced that God had it in for Hardy, which was only fair. Hardy was especially worried whenever he travelled by sea, in case God sank the boat. So before travelling, he would send his colleagues a telegram: 'HAVE PROVED RIEMANN HYPOTHESIS. HARDY.' He would then retract this claim on arrival.

As just discussed, the Riemann Hypothesis is the most famous unsolved problem in mathematics, and one of the most important. And so it was in Hardy's day, too. When his colleagues asked him why he sent such telegrams, he explained that God would never let him die if that would give him credit – however controversial – for proving the Riemann Hypothesis.

• •

Disproof of the Riemann Hypothesis

Consider the following logical argument:

- Elephants never forget.
- No creature that has ever won *Mastermind* has possessed a trunk.
- A creature that never forgets will always win *Mastermind* provided it takes part in the competition.
- A creature lacking a trunk is not an elephant.
- In 2001 an elephant took part in *Mastermind*.

Therefore:

- The Riemann Hypothesis is false.

Is this a correct deduction?

Answer on page 302

● ●

Murder in the Park

This puzzle – like several in this book – goes back to the great English puzzlist Henry Ernest Dudeney. He called it 'Ravensdene Park'. I've made a few trivial changes.

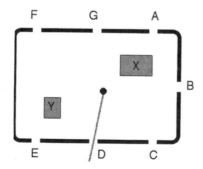

Ravensdene Park.

Soon after a heavy fall of snow, Cyril Hastings entered Ravensdene Park at gate D, walked straight to the position marked with a black dot, and was stabbed through the heart. His body was found the next morning, along with several tracks in the snow. The police immediately closed the park.

Their subsequent investigations revealed that each track had been made by a different, very distinctive shoe. Witnesses placed four individuals, other than Hastings, in the park during the period concerned. So the murderer had to be one of them. Examining their shoes, the police deduced that:

- The butler – who could prove he had been in the house X at the time of the murder – had entered at gate E and gone to X.
- The gamekeeper – who had no such alibi – had entered at gate A and gone to his lodge at Y.
- A local youth had entered at gate G and left by gate B.
- The grocer's wife had entered at gate C and left by gate F.

None of these individuals entered or left the park more than once.

It had been foggy as well as snowy, so the routes these people took were often rather indirect. The police did notice that no two paths crossed. But they failed to make a sketch of the routes before the snow melted and they disappeared.

So who was the murderer?

Answer on page 303

. . . .

The Cube of Cheese

An oldie, but none the worse for that. Marigold Mouse has a cube of cheese and a carving-knife. She wishes to slice the cheese along a flat plane, to obtain a cross-section that is a regular hexagon. Can she do this, and if so, how?

Answer on page 304

. . . .

The Game of Life

In the 1970s John Conway invented the Game of Life. Strange black creatures scuttle across a grid of white tiles, changing shape, growing, collapsing, freezing and dying. The best way to play Life (as it's commonly known) is to download suitable software. There are several excellent free programs for Life on the web, easily located by searching. A Java version, which is easy to use and will give hours of pleasure, can be found at: www.bitstorm.org/gameoflife/

Life is played with black counters on a potentially infinite grid of square cells. Each cell holds either one counter or none. At each stage, or *generation*, the set of counters defines a *configuration*. The initial configuration at generation 0 evolves at successive stages according to a short list of rules. The rules are illustrated in the diagrams below. The *neighbours* of a given cell are the eight cells immediately adjacent to it, horizontally, vertically or diagonally. All births and deaths occur simultaneously: what happens to each counter or empty cell in generation $n+1$ depends only on its neighbours in generation n.

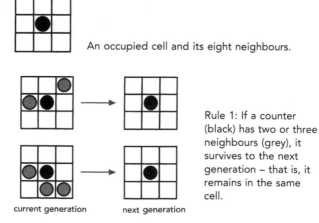

An occupied cell and its eight neighbours.

Rule 1: If a counter (black) has two or three neighbours (grey), it survives to the next generation – that is, it remains in the same cell.

current generation   next generation

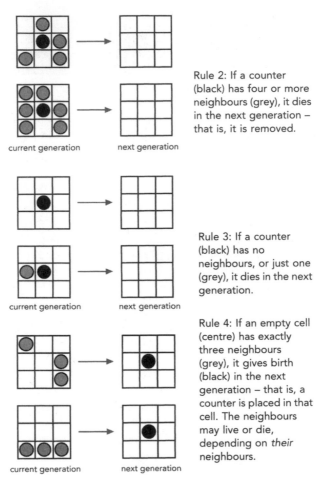

Rule 2: If a counter (black) has four or more neighbours (grey), it dies in the next generation – that is, it is removed.

current generation   next generation

Rule 3: If a counter (black) has no neighbours, or just one (grey), it dies in the next generation.

current generation   next generation

Rule 4: If an empty cell (centre) has exactly three neighbours (grey), it gives birth (black) in the next generation – that is, a counter is placed in that cell. The neighbours may live or die, depending on *their* neighbours.

current generation   next generation

Starting from any given initial configuration, the rules are applied repeatedly to produce the life history of its succeeding generations. For example, here is the life history of a small triangle built from four counters:

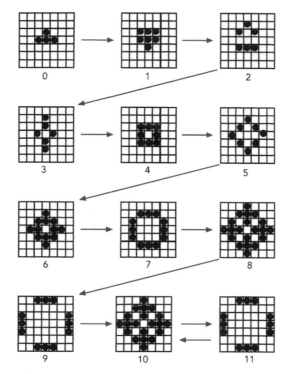

Life history of a configuration. Configurations 8 and 9 alternate periodically.

Even this simple examples shows that the rules for Life can generate complex structures from simpler ones. Here the sequence of generations becomes *periodic*: at generation 10 the configuration is the same as at generation 8, and thereafter configurations 8 and 9 alternate, a sequence known as *traffic lights*.

One of the fascinations of Life is the astonishing variety of life histories, and the absence of any *obvious* relation between the initial configuration and what it turns into. The system of rules is entirely deterministic – the entire infinite future of the system is implicit in its initial state. But Life dramatically demonstrates the difference between determinism and predictability. This is where the name 'Life' comes from.

From a mathematical point of view, it is natural to classify Life configurations according to their long-term behaviour. For example, configurations may:

(1) disappear completely (die)

(2) attain a steady state (stasis)

(3) repeat the same sequence over and over again (periodicity)

(4) repeat the same sequence over and over again but end up in a new location

(5) behave chaotically

(6) exhibit computational behaviour (universal Turing machine).

Among the common periodic configurations are the *blinker* and traffic lights, with periods 3 and 8, respectively:

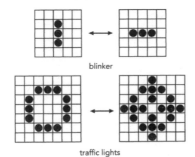

blinker

traffic lights

Two periodic configurations.

The outcome of a game of Life is extraordinarily sensitive to the precise choice of the initial state. A difference of one cell can totally change the state's future. Moreover, simple initial configurations can sometimes develop into very complicated ones. This behaviour – to some extent 'designed in' by the choice of update rules – motivates the game's name.

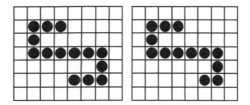

The left-hand S-shaped configuration eventually settles down after 1,405 generations, by which time it has given birth to 2 gliders, 24 blocks, 6 ponds, 4 loaves, 18 beehives and 8 blinkers. If you delete just one cell to give the right-hand configuration, everything dies completely after 61 generations.

The prototype mobile state is the *glider*, which moves one cell diagonally every four moves:

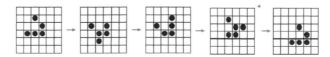

Motion of a glider.

Three *spaceships* (lightweight, middleweight, heavyweight) repeat cycles that cause them to move horizontally, throwing off sparks that vanish immediately. Longer spaceships do not work on their own – they break apart in complicated ways – but they can be supported by flotillas of accompanying smaller spaceships.

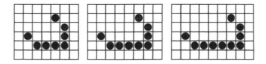

Spaceships.

One of the earliest mathematical questions about Life was whether there can exist a finite initial configuration whose future configurations are *unbounded* – that is, it will become as large as we wish if we allow enough time to pass. This question was answered in the affirmative by Bill Gosper's invention of a *glider gun*. The configuration shown below in black oscillates with period 30, and repeatedly fires gliders (the first two are shown in grey). The stream of gliders grows unboundedly.

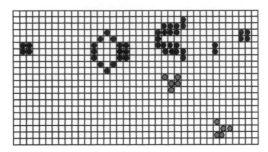

The glider gun, with two gliders it has spat out.

It turns out that the Game of Life has configurations which act like computers, able in principle to calculate anything that a computer program can specify. For example, such a configuration can compute π to as many decimal places as we want. In practice, such computations run incredibly slowly, so don't throw away your PC just yet.

In fact, even simpler 'games' of the same kind – known as cellular automata (page 239) – living on a line of squares instead of a two-dimensional array, can behave as universal computers. This automaton, known as 'rule 110', was suggested by Stephen Wolfram in the 1980s, and Matthew Cook proved its universality in the 1990s. It illustrates, in a very dramatic way, how astonishingly complex behaviour can be generated by very simple rules. See mathworld.wolfram.com/Rule110.html

• •

Two-Horse Race

Every whole number can be obtained by multiplying suitable primes together. If this requires an even number of primes, we say that the number is of *even type*. If it requires an odd number of primes, we say that the number is of *odd type*. For instance,

$$96 = 2 \times 2 \times 2 \times 2 \times 2 \times 3$$

uses six primes, so is of even type. On the other hand,

$$105 = 3 \times 5 \times 7$$

uses three primes, so is of odd type. By convention, 1 is of even type.

For the first ten whole numbers, 1–10, the types are:

Odd		2	3		5		7	8		
Even	1			4		6			9	10

A striking fact emerges: in general, odd types occur at least as frequently as even types. Imagine two horses, Odd and Even, racing. Start them level with each other, and read along the sequence of numbers: $1, 2, 3, \ldots$ At each stage, move Odd forward one step if the next number has odd type; move Even forward one step if the next number has even type. So:

> After 1 step, Even is ahead.
> After 2 steps, Odd and Even are level.
> After 3 steps, Odd is ahead.
> After 4 steps, Odd and Even are level.
> After 5 steps, Odd is ahead.
> After 6 steps, Odd and Even are level.
> After 7 steps, Odd is ahead.
> After 8 steps, Odd is ahead.
> After 9 steps, Odd is ahead.
> After 10 steps, Odd and Even are level.

Odd always seems to be level, or ahead. In 1919 George Pólya conjectured that Odd *never* falls behind Even, except right at the start, step 1. Calculations showed that this is true for the first million steps. Given this weight of favourable evidence, surely it has to be true for any number of steps?

Without a computer you can waste a lot of time on this question, so I'll tell you the answer. Pólya was *wrong*! In 1958, Brian Haselgrove proved that at some (unknown) stage Odd falls behind Even. Once reasonably fast computers were available, it was easy to test ever larger numbers. In 1960 Robert Lehman discovered that Even is in the lead at step 906,180,359. In 1980

Minoru Tanaka proved that Even *first* takes the lead at step 906,150,257.

This kind of thing is what makes mathematicians insist upon proofs. And it shows that even a number like 906,150,257 can be interesting and unusual.

• •

Drawing an Ellipse – and More?

It is well known that an easy way to draw an ellipse is to fix two pins through the paper, tie a loop of string round them, and place your pencil so that the string stays taut. Gardeners sometimes use this method to map out elliptical flowerbeds. The two pins are the *foci* (plural of *focus*, and pronounced 'foe-sigh') of the ellipse.

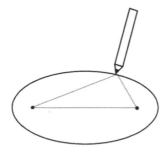

How to draw an ellipse.

Suppose that you use three pegs, in a triangle. It need not be an equilateral or isosceles triangle.

Why isn't this interesting too?

That ought to give some interesting new kinds of curves. So why don't the mathematics books mention them?

Answer on page 304

· ·

Mathematical Jokes 3

Two mathematicians in a cocktail bar are arguing about how much maths the ordinary person knows. One thinks they're hopelessly ignorant; the other says that quite a few people actually know a lot about the subject.

'Bet you twenty pounds I'm right,' says the first, as he heads for the gents. While he is gone, his colleague calls the waitress over.

'Listen, there's ten pounds in it for you if you come over when my friend gets back and answer a question. The answer is "one-third *x* cubed." Got that?'

'Ten pounds for saying "One thirdex cue?"'

'No, one-third *x* cubed.'

'One thir dex cubed?'

'Yeah, that'll do.'

The other mathematician comes back, and the waitress comes over.

'Hey – what's the integral of *x* squared?'

'One third *x* cubed,' says the waitress. As she walks away, she adds, over her shoulder, 'Plus a constant.'

· ·

The Kepler Problem

Mathematicians have learned that apparently simple questions are often hard to answer, and apparently obvious facts may be false, or may be true but extremely hard to prove. The Kepler problem is a case in point: it took nearly three hundred years to solve it, even though everyone knew the correct answer from the start.

It all began in 1611 when Johannes Kepler, a mathematician and astrologer (yes, he cast horoscopes; lots of mathematicians did at that time – it was a quick way to make money) wanted to give his sponsor a New Year's gift. The sponsor rejoiced in the name Johannes Mathäus Wacker of Wackenfels, and Kepler wanted to say 'thanks for all the cash' without actually spending any of it. So he wrote a book, and presented it to his sponsor. Its title (in Latin) was *The Six-Cornered Snowflake*. Kepler started with the curious shapes of snowflakes, which often form beautiful sixfold symmetric crystals, and asked why this happened.

A typical 'dendritic' snowflake.

It is often said that 'no two snowflakes are alike'. The logician in me objects 'How can you tell?' but a back-of-the-envelope calculation suggests that there are so many features in a 'dendritic' snowflake, of the kind illustrated, that the chance of two being identical is pretty much zero.

No matter. What matters here is that Kepler's analysis of the snowflake led him to the idea that its sixfold symmetry arises because that's the most efficient way to pack circles in a plane.

Take a lot of coins, of the same denomination – pennies, say. If you lay them on a table and push them together tightly, you quickly discover that they fit perfectly into a honeycomb pattern, or 'hexagonal lattice':

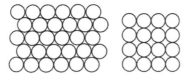

(Left) The closest way to pack circles, and (right) a less efficient lattice packing.

And this is the *closest* packing – the one that fills space most efficiently, in the ideal case of infinitely many circles arranged on a plane. Alternatives, such as the square lattice on the right, are less efficient.

Mind you, this innocent assertion wasn't proved until 1940, when László Fejes Tóth managed it. (Axel Thue sketched out a proof in 1892, and gave more details in 1910, but he left some gaps.) Tóth's proof was quite hard. Why the difficulty? We don't know, to begin with, that the most efficient packing forms a regular lattice. Maybe something more random could work better. (For *finite* packings, say inside a square, this can actually happen – see the next puzzle, about a milk crate.)

Along the way, Kepler came very close indeed to the idea that all matter is made from tiny indivisible components, which we now call 'atoms'. This is impressive, given that he did no experiments in the course of writing his book. Atomic theory, introduced by the Greek Democritus, was not established experimentally until about 1900.

Kepler had his eye on something a bit more complicated, though: the closest way to pack identical spheres in space. He was aware of three regular 'lattice' packings, which we now call the *hexagonal, cubic* and *face-centred cubic* lattices. The first of these is formed by stacking lots of honeycomb layers of spheres on top of one another, with the centres of corresponding spheres forming a vertical line. The second is made from square-lattice layers, also stacked vertically. For the third, we stack hexagonal layers, but fit the spheres in any given layer into the hollows in the one below.

You can get the same result, though tilted, by similarly stacking square-lattice layers so that the spheres in any given layer fit into the hollows in the one below – this isn't entirely obvious, and – like the milk crate puzzle – it shows that intuition may not be a good guide in this area. The picture shows how this happens: the horizontal layers are square, but the slanting layers are hexagonal.

Part of a face-centred cubic lattice.

Now, every greengrocer knows that the way to stack oranges is to use the face-centred cubic lattice.* By thinking about pomegranate seeds, Kepler was led to the casual remark that with the face-centred cubic lattice, 'the packing will be the tightest possible'.

That was in 1611. The proof that Kepler was right had to wait until 1998, when Thomas Hales announced that he had achieved this with massive computer assistance. Basically, Hales considered all possible ways to surround a sphere with other spheres, and showed that if the arrangement wasn't the one found in the face-centred cubic lattice, then the spheres could be shoved closer together. Tóth's proof in the plane used the same ideas, but he only had to check about forty cases.

Hales had to check thousands, so he rephrased the problem in terms that could be verified by a computer. This led to a huge computation – but each step in it is essentially trivial. Almost all of the proof has been checked independently, but a very tiny level of doubt still remains. So Hales has started a new computer-

* They don't say it that way, but they stack it that way.

based project to devise a proof that can be verified by standard proof-checking software. Even then, a computer will be involved in the verifications, but the software concerned does such simple things – in principle – that a human can check that the software does what it is supposed to. The project will probably take 20 years. You can still object on philosophical grounds if you wish, but you'll be splitting logical hairs very finely indeed.

What makes the problem so hard? Greengrocers usually start with a square box that has a flat base, so they naturally pack their oranges in layers, making each layer a square lattice. It is then natural to make the second layer fill the gaps in the bottom one, and so on. If by chance they start with a hexagonal layer instead, they get the same packing anyway, except for a tilt. Gauss proved in 1831 that Kepler's packing is the tightest *lattice* packing. But the mathematical problem here is to prove this, without assuming at the start that the packing forms flat layers. The mathematician's spheres can hover unsupported in space. So the greengrocer's 'solution' involves a whole pile of assumptions – well, actually, oranges. Since experiments aren't proofs, and here even the experiment is dodgy, you can see that the problem could be harder than it seems.

• •

The Milk Crate Problem

Here's a simpler question of the same kind. A milkman wishes to pack identical bottles, with circular cross-section, into a square crate. To him, it's obvious that for any given *square* number of bottles – 1, 4, 9, 16, and so on – the crate can be made as small as possible by packing the bottles in a regular square array. (He can see that with a non-square number of bottles, there are gaps and maybe the bottles can be jiggled around to shrink the crate.)

Is he right?

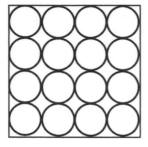

How the milkman fits
16 bottles into the
smallest possible
square crate.

Answer on page 305

Equal Rights

One of the leading female mathematicians of the early twentieth century was Emmy Noether, who studied at the University of Göttingen. But after she completed her doctorate the authorities refused to allow her to proceed to the status of *Privatdozent*, which would allow her to charge students fees for tuition. Their stated reason was that women were not permitted to attend faculty meetings at the university senate. The head of the mathematics department, the great David Hilbert, is said to have remarked: 'Gentlemen! There is nothing wrong with having a woman in the senate. Senate is not a public bath.'

Road Network

Four towns – Aylesbury, Beelsbury, Ceilsbury and Dealsbury – lie at the corners of a square, of side 100 km. The highways department wishes to connect them all together using the shortest possible network of roads.

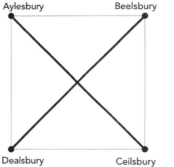

Not like this.

'We can run roads straight from Aylesbury to Beelsbury to Ceilsbury to Dealsbury,' said the assistant town planner. 'That's 300 km of roads.'

'No, we can do better than that!' his boss replied. 'Two diagonals, which, if you recall your Pythagoras, amount to 200 $\sqrt{2}$ km – about 282 km.'

What is the shortest network? Using the diagonals of the square is *not* the answer.

Answer on page 305

• •

Tautoverbs

In Terry Pratchett's *Discworld* series of fantasy novels, the members of the Order of Wen the Eternally Surprised, better known as the History Monks, are greatly impressed by the homespun wisdom of Mrs Marietta Cosmopilite. They have never before heard her homely homilies (such as 'I haven't got all day, you know'), so to those monks who follow the Way of Mrs Cosmopilite, her simple assertions are marvellous new philosophical insights.

Mathematicians take a more jaundiced view of folk wisdom, and habitually revise proverbs to make them more logical. Indeed tautological – trivially true. Thus the proverb 'Penny wise, pound foolish' becomes the *tautoverb* 'Penny wise, wise about pennies,' which makes more sense and is difficult to dispute. And 'Look after the pennies and the pounds will look after themselves' is more convincing in the form 'Look after the pennies and the pennies will be looked after.'

As a kid I was always vaguely bothered that in their original form these two proverbs contradict each other, though I now see this as the default mechanism whereby folk wisdom ensures that it is perceived to be wise. The revised versions do *not* conflict – clear evidence of their superiority. I'll give you a couple of other examples to get you started, and then turn you lose on the opening words of several proverbs. Your job is to complete them to make tautoverbs. The first example is simple and direct, the other more baroque. Both forms are permissible. So is helpful commentary, preferably blindingly obvious. Logical quibbles are actively encouraged – the more pedantic, the better.

- He who fights and runs away will live to run away again.
- A bird in the bush is worth two in the hand, because free-range produce is always expensive.

OK, now it's your turn. In the same spirit, complete the following tautoverbs:

- No news is—
- The bigger they are—
- Nothing ventured—
- Too many cooks—
- You cannot have your cake—
- A watched pot—
- If pigs had wings—

If you have enjoyed this game, psychiatric help is recommended, but until it arrives you can find a lot more proverbs to work on at www.manythings.org/proverbs/index.html

Answers on page 306

Complexity Science

Complexity science, or the theory of complex systems, came to prominence with the founding of the Santa Fe Institute (SFI) in 1984, by George Cowan and Murray Gell-Mann. This was, and still is, a private research centre for interdisciplinary science, with emphasis on 'the sciences of complexity'. You might think that 'complexity' refers to anything complicated, but the SFI's main objective has been to develop and disseminate new mathematical techniques that could shed light on systems in which very large numbers of agents or entities interact with one another according to relatively simple rules. A key phenomenon is what is called *emergence*, in which the system as a whole behaves in ways that are not available to the individual entities.

An example of a real-world complex system is the human brain. Here the entities are nerve cells – neurons – and the emergent features include intelligence and consciousness. Neurons are neither intelligent nor conscious, but when enough of them are hooked together, these abilities emerge. Another example is the world's financial system. Now the entities are bankers and traders, and emergent features include stock-market booms and crashes. Other examples are ants' nests, ecosystems and evolution. You can probably work out what the entities are for each of these, and think of some emergent features. Anyone can play this game.

What's harder, and what SFI was, and still is, all about, is to model such systems mathematically in a way that reflects their underlying structure as an interacting system of simple components. One such modelling technique is to employ a cellular automaton – a more general version of John Conway's Game of Life. This is like a computer game played on a square grid. At any given instant, each square exists in some state, usually represented by what colour it is. As time ticks to the next instant, each square changes colour according to some list of rules. The rules involve the colours of neighbouring squares, and might be something like this: 'a red square changes to green if it has between two and six blue neighbours'. Or whatever.

Three types of pattern formed by a simple cellular automaton: static (blocks of the same colour), structured (the spirals), and chaotic (for example the irregular patch at bottom right).

It might seem unlikely that such a rudimentary gadget can achieve anything interesting, let alone solve deep problems of complexity science, but it turns out that cellular automata can behave in rich and unexpected ways. In fact their earliest use, by John von Neumann in the 1940s, was to prove the existence of an abstract mathematical system that could self-replicate – make copies of itself.* This suggested that the ability of living creatures to reproduce is a logical consequence of their physical structure, rather than some miraculous or supernatural process.

Evolution, in Darwin's sense, offers a typical example of the complexity-theory approach. The traditional mathematical model of evolution is known as population genetics, which goes

---

* There is now a lot of interest in doing the same with real machines, using nanotechnology. There are many science fiction stories about 'Von Neumann machines', often employed by aliens or machine cultures to invade planets, including our own. The techniques used to pack millions of electronic components on to a tiny silicon chip are now being used to build extremely tiny machines, 'nanobots', and a true replicating machine may not be so far away. Alien invasions are not a current cause for concern, but the possibility of a mutant Von Neumann machine turning the Earth into 'grey goo' has raised issues about the safety, and control, of nanotechnology. See en.wikipedia.org/wiki/Grey_goo

back to the British statistician Sir Ronald Fisher, around 1930. This approach views an ecosystem – a rainforest full of different plants and insects, or a coral reef—as a vast pool of genes. As the organisms reproduce, their genes are mixed together in new combinations.

For example, a hypothetical population of slugs might have genes for green or red skins, and other genes for a tendency to live in bushes or on bright red flowers. Typical gene combinations are green–bush, green–flower, red–bush, and red–flower. Some combinations have greater survival value than others. For example red–bush slugs would easily be seen by birds against the green bushes they live in, whereas red–flower slugs would be less visible.

As natural selection weeds out unfit combinations, the combinations that allow organisms to survive better tend to proliferate. Random genetic mutations keep the gene pool simmering. The mathematics centres on the *proportions* of particular genes in the population, and works out how those proportions change in response to selection.

A complexity model of slug evolution would be very different. For instance, we could set up a cellular automaton, assigning various environmental characteristics to each cell. For example, a cell might correspond to a piece of bush, or a flower, or whatever. Then we choose a random selection of cells and populate them with 'virtual slugs', assigning a combination of slug genes to each such cell.

Other cells could be 'virtual predators'. Then we specify rules for how the virtual organisms move about the grid and interact with one another. For example, at each time-step a slug must either stay put or move to a random neighbouring cell. On the other hand, a predator might 'see' the nearest slug and move five cells towards it, 'eating' it if it reaches the slug's own cell – so that particular virtual slug is removed from the computer's memory.

We would set up the rules so that green slugs are less likely to be 'seen' if they are on bushes rather than flowers. Then this mathematical computer game would be allowed to run for a few million time-steps, and we would read off the proportions of various surviving slug gene combinations.

Complexity theorists have invented innumerable models in the same spirit: building in simple rules for interactions between many individuals, and then simulating them on a computer to see what happens. The term 'artificial life' has been coined to describe such activities. A celebrated example is Tierra, invented by Tom Ray around 1990. Here, short segments of computer code compete with one another inside the computer's memory, reproducing and mutating (see www.nis.atr.jp/~ray/tierra/). His simulations show spontaneous increases in complexity, rudimentary forms of symbiosis and parasitism, lengthy periods of stasis punctuated by rapid changes – even a kind of sexual reproduction. So the message from the simulations is that all these puzzling phenomena are entirely natural, provided they are seen as emergent properties of simple mathematical rules.

The same difference in working philosophy can be seen in economics. Conventional mathematical economics is based on a model in which every player has complete and instant information. As the Stanford economist Brian Arthur puts it, the assumption is that 'If two businessmen sit down to negotiate a deal, in theory each can instantly foresee all contingencies, work out all possible ramifications, and effortlessly choose the best strategy.' The goal is to demonstrate mathematically that any economic system will rapidly home in on an equilibrium state, and remain there. In equilibrium, every player is assured of the best possible financial return for themselves, subject to the overall constraints of the system. The theory puts formal flesh on the verbal bones of Adam Smith's 'invisible hand of the market'.

Complexity theory challenges this cosy capitalist utopia in a number of ways. One central tenet of classical economic theory is the 'law of diminishing returns', which originated with the English economist David Ricardo around 1820. This law asserts that any economic activity that undergoes growth must eventually be limited by constraints. For example, the plastics industry depends upon a supply of oil as raw material. When oil is cheap, many companies can move over from, say, metal components to plastic ones. But this creates increasing demand for oil, so the price goes up. At some level, everything balances out.

Modern hi-tech industries, however, do not follow this pattern at all. It costs perhaps a billion dollars to set up a factory to make the latest generation of computer memory chips, and until the factory begins production, the returns are zero. But once the factory is in operation, the cost of producing chips is tiny. The longer the production run, the cheaper chips are to make. So here we see a law of *increasing* returns: the more goods you make, the less it costs you to do so.

From the point of view of complex systems, the market is not a simple mathematical equilibrium-seeker, but a 'complex adaptive system', where interacting agents modify the rules that govern their own behaviour. Complex adaptive systems often settle into interesting patterns, strangely reminiscent of the complexities of the real world. For example, Brian Arthur and his colleagues have set up computer models of the stock market in which the agents search for patterns – genuine or illusory – in the market's behaviour, and adapt their buying and selling rules according to what they perceive. This model shares many features of real stock markets. For example, if many agents 'believe' that the price of a stock will rise, they buy it, and the belief becomes self-fulfilling.

According to conventional economic theory, none of these phenomena should occur. So why do they happen in complexity models? The answer is that the classical models have inbuilt mathematical limitations, which preclude most kinds of 'interesting' dynamics. The greatest strength of complexity theory is that it resembles the untidy creativity of the real world. Paradoxically, it makes a virtue of simplicity, and draws far-ranging conclusions from models with simple – but carefully chosen – ingredients.

• •

Scrabble Oddity

The letter scores at Scrabble are:

Score 1	A, E, I, L, N, O, R, S, T, U
Score 2	D, G
Score 3	B, C, M, P
Score 4	F, H, V, W, Y
Score 5	K
Score 8	J, X
Score 10	Q, Z

Which positive integer is equal to its own Scrabble score when spelt out in full?

Answer on page 306

Dragon Curve

The picture shows a sequence of curves, called dragon curves (look at the last one). The sequence can be continued indefinitely, getting ever more complicated curves.

What is the rule for making them? Ignore the 'rounding' of the corners by the short lines, which is done so that later curves in the sequence remain intelligible.

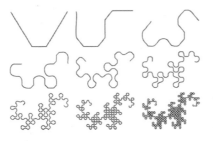

The first nine dragon curves.

Answer on page 307

Counterflip

Make some circular counters out of card, each black on one side and white on the other (the precise number doesn't matter, but 10 or 12 is about right). Arrange them in a row, with a random choice of colours facing upwards.

Your task is now to remove all the counters, by making a series of moves. Each move involves choosing a black counter, removing it, and flipping any neighbouring counters over to change their colours. Counters are 'neighbours' if they are next to each other in the original row of counters; removing any counter creates a gap. As the game progresses, a counter may have two, one or no neighbours.

Here is a sample game in which the player succeeds in removing all the counters:

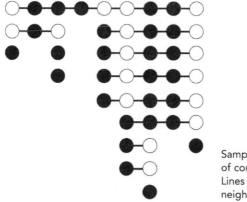

Sample game of counterflip. Lines show neighbours.

The key to this puzzle is simple, but far from obvious: with correct play, you can always succeed if the initial number of black counters is odd. If it is even, there is no solution.

You can play the game for fun, without analysing its mathematical structure. If you feel ambitious, you can look for a winning strategy – and explain why there is no way to win when the initial number of black counters is even.

Answer on page 307

Spherical Sliced Bread

Araminta Ponsonby took her two sets of quins to the Archimedes bakery, which makes spherical loaves. She likes to go there because each loaf is cut into ten slices, of equal thickness, so each child can have one slice of bread. They have different appetites – which is fortunate because some slices have smaller volume than others. But, being extremely well-behaved, all ten children love the crust, and want as much as they can get.

Which slice has the most crust?

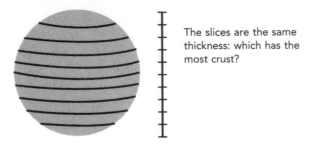

The slices are the same thickness: which has the most crust?

Assume that the loaf is a perfect sphere, the slices are formed by parallel equally spaced planes, and the crust is infinitely thin – so the amount of crust on each slice is equal to the *area* of the corresponding part of the sphere's surface.

Answer on page 309

● ●

Mathematical Theology

It is said that during Leonhard Euler's second stint at the Court of Catherine the Great, the French philosopher Denis Diderot was trying to convert the Court to atheism. Since royalty generally claims to have been appointed by God, this didn't go down terribly well. At any rate, Catherine asked Euler to put a spoke in Diderot's wheel. So Euler told the Court that he knew an algebraic proof of the existence of God. Facing Diderot, he declaimed: 'Sir,

$a + b^n/n = x$, hence God exists – reply!' Diderot had no answer, and left the Court to widespread laughter, humiliated.

Yes, well ... There are some little problems with this anecdote, which seems to have originated with the English mathematician Augustus De Morgan in his *Budget of Paradoxes*. As the historian Dirk Struik pointed out in 1967, Diderot was an accomplished mathematician who had published work on geometry and probability, and would have been able to recognise nonsense when he heard it. Euler, an even better mathematician, would not have expected something that simple-minded to work. The formula is a meaningless equation unless we know what a, b, n and x are supposed to be. As Struik remarks, 'No reason exists to think that the thoughtful Euler would have behaved in the asinine way indicated.'

Euler was a religious man, who apparently considered the Bible to be literal truth, but he also believed that knowledge stems, in part, from rational laws. In the eighteenth century there was occasional talk about the possibility of an algebraic proof of the Deity's existence, and Voltaire mentions one by Maupertuis in his *Diatribe*.

A much better attempt was found among Kurt Gödel's unpublished papers. Naturally, it is formulated in terms of mathematical logic, and for the record here it is in its entirety:

Ax.1 $\Box\forall x[\phi(x) \to \psi(x)] \land P(\phi) \to P(\psi)$
Ax.2 $P(\neg\phi)\sqrt{}\neg P(\phi)$
Th.1 $P(\phi) \to \Diamond\exists x[\phi(x)]$
Df.1 $G(x) \Leftrightarrow \forall\phi[P(\phi) \to \phi(x)]$
Ax.3 $P(G)$
Th.2 $\Diamond\exists x G(x)$
Df.2 $\phi \operatorname{ess} x \Leftrightarrow \phi(x) \land \forall\Psi\Psi(x) \to \Box\forall x[\phi(x) \to \Psi(x)]$
Ax.4 $P(\phi) \to \Box P(\phi)$
Th.3 $G(x) \to G\operatorname{ess} x$
Df.3 $E(x) \Leftrightarrow \forall\phi[\phi\operatorname{ess} x \to \Box\exists x\phi(x)]$
Ax.5 $P(E)$
Th.4 $\Box\exists x G(x)$

The symbolism belongs to a branch of mathematical logic called *modal logic*. Roughly speaking, the proof works with 'positive properties', denoted by P. The expression $P(\phi)$ means that ϕ is a positive property. The property 'being God' is defined (Df.1) by requiring God to have *all* positive properties. Here $G(x)$ means 'x has the property of being God', which is a fancy way of saying 'x is God'. The symbols $\Box$ and $\Diamond$ denote 'necessary truth,' and 'contingent truth,' respectively. The arrow $\rightarrow$ means 'implies', $\forall$ is 'for all' and $\exists$ is 'there exists'. The symbol $\neg$ means 'not', $\wedge$ is 'and', and $\leftrightarrow$ and $\Leftrightarrow$ are subtly different versions of 'if and only if'. The symbol 'ess' is defined in Df.2. The axioms are Ax.1–5. The theorems (Th.1–4) culminate in the statement 'there exists x such that x has the property of being God' – that is, God exists.

The distinction between necessary and contingent truth is a key novelty of modal logic. It distinguishes statements that *must* be true (such as '$2 + 2 = 4$' in a suitable axiomatic treatment of mathematics) from those that conceivably might be false (such as 'it is raining today'). In conventional mathematical logic, the statement 'If A then B' is always considered to be true when A is false. For instance '$2 + 2 = 5$ implies $1 = 1$' is true, and so is '$2 + 2 = 5$ implies $1 = 42$'. This may seem strange, but it is possible to prove that $1 = 1$ starting from $2 + 2 = 5$, and it is also possible to prove that $1 = 42$ starting from $2 + 2 = 5$. So the convention makes good sense. *Can you find any such proofs?*

If we extend this convention to human activities, then the statement 'If Hitler had won World War II then Europe would now be a single nation' is trivially true, because Hitler did *not* win World War II. But 'If Hitler had won World War II then pigs would now have wings' is *also* trivially true, for the same reason. In modal logic, however, it would be sensible to debate the truth or falsity of the first of these statements, depending on how history might have changed if the Nazis had won the war. The second would be false, because pigs don't have wings.

Gödel's sequence of statements turns out to be a formal version of the ontological argument put forward by St Anselm of Canterbury in his *Proslogion* of 1077–78. Defining 'God' as 'the greatest conceivable entity', Anselm argued that God is conceivable. But if he is not real, we could conceive of Him being greater by existing in reality. Therefore, God must be real.

Aside from deep issues of what we mean by 'greatest' and so forth, there is a basic logical flaw here, one that every mathematician learns at his mother's knee. Before we can deduce any property of some entity or concept from its definition, we must first prove that something satisfying the definition exists. Otherwise the definition might be self-contradictory. For instance, suppose we define n to be 'the largest whole number'. Then we can easily prove that $n = 1$. For if not, $n^2 > n$, contradicting the definition of n. Therefore 1 is the largest whole number. The flaw is that we cannot use any properties of n until we know that n exists. As it happens, it doesn't – but even if it did, we would have to *prove* that it did before proceeding with the deduction.

In short: in order to prove that God exists by Anselm's line of thinking, we must first establish that God exists (by some other line of reasoning, or else the logic is circular). Of course I've simplified things here, and later philosophers tried to remove the flaw by being more careful with the logic or the philosophy. Gödel's proof is essentially a formal version of one proposed by Leibniz. Gödel never published his proof because he was worried that it might be seen as a rigorous demonstration of the existence of God, whereas he viewed it as a formal statement of Leibniz's tacit assumptions, which would help to reveal potential logical errors. For further analysis see en.wikipedia.org/wiki/G%C3%B6del's_ontological_proof and for a detailed discussion of modal logic and its use in the proof see www.stats.uwaterloo.ca/~cgsmall/ontology1.html

Answers on page 310

Professor Stewart's Cunning Crib Sheet

Wherein the discerning or desperate reader may locate answers to those questions that are currently known to possess them ... with occasional supplementary facts for their further edification.

Alien Encounter

Alfy is a Veracitor, whereas Betty and Gemma are Gibberish.

There are only eight possibilities, so you can try each in turn. But there's a quicker way. Betty said that Alfy and Gemma belong to the same species, but they have given different answers to the same question, so Betty is Gibberish. Alfy said precisely that, making him a Veracitor. Gemma said the opposite, so she must be Gibberish.

Curious Calculations

$$1 \times 1 = 1$$
$$11 \times 11 = 121$$
$$111 \times 111 = 12,321$$
$$1,111 \times 1,111 = 1,234,321$$
$$11,111 \times 11,111 = 123,454,321$$

If you know how to do 'long multiplication', you can see why this striking pattern occurs. For instance,

$$111 \times 111 =$$
$$11,100 +$$
$$1,110 +$$
$$111$$

We find one '1' in the units column, two in the tens column, three in the hundreds; then the numbers shrink again, with two in the thousands and one in the ten thousands. So the answer must be 12,321.

The pattern does continue – but your calculator may run out of digits. In fact,

$$111,111 \times 111,111 = 12,345,654,321$$
$$1,111,111 \times 1,111,111 = 1,234,567,654,321$$
$$11,111,111 \times 11,111,111 = 123,456,787,654,321$$
$$111,111,111 \times 111,111,111 = 12,345,678,987,654,321$$

After this the pattern breaks down, because digits 'carry' and spoil it.

$$142,857 \times 2 = 285,714$$
$$142,857 \times 3 = 428,571$$
$$142,857 \times 4 = 571,428$$
$$142,857 \times 5 = 714,285$$
$$142,857 \times 6 = 857,142$$
$$142,857 \times 7 = 999,999$$

When we multiply 142,857 by 2, 3, 4, 5 or 6, we get the same sequence of digits in cyclic order, but starting at a different place. The 999,999 is a bonus.

This curious fact is not an accident. Basically, it happens because 1/7 in decimals is $0.142\,857\,142\,857\ldots$, repeating for ever.

Triangle of Cards

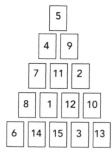

The 15-card difference triangle.

Turnip for the Books

Hogswill started with 400 turnips.

The way to solve this kind of puzzle is to work backwards.

Suppose that at the start of hour 4, Hogswill has x turnips. By the end of the hour he has sold $\frac{6x}{7} + \frac{1}{7}$ turnips, and since none are left, this equals x. So $x - \frac{6x}{7} + \frac{1}{7} = \frac{x-1}{7} = 0$, and $x = 1$. Similarly, if he had x turnips at the start of hour 3, then $\frac{x-1}{7} = 1$, so $x = 8$. If he had x turnips at the start of hour 2, then $\frac{x-1}{7} = 8$, so $x = 57$. Finally, if he had x turnips at the start of hour 1, then $\frac{x-1}{7} = 57$, so $x = 400$.

The Four-Colour Theorem

Here are the four counties for which each is adjacent to all of the others. The middle one is West Midlands – which coincidentally is where I live – and the three surrounding it are Staffordshire, Warwickshire and Worcestershire, clockwise from the top.

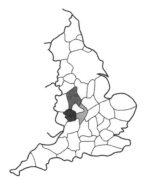

These counties imply that we need at least four colours.

Shaggy Dog Story

First, the dodgy arithmetic. The method 'works' because the will's terms are inconsistent. The fractions do not add up to 1. In fact,

$$\frac{1}{2} + \frac{1}{3} + \frac{1}{9} = \frac{17}{18}$$

which should make the trick obvious.

Whoever first designed this puzzle was clever – there are very few numbers that work, and this choice disguises the inconsistency very neatly. I mean, how would you feel about a puzzle where the uncle has 1,129 dogs, the sons are bequeathed $\frac{4}{7}$, $\frac{3}{11}$ and $\frac{2}{15}$ of them, and Lunchalot rides to the rescue with 26 extra dogs?

However, there is another neat possibility: exactly the same, except that the third son gets one-seventh of the dogs. If the same trick works, how many dogs were there?

Answer to the Answer

The clue is that

$$\frac{1}{2} + \frac{1}{3} + \frac{1}{7} = \frac{41}{42}$$

So there were 41 dogs.

Answer Continued

Oops, I nearly forgot the actual *question*: what did Gingerbere say to Ethelfred that so offended Sir Lunchalot?

It was this: 'Surely you wouldn't send a knight out on a dog like this?'

I said it was a shaggy dog story.

Confession

The shaggy dog story is inspired, in part, by the science fiction short story 'Fall of knight' by A. Bertram Chandler, which appeared in *Fantastic Universe* magazine in 1958.

Rabbits in the Hat

Nothing is wrong with the calculation, but its interpretation is nonsense. When the various probabilities are combined, we are working out the probability of extracting a black rabbit, over *all possible combinations* of rabbits. It is fallacious to imagine that this probability is valid for any specific combination. The fallacy is glaring if there is only one rabbit in the hat. With one rabbit, a similar argument (ignoring adding and removing a black rabbit, which changes nothing essential) goes like this: the hat contains either B or W, each with probability $\frac{1}{2}$. The probability of extracting a black rabbit is therefore

$$\frac{1}{2} \times 1 + \frac{1}{2} \times 0$$

which is $\frac{1}{2}$. Therefore (*really?*) half the rabbits in the hat are black, and half are white.

But there's only one rabbit in the hat ...

River Crossing 1 – Farm Produce

There are two solutions. One is:

(1) Take the goat across.

(2) Come back with no cargo, pick up the wolf, and take that across.

(3) Bring the goat back, but leave the wolf.

(4) Drop off the goat, pick up the cabbage, cross the river, leave the cabbage.

(5) Come back with no cargo, pick up the goat, take it across.

In the other, the roles of wolf and cabbage are exchanged.

I like to solve this geometrically, using a picture in *wolf–goat–cabbage space*. This consists of triples (w, g, c) where each symbol is either 0 (on this side of the river) or 1 (on the far side). So, for instance, $(1, 0, 1)$ means that the wolf and cabbage are on the far side but the goat is on this side. The problem is to get from $(0, 0, 0)$ to $(1, 1, 1)$ without anything being eaten. We don't need to say where the farmer is, since he always travels in the boat during river crossings.

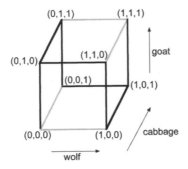

Wolf–goat–cabbage space: now it's obvious.

There are eight possible triples, and they can be thought of as the vertices of a cube. Because only one item can accompany the farmer on each trip, the permissible moves are the edges of the cube.

However, four edges (shown in grey) are not permitted, because things get eaten. The remaining edges (black) do not cause mayhem.

So the puzzle reduces to a geometric one: find a route along the black edges, from $(0, 0, 0)$ to $(1, 1, 1)$. The two solutions are immediately evident.

More Curious Calculations

(1) $13 \times 11 \times 7 = 1,001$, and this is why the trick works. If you multiply a three-digit number abc by 1,001 the result is $abcabc$. Why? Well, multiplying by 1,000 gives $abc000$. Then you add a final abc to multiply by 1,001.

(2) For four-digit numbers, everything is similar, but we have to multiply by 10,001. This can be done in two stages – multiply by 73 and then by 137 – because $73 \times 137 = 10,001$.

(3) For five-digit numbers we have to multiply by 100,001. This can be done in two stages – multiply by 11 and then by 9,091 – because $11 \times 9,091 = 100,001$. As a party trick, this is a bit contrived, though.

(4) We get 471,471,471,471 – the same three digits repeated four times. Why? Because

$$7 \times 11 \times 13 \times 101 \times 9,901 = 1,001,001,001$$

(5) Adding the final 128 leads to 128,000,000 – a million times the original number. This trick works for all three-digit numbers, and it does so because

$$3 \times 3 \times 3 \times 7 \times 11 \times 13 \times 37 = 999,999$$

Add 1 and you get a million.

You can turn all these tricks into party magic tricks. For instance, the trick that turns 471,471 into 471 could be presented like this. The magician, with eyes blindfolded, asks a member of the audience to write down a three-digit number (say 471) on a blackboard or a sheet of paper. A second person then writes it down twice (471,471). A third, armed with a calculator, divides that by 13 (getting 36,267). A fourth divides the result by 11 (getting 3,297). While this is going on, the magician makes a lot of fuss about how unlikely it is that either of these numbers divides without remainder. Then she asks what the final result is, and instantly announces that the original number was 471.

To work this out, she mentally divides 3,297 by 7. OK, you have to be able to do that, but if you know your seven times table it's easy.

Extracting the Cherry

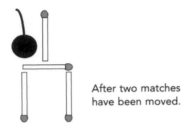

After two matches have been moved.

Make Me a Pentagon

Tie a knot in the strip, and flatten it – carefully.

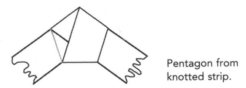

Pentagon from knotted strip.

An interesting challenge is to prove that the result really is a regular pentagon – in an idealised Euclidean version of the problem. I'll leave that for anyone who is interested.

Empty Glasses

Pick up the second glass from the left, pour its contents into the fifth glass, and replace the second glass.

Three Quickies

(1) If you and your partner hold all the spades, your opponents

hold none, and vice versa. So the likelihood is the same in each case.

(2) Three. You *took* three, so that's how many you have.

(3) Zero. If five are in the right envelope, so is the sixth.

Knight's Tours

There is no closed tour on the 5 × 5 square. Imagine colouring the squares black and white in the usual chessboard fashion. Then the knight changes colour at each move. A closed tour must then have equal numbers of black and white squares. But 5 × 5 = 25 is odd. The same argument rules out closed tours on all squares with odd sides.

There is no tour on the 4 × 4 square. The main obstacle is that each corner square connects to only two other squares, and the diagonally opposite corner *also* connects to those two squares. A little thought proves that if a tour of all 16 squares exists, it must start at one corner and finish in an adjacent corner. Systematic consideration of possibilities shows that this is impossible.

However, there is a tour that visits 15 of the 16 squares (showing that the situation regarding a complete tour is delicate):

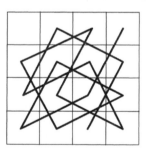

How the knight
can visit 15 squares.

White-Tailed Cats

Suppose that there are c cats, of which w have white tails. There are c $(c-1)$ ordered pairs of distinct cats, and $w(w-1)$ ordered pairs of white-tailed cats. (You can choose the first cat of the pair in c ways, but the second in only $c-1$ ways since you've used up one cat. Ditto for white-tailed cats. By 'ordered' I mean that choosing first cat A and then cat B is considered to be different from B first and then A. If you don't like that, then both formulas have to be halved – with the same result.)

This means that the probability of *both* cats having white tails is

$$\frac{w(w-1)}{c(c-1)}$$

and this must be $\frac{1}{2}$. Therefore

$$c(c-1) = 2w(w-1)$$

with c and w whole numbers. The smallest solution is $c = 4$, $w = 3$. The next smallest turns out to be $c = 21$, $w = 15$. Since Ms Smith has fewer than 20 cats, she must have four cats, of which three have white tails.

Perpetual Calendar

Each cube must include 1 and 2 so that 11 and 22 can be represented. If only one cube bears a 0, then at most six of the nine numbers 01–09 can be represented, so both must bear a 0 as well. That leaves six spare faces for the seven digits 3–9, so the puzzle looks impossible ... until you realise that the cube bearing the number 6 can be turned upside down to represent 9. So the white cube bears the numbers 0, 1, 2, 6 (also 9), 7 and 8, and the grey cube bears the numbers 0, 1, 2, 3, 4 and 5. (Note that I've shown a 5 on my grey cube, and that tells us which cube is which.)

Deceptive Dice

There is no best dice. If Innumeratus plays, and Mathophila chooses correctly (as she will, because she's like that), then he will lose, in the long run. The odds will always favour Mathophila.

How come? Mathophila has constructed her dice so that on average, the yellow one beats the red one, the red one beats the blue one – and the blue one beats the yellow one! At first sight this seems impossible, so let me explain why it's true.

Each number occurs twice on each of the dice, so the chance of rolling any particular number is always $\frac{1}{3}$. So I can make a table of the possibilities, and see who wins for which combinations of numbers thrown. Each combination has the same probability, $\frac{1}{9}$.

Yellow versus red

	1	5	9
3	Red	Yellow	Yellow
4	Red	Yellow	Yellow
8	Red	Red	Yellow

Here yellow wins five times out of nine, red wins only four times.

Red versus blue

	3	4	8
2	Red	Red	Red
6	Blue	Blue	Red
7	Blue	Blue	Red

Here red wins five times out of nine, blue wins only four times.

Blue versus yellow

	2	6	7
1	Blue	Blue	Blue
5	Yellow	Blue	Blue
9	Yellow	Yellow	Yellow

Here blue wins five times out of nine, yellow wins only four times.

So yellow beats red $\frac{5}{9}$ of the time, red beats blue $\frac{5}{9}$ of the time, and blue beats yellow $\frac{5}{9}$ of the time.

This gives Mathophila an advantage if she chooses *second*, which she has cunningly arranged. If Innumeratus chooses the red dice, she should choose yellow. If he chooses the yellow dice, she should choose blue. And if he chooses the blue dice, she should choose red.

It may not be a huge advantage – five chances out of nine of winning, compared with four out of nine – but it's still an advantage.

In the long run, Innumeratus will lose his pocket money. If he wants to play, then a gentlemanly 'No, *you* choose' would be a good idea.

It may seem impossible to have yellow 'better than' red, and red 'better than' blue – but not to have yellow 'better than' blue. What's happening is that the meaning of 'better than' depends on which dice are being used. It's a bit like three football teams:

- The Reds have a good goalkeeper and a good defence, but a poor attack. They win if and only if the opposing goalie is poor.
- The Yellows have a poor goalie, a good defence, and a good attack. They win if and only if the opposing defence is poor.
- The Blues have a good goalie, a poor defence, and a good attack. They win if and only if the opposing attack is poor.

Then (check this!) the Reds always beat the Yellows, the Yellows always beat the Blues, and the Blues always beat the Reds.

Dice like this are said to be *intransitive*. ('Transitive' means that if A beats B and B beats C then A beats C. That doesn't happen here.) On the practical side, the existence of intransitive dice tells us that some 'obvious' assumptions about economic behaviour are actually wrong.

An Age-Old Old-Age Problem

Scrumptius was 69. There was no year 0 between BC dates and AD dates. (If you decided that he might be 68 if he died earlier in the day than he was born, you get a point for ingenuity. But you lose two for pedantry, because it is usual to increase a person's age by one year as soon as their birthday begins, immediately after midnight.)

Heron Suit

The deduction is incorrect. Consider a cat with blunt claws that plays with a gorilla, does not wear a heron suit, has a tail, has no whiskers and is unsociable. The first five statements are all true, but the sixth is not.

I'd explain about the heron suit, but my cat has refused permission on the grounds that it might incriminate itself.

How to Unmake a Greek Cross

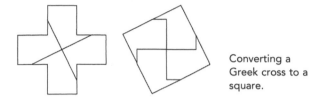

Converting a Greek cross to a square.

Euler's Pentagonal Holiday

Here's a solution to (b), which is automatically a solution to (a) as well. There are others. But they all have to start and end at the two vertices with valency 3, and a mirror-symmetric one must always have the bottom edge of the pentagon in the middle of the tour.

A solution with left-right symmetry.

Ouroborean Rings

One possible ouroborean ring for quadruplets is

1111000010100110

There are others. The topic has a long history, going back to Irving Good in 1946. Ouroborean rings exist for all m-tuples of n digits: for example, in this one

000111222121102202101201002

each triple of the three digits 0, 1, 2 occurs exactly once.

How many ouroborean rings are there? In 1946 Nicholas de Bruijn proved that for m-tuples formed from the two digits 0 and 1, this number is $2^{2^{m-1}-m}$, which grows extremely fast. Here rings obtained by rotating a given one are considered to be the same.

m	Number of ouroborean rings
2	1
3	2
4	16
5	2,048
6	67,108,864
7	144,115,188,075,855,872

The Ourotorus

There is a unique solution, except for various symmetry transformations – rotation, reflection and translations horizontally or vertically. Bear in mind the 'wrap round' convention. So you can, for instance, cut off the four pieces on the right and move them to the left.

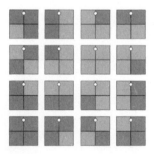

Solution to the ourotorus puzzle.

A Constant Bore

The only reason for including this kind of question in this kind of book is if something surprising happens, and the only surprising thing that makes much sense is that the answer does *not* depend on the radius of the sphere.

That sounds crazy – suppose the sphere were the Earth? But to make the hole only 1 metre long, you have to remove almost the entire planet, leaving only a *very* thin band round the equator, one metre wide. So just maybe . . .

Here comes the easy bit. Assuming that the radius really does not matter, we can work out the answer by considering the special case when the hole is very narrow – in fact, when its width is zero.

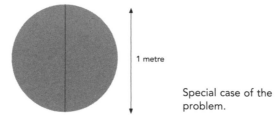

1 metre

Special case of the problem.

Now the volume of copper is equal to that of the entire sphere, and the diameter of the sphere is 1 metre. So its radius is $r = \frac{1}{2}$, and its volume is given by the famous formula

$$V = \frac{4}{3}\pi r^3$$

which equals $\pi/6$ when $r = \frac{1}{2}$.

Ah, but how do we *know* that the answer doesn't depend on the radius? That's a bit more complicated, and it uses more geometry. (Or you can do it by calculus, if you know how.)

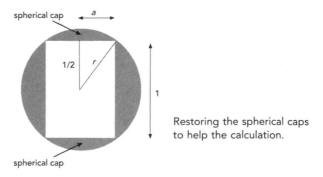

spherical cap

a

1/2 r

1

spherical cap

Restoring the spherical caps to help the calculation.

Put back the missing 'spherical caps' on the top and bottom. Suppose that the radius of the sphere is r, and the radius of the cylindrical hole is a. Then Pythagoras's Theorem applied to the small triangle at the top right tells us that

$$r^2 = a^2 + (\tfrac{1}{2})^2$$

so

$$a^2 = r^2 - \tfrac{1}{4}.$$

Now we need three volume formulas:

- The volume of a sphere of radius r is $\tfrac{4}{3}\pi r^3$.
- The volume of a cylinder of base radius a and height h is $\pi a^2 h$.
- The volume of a spherical cap of height k in a sphere of radius r is $\tfrac{1}{3}\pi k^2(3r - k)$.

Don't worry, I had to look that last one up myself.

The volume of copper required is the volume of the sphere, minus that of the cylinder, minus that of two spherical caps, which is

$$\tfrac{4}{3}\pi r^3 - \pi a^2 h - \tfrac{2}{3}\pi k^2(3r - k)$$

since there are two spherical caps. But $h = 1, k = r - \tfrac{1}{2}$ and $a^2 = r^2 - \tfrac{1}{4}$, so the volume is

$$\tfrac{4}{3}\pi r^3 - \pi\left(r^2 - \tfrac{1}{4}\right) - \tfrac{2}{3}\pi\left(r - \tfrac{1}{2}\right)^2\left[2r + \left(\tfrac{1}{2}\right)\right]$$

Doing the algebra, almost everything miraculously cancels, and all that remains is $\pi/6$.

Digital Century

$$123 - 45 - 67 + 89 = 100$$

This solution was found by the great English puzzle-creator, Henry Ernest Dudeney, and can be found in his book *Amusements in Mathematics*. There are lots of answers if you use four or more arithmetical symbols.

Squaring the Square

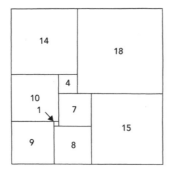

How Morón's tiles make a rectangle.

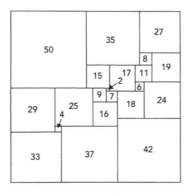

How Duijvestijn's tiles make a square.

You can also rotate or reflect these arrangements.

Ring a-Ring a-Ringroad

The difference is 20π metres, or roughly 63 metres, for roads on the flat. It doesn't depend on the length of the motorway, or how wiggly it is, provided the curvature is gradual enough for '10 metres distance between lanes' to be unambiguous.

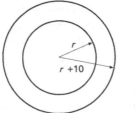

Data for a
circular M25.

Let's start with an idealised version, where the M25 is a perfect circle. If the anticlockwise lane has radius r, then the clockwise one has radius $r + 10$. Their circumferences are then $2\pi r$ and $2\pi(r + 10)$. The difference is

$$2\pi(r + 10) - 2\pi r = 20\pi$$

which is independent of r.

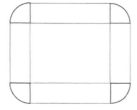

A rectangular
motorway also creates
an excess
of 20.

However, the M25 is not circular. For argument's sake, try a rectangle. Now the outer lane consists of four straight bits, which match the inner lane exactly, plus four quarter-circles at the corners. These extra arcs fit together to make a single circle of radius 10. Again, we get an excess of exactly 20π.

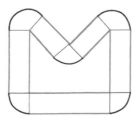

A non-convex polygon
gives 20π as well.

The same point holds for any 'polygonal' road – one composed of straight lines, plus arcs of circles at corners. The straight-line parts match; the arcs add up to one complete circle of radius 10. This is true even when the polygon is not convex, such as the M-shape shown above.* Now the outer lane has arcs that add up to one and a quarter circles, and the inner lane has a quarter circle of its own. But this quarter-turn is of opposite curvature, so it cancels out the excess quarter-turn in the outer lane. The point is that any sufficiently smooth curve can be approximated as closely as we wish by polygons, so the excess is 20π in *all* cases.

The same argument applies to runners on a curved track. In the 400 metres, runners start from 'staggered' positions, to make the overall distance the same in each lane. The stagger between adjacent lanes must be 2π times the width of a lane. This width is usually 1.22 metres, so the stagger should be 7.66 metres per lane – provided it is applied on a straight section of the track. In practice the region where the athletes start often includes part of a bend, so the numbers are a bit different. The easy way to calculate them is to make sure that each runner goes exactly
400 metres, which is what the rules actually state.

Magic Hexagon

The only solution (apart from rotations and reflections of it) is

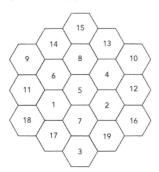

The only non-trivial magic hexagon.

This magic hexagon was found independently by several people

---

* It's not true if the polygon crosses itself, as does the Suzuka racing circuit in Japan. But for some reason figure-of-eight orbital motorways don't seem to have caught on.

between 1887 and 1958. If we try similar patterns of hexagons with n cells along the edge instead of 3, then the only other case where a magic hexagon (using consecutive numbers $1, \ldots, n$) exists is the trivial pattern when $n = 1$: a single hexagon containing the number 1. Charles W. Trigg explained why in 1964, by proving that the magic constant must be

$$\frac{9(n^4 - 2n^3 + 2n^2 - n) + 2}{2(2n - 1)}$$

which is an integer only when $n = 1$ or 3.

Pentalpha

The star shape is designed to mislead. The important aspect of the structure is which circles are two steps away from which, because these are where each new counter starts and finishes. By focusing on this we can draw a much simpler diagram:

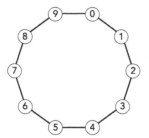

A transformed version of the puzzle.

The rule for placing counters is now: place each new counter on an empty circle and slide it to an adjacent empty circle. It is now obvious how to cover nine circles. For example, place a counter on 1 and slide it to 0. Then place a counter on 2 and slide it to 1. Then place a counter on 3 and slide it to 2. Continue in this way, placing each new counter two empty dots away from the existing string of counters.

Copy these moves on the original diagram to solve the puzzle.

In the second diagram, you can add new counters at either end, so there are lots of solutions. But you can't create more than one connected chain of counters at any stage, because there are then at least two gaps where no counters exist, and each gap leads to at least one circle that can't be covered.

How Old Was Diophantus?

Diophantus was 84 when he died. Let x be his age. Then

$$\frac{x}{6} + \frac{x}{12} + \frac{x}{7} + 5 + \frac{x}{2} + 4 = x$$

So

$$\frac{9}{84}x = 9$$

and $x = 84$.

The Sphinx is a Reptile

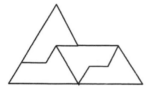

Four sphinxes make a bigger sphinx.

Langford's Cubes

Langford's cubes with four colours.

Magic Stars

This arrangement – possibly rotated or reflected – is the only solution.

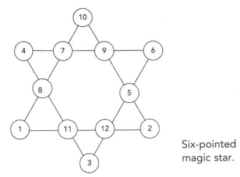

Six-pointed
magic star.

Curves of Constant Width

Surprisingly, the circle isn't the only curve of constant width. The simplest curve of constant width that is not a circle is an equilateral triangle with rounded edges:

(Left) Constant-width triangle.

(Right) Twenty-pence coin.

Each edge is an arc of a circle, with centre at the opposite vertex. Two British coins, the 20p and 50p, are 7-sided curves of constant width; this shape was chosen because it makes the coins suitable for use in slot machines, but distinguishes them from other circular coins worth different amounts – which is especially useful for the visually impaired.

Connecting Cables

The main point is *not* to connect the dishwasher first, with a straight cable. This isolates each of the other appliances from its socket, and makes a solution impossible. If you connect the fridge and cooker first, it's then obvious how to hook up the dishwasher.

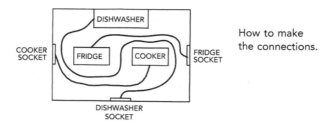

How to make the connections.

Coin Swap

One solution is to successively swap the following pairs: HK, HE, HC, HA, IL, IF, ID, KL, GJ, JA, FK, LE, DK, EF, ED, EB, BK. There are many others.

The Stolen Car

Fenderbender paid out £900 for the car and an extra £100 to the clergyman as change. He counted all his outgoings but forgot to include the corresponding income. All other transactions cancel out, so he lost £1,000.

Compensating Errors

The numbers were 1, 2 and 3. Then $1 + 2 + 3 = 6 = 1 \times 2 \times 3$. This is the only solution for three positive whole numbers.

With two numbers, the only possibility is $2 + 2 = 4 = 2 \times 2$. With four numbers, the only solution is $1 + 1 + 2 + 4 = 8 = 1 \times 1 \times 2 \times 4$.

With more numbers, there are usually lots of solutions, but in some exceptional cases there is just one solution. If the sum of k positive whole numbers is equal to their product, and only *one* set of k numbers has that property, then k is one of the numbers 2, 3, 4, 6, 24, 114, 174 and 444, or it is at least 13, 587, 782, 064. No examples greater than that are known, but their possible existence remains open.

River Crossing 2 – Marital Mistrust

A graphical solution is a bit messy to draw because it involves a 6-dimensional hypercube in $husband_1$–$husband_2$–$husband_3$–$wife_1$–

wife$_2$–wife$_3$ space. Fortunately there's an alternative. Eliminating unsuitable moves and using a bit of logic leads to a solution in 11 moves, which is the smallest possible number. Here the husbands are *A B C* and the corresponding wives are *a b c*.

This bank	In boat	Direction	Far bank
A C a c	*B b*	→	—
A C a c	*B*	←	*b*
A B C	*a c*	→	*b*
A B C	*a*	←	*b c*
A a	*B C*	→	*b c*
A a	*B b*	←	*C c*
a b	*A B*	→	*C c*
a b	*c*	←	*A B C*
b	*a c*	→	*A B C*
b	*B*	←	*A C a c*
—	*B b*	→	*A C a c*

There are minor variations on this solution in which various couples are interchanged.

Wherefore Art Thou Borromeo?

In the second pattern, the two lower rings are linked. In the third pattern, all three pairs are linked. In the fourth pattern, the top ring is linked to the left one which in turn is linked to the right one.

There are lots of four-ring versions. Here's one:

A set of four 'Borromean' rings.

Analogous arrangements exist for any finite number of rings. It has been proved that the Borromean property can't be obtained using perfectly circular (and therefore flat) rings. This is a topological phenomenon.

Percentage Play

The profit and loss do not cancel out. The bicycle he sold to Bettany cost him £400 (he lost £100, which is 25% of £400). The one he sold to Gemma cost him £240 (he gained £60, which is 25% of £240). Overall, he paid £640 and received £600, so he lost £40.

New Merology

Assign the values

E	F	G	H	I	L	N	O	R	S	T	U	V	W	X	Z
3	9	6	1	−4	0	5	−7	−6	−1	2	8	−3	7	11	10

Then

$$Z + E + R + O = 0$$
$$O + N + E = 1$$
$$T + W + O = 2$$
$$T + H + R + E + E = 3$$
$$F + O + U + R = 4$$
$$F + I + V + E = 5$$
$$S + I + X = 6$$
$$S + E + V + E + N = 7$$
$$E + I + G + H + T = 8$$

$$N + I + N + E = 9$$
$$T + E + N = 10$$
$$E + L + E + V + E + N = 11$$
$$T + W + E + L + V + E = 12$$

Spelling Mistakes

There are four *spelling* mistakes, in the words 'there', 'mistakes', 'in' and 'sentence'. The fifth mistake is the claim that there are five mistakes, when there are really only four.

But ... this means that if the sentence is true, it has to be false, but if it's false, it has to be true. Oops.

Expanding Universe

Perhaps surprisingly, the *Indefensible* actually *does* get to the edge of the universe ... but it takes about 10^{434} years to do so. By then the universe has grown to a radius of about 10^{437} light years.

Let's see why.

At each stage, when the universe expands, the *fraction* of the distance that the *Indefensible* has already covered doesn't change. That suggests that if we think about the fractions, we ought to be able to find the answer more easily.

In the first year the ship travels 1/1,000 of the distance to the edge. In the next year it travels 1/2,000 of the distance. In the third year it travels 1/3,000 of the distance, and so on. In the nth year it travels $1/1,000n$ of the distance. So the total fraction travelled after n years is

$$\frac{1}{1,000}\left(1 + \frac{1}{2} + \frac{1}{3} + \frac{1}{4} + \ldots + \frac{1}{n}\right) = \frac{1}{1,000}H_n$$

which is why harmonic numbers are relevant. In particular, the number of years required to reach the edge is whatever value of n first makes this fraction bigger than 1 – that is, makes H_n bigger than 1,000. There is no known formula for the value of H_n in terms of n, and it grows very slowly as n increases. However, it can be proved that by making n large enough, H_n can be made as large as we wish – and in particular, greater than 1. So the *Indefensible* does get to the edge if n is sufficiently large.

To find out how large, we use the hint. To make $H_n > 1,000$ we require $\log n + \gamma > 1,000$, so that $n > e^{1,000-\gamma}$. So the number of years required to reach the edge of the universe is very close to $e^{999.423}$, which is 10^{434} in round numbers. By then the universe will have grown to $n + 1$ thousand light years, which is near enough 10^{437} light years.

Initially the remaining distance keeps increasing each year, but eventually the ship starts to catch up with the expanding edge of the universe. Its 'share' of the expansion grows as it gets farther out, and in the long run this beats the fixed expansion rate of 1,000 light years per year of the edge of the universe. The 'long run' here is very long: it takes about $e^{999-\gamma} = 10^{433.61}$ years before the remaining distance starts to decrease – roughly the first third of the voyage.

Family Occasion

The smallest possible number of party guests is seven: two small girls and one boy, their father and mother, and their father's father and mother.

Don't Let Go!

Your body plus the rope forms a closed loop. It is a topological theorem that a knot cannot be created in a closed loop by deforming it continuously, so the problem can never be solved if you pick up the rope in the obvious 'normal' way. Instead, you must first tie a knot *in your arms*. This may sound difficult, but anyone can do it: just fold them across your chest. Now lean forward so that the hand that is on top of an arm can reach over the arm to pick up one end of the rope, and pick up the other end with the other hand. Unfold your arms, and the knot appears.

Möbius and His Band

If you cut a Möbius band along the middle, it stays in one piece – see the second limerick. The resulting band has a 360° twist.

If you cut a Möbius band one-third of the way across, you get two linked bands. One is a Möbius band, the other (longer) one has a 360° twist.

If you cut a band with a 360° twist along the middle, you get two linked bands with 360° twists.

Three More Quickies

(1) Five days. (Each dog digs a hole in five days.)

(2) The parrot is deaf.

(3) The usual answer is that one hemisphere of the planet is land, and the other is water, so the continent and the island are

identical. But puzzles like this are easily 'cooked' by finding loopholes in the conditions. For instance, maybe Nff lives on the continent but its house is on the island, and Pff eats houses for breakfast. Or on Nff-Pff, the land moves – after all, who knows what happens on an alien world? *Or* ...

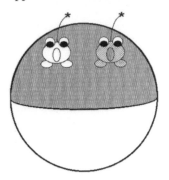

Nff and Pff on their home planet of Nff-Pff.

Miles of Tiles

I forgot to add an extra condition: the tiles should meet at their corners. The corners of some might meet the edges of others. This doesn't change the answer, but it complicates the proof a little.

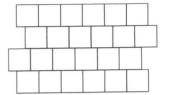

I forgot this kind of thing.

Après-le-Ski

The cables cross at a height of 240 metres.

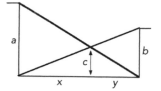

More generally...

It's simpler to tackle a more general problem, where the lengths are as shown. By similar triangles,

$$\frac{x+y}{a} = \frac{y}{c} \quad \text{and} \quad \frac{x+y}{b} = \frac{x}{c}$$

Adding, we get

$$(x+y)\left(\frac{1}{a} + \frac{1}{b}\right) = \frac{x+y}{c}$$

Dividing by $x+y$, we obtain

$$\frac{1}{a} + \frac{1}{b} = \frac{1}{c}$$

leading to

$$c = \frac{ab}{a+b}$$

We notice that c does not depend on x or y, which is a good job since the puzzle didn't tell us those. We know that $a = 600$, and $b = 400$, so

$$c = \frac{600 \times 400}{1000} = 240$$

Pick's Theorem

The lattice polygon illustrated has $B = 21$ and $I = 5$, so its area is $14\frac{1}{2}$ square units.

Paradox Lost

I don't think this one stands up to scrutiny. Both litigants are doing a pick-and-mix – at one moment assuming that the agreement is valid, but at another, assuming that the court's decision can override the agreement. But why do you take an issue like this to court? Because the court's job is to resolve any claimed ambiguities in the contract, *override the contract if need be*, and tell you what to do. So if the court orders the student to pay up, then he has to, and if the court says that he doesn't have to pay up, then Protagoras doesn't have a leg to stand on.

Six Pens

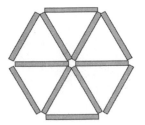

12 panels
making 6 pens.

Hippopotamian Logic

Therefore oak trees grow in Africa.

Why? Suppose, on the contrary, that oak trees *don't* grow in Africa. Then squirrels hibernate in the winter, and hippos eat acorns. Therefore I'll eat my hat. But I won't eat my hat, a contradiction. Therefore (*reductio ad absurdum*) my assumption that oak trees don't grow in Africa must be false. So oak trees grow in Africa.

Pig on a Rope

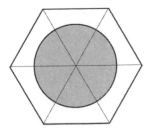

Six copies simplify
the geometry.

To simplify the problem, make six copies of the field, with six copies of the (shaded) region accessible to the pig. Then we want the shaded circle to have half the area of the hexagon. The area of the circle is πr^2 where r is the radius. The hexagon has sides 100 metres long, so its area is $10,000 \times 3\sqrt{3}/2$, or $15,000\sqrt{3}$. So $\pi r^2 = 7,500\sqrt{3}$, and

$$r = \sqrt{\left(\frac{7,500\sqrt{3}}{\pi}\right)}$$

which is about 64.3037 metres.

The Surprise Examination

I think that the Surprise Examination Paradox is a very interesting case of something that looks like a paradox but isn't. My reason is that there is a logically equivalent statement, which is obviously true – but totally uninteresting.

Suppose that every morning the students announce confidently, 'The test will be today.' Then eventually they will do so on the day of the actual test, at which point they will be able to claim that the test was not a surprise.

I don't see any logical objection to this technique, but it's obviously a cheat. If you expect something to happen every day, then of course you won't be surprised when it does. My view – and I've argued with enough mathematicians who didn't agree with me, let alone anybody else, so I'm aware that there's room for differences of opinion – is that the paradox is bogus. It is nothing more than this obvious strategy, dressed up to look mysterious. The cheat is not entirely obvious, because everything is intuited instead of being acted upon, but it's the *same* cheat.

Let me sharpen the conditions by requiring the students to state, each morning before school begins, whether they think the test will be held that day. With this condition, in order for the students to *know* that it can't be on Friday, they have to leave themselves the option of announcing on Friday morning that 'It will be today.' And the same goes for Thursday, Wednesday, Tuesday and Monday. So they have to say 'It will be today' five times in all – once per day. This makes sense: if the students are allowed to revise their prediction each day, then eventually they'll be right.

If we demand even the tiniest bit more, though, their strategy falls to bits. For example, suppose that they're allowed only one such announcement. If Friday arrives and they haven't used up their guess, then they can make the announcement then. But if they *have* used up their guess, they're in trouble. Worse, they *can't* wait until Friday to use their guess, because the test might be on Monday, Tuesday, Wednesday or Thursday.

In fact, if they are allowed *four* guesses, they're still sunk. Only if they are permitted five guesses can they guarantee to predict the correct day. But any fool could do that.

I'm proposing two things here. The less interesting one is that the paradox hinges on what we mean by 'surprise'. The more interesting proposal is that *whatever* we mean by 'surprise', there are two logically equivalent ways to state the students' prediction strategy. One – the usual presentation – seems to indicate a genuine paradox. The other – describe the strategy in terms of actual actions, not hypothetical ones – turns it into something correct but unsurprising, destroying the element of paradox.

Equivalently, we can up the ante by letting the teacher add another condition. Suppose that the students have poor memories, so that any work they do on a given evening to prepare for the test is forgotten by the next evening. If, as the students claim, the test is not going to be a surprise, then they ought to be able to get away with very little homework: just wait until the evening before the test, then cram, pass and forget. But the teacher, in her wisdom, knows that this won't work. If they don't do their homework on Sunday evening, the test could be on Monday, and if it is, they'll fail. Ditto Tuesday through Friday. So despite claiming never to be surprised by the test, the students have to do five evenings of homework.

Antigravity Cone

The uphill motion is an illusion. As the cone moves in the 'uphill' direction, its centre of gravity moves downwards, because the slope widens out and the cone is supported nearer to its two ends.

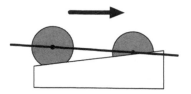

Side view: as the cone follows the arrow, which points *up* the slope, its centre of gravity moves *down* (black line).

What Shape is a Crescent Moon?

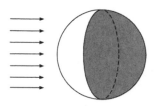

Geometry of an illuminated sphere.

The left curve of the crescent is a semicircle, but the other edge is not an arc of a circle. It is a 'semiellipse' – an ellipse cut in half along its longest axis. The diagram shows the rays of the Sun, which we are assuming to be parallel. In this view the Sun must be positioned some distance *behind* the plane of the page to create the crescent. The light and dark portions of the Moon are *hemispheres*, so the boundary between them is a circle; in fact, it is where a plane at right angles to the Sun's rays cuts the sphere. We observe this circle at an angle. A circle viewed at an angle is an ellipse – its hidden edge is drawn dotted, and we see only the front half. (I've used grey shading to show the dark part of the moon.)

In reality the illumination becomes very faint near the boundary between light and dark, and the Moon is a bit bumpy. So the shape is not as clearly defined as this discussion suggests. You can also quibble about how the circle is projected on to the retina if you so desire.

The crescent shape formed by two circular arcs *can* sometimes be seen in the sky – most dramatically during an eclipse of the Sun, when the Moon partially overlaps the Sun's disk. But now it is the Sun, not the Moon, that looks like a crescent.

Famous Mathematicians

The odd famous mathematician out is Carol Vorderman – see below.

Pierre Boulez Modernist composer and conductor. Studied mathematics at the University of Lyon but then switched to music.

Sergey Brin Co-founder of Google™, with Larry Page. Computer Science and Mathematics degree from University of Maryland. Net worth estimated at $16.6 billion in 2007, making him the 26th-richest person in the world. The Google search engine is based on mathematical principles.

Lewis Carroll Pseudonym of Charles Lutwidge Dodgson. Author of *Alice in Wonderland*. Logician.

J. M. Coetzee South African author and academic, winner of the 2003 Nobel Prize in Literature. BA in Mathematics at the University of Cape Town in 1961. Also BA in English, Cape Town, 1960.

Alberto Fujimori President of Peru, 1990–2000. Holds a master's degree in mathematics from the University of Wisconsin-Milwaukee.

Art Garfunkel Singer. Master's in mathematics from Columbia University. Started on his PhD, but stopped to pursue a career in music.

Philip Glass Modern composer, 'minimalist' (now 'post-minimalist') in style. Accelerated college programme in Mathematics and Philosophy, University of Chicago, at the age of fifteen.

Teri Hatcher Actor. Played Lois Lane in *The New Adventures of Superman* and also starred in *Desperate Housewives*. Mathematics and engineering major at DeAnza Junior College.

Edmund Husserl Philosopher. Mathematics PhD from Vienna in 1883.

Michael Jordan Basketball player. Started as a mathematics student at university but changed subject after his second year.

Theodore Kaczynski PhD in mathematics from the University of Michigan. Retreated to the Montana foothills and became the notorious 'Unabomber.' Sentenced to life imprisonment, with no possibility of parole, for murder.

John Maynard Keynes Economist. MA and 12th Wrangler in mathematics, Cambridge University.

Carole King Prolific pop songwriter of the 1960s, later also became a singer. Dropped out after one year of a mathematics degree to develop her musical career.

Emanuel Lasker Chess grandmaster, world chess champion 1894–1921. Mathematics professor at Heidelberg University.

J. P. Morgan Banking, steel and railroad magnate. He was so good at mathematics that the faculty of Göttingen University tried to persuade him to become a professional.

Larry Niven Author of *Ringworld* and numerous other science fiction bestsellers. Majored in mathematics.

Alexander Solzhenitsyn Winner of the 1970 Nobel Prize in literature. Author of *The Gulag Archipelago* and other influential literary works. Degree in mathematics and physics from the University of Rostov.

Bram Stoker Author of *Dracula*. Mathematics degree from Trinity College, Dublin.

Leon Trotsky Revolutionary. Studied mathematics at Odessa in 1897. Mathematical career terminated by imprisonment in Siberia.

Eamon de Valera Prime Minister and later President of the Republic of Ireland. Taught mathematics at university before Irish independence.

Carol Vorderman Highly numerate co-presenter of television series *Countdown*. Actually studied Engineering, so strictly speaking does not belong in this list.

Virginia Wade Tennis player, winner of the 1977 Wimbledon ladies' singles title. Degree in mathematics and physics from Sussex University.

Ludwig Wittgenstein Philosopher. Studied mathematical logic with Bertrand Russell.

Sir Christopher Wren Architect, in particular of St Paul's Cathedral. Science and mathematics at Wadham College, Oxford.

A Puzzling Dissection

The area can't change when the pieces are reassembled in a different way. When we form the rectangle, the pieces don't quite fit, and a long, thin parallelogram is missing – I've exaggerated the effect to show you what I mean.

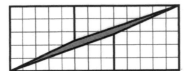

Why the area isn't 65.

In fact, if we calculate the slopes of the slanting lines, the top-left line has a slope of 2/5 = 0.4 and the top-right line has a slope of 3/8 = 0.375. These are different, and the first is slightly larger, so the top-left line is slightly steeper than the top-right one. In particular, they are not two pieces of the same straight line.

The key lengths in this puzzle are 5, 8 and 13 – consecutive Fibonacci numbers (page 98). You can create a similar puzzle using other sets of consecutive Fibonacci numbers.

Nothing Up My Sleeve ...

The topological point is that because your jacket has holes,* the string is not actually linked to your body and the jacket. It just looks

---

* Armholes, not moth holes.

that way. To see that the string is not linked to your body or the jacket, imagine shrinking your body down to the size of a walnut, so that it slides down your sleeve and into your pocket. Now you can obviously pull the loop away, because your wrist is no longer blocking the gap between sleeve and pocket. However, this method is impractical, so we need a substitute.

Here's how.

Begin by pulling the end of the loop up the outside of your arm inside the jacket sleeve, as shown by the arrow in the diagram on the left. Pull out a loop at the top and draw it over your head to reach the position shown in the right-hand diagram. Then pull the loop down the outside of the other arm, inside the sleeve, as shown by the arrow in the right-hand diagram. Pull it over your hand and then back up through the sleeve. Now take hold of the string where it passes in front of your head and push it down inside the front of the jacket. The string pulls through the jacket armholes, and after a few wiggles it drops down around your ankles and you can step out of it.

Nothing Down My Leg ...

After the moves that solve the previous problem, the string ends up looped around your waist, and it is still looped around your arm. So you now follow a similar sequence of moves again, with the trousers instead of the jacket: pass a loop down the trouser leg on the side opposite the pocket with the hand in it, over the foot, back up the trouser leg – and finally remove the string down the other trouser leg. All of this is enormously undignified, and thus highly amusing to spectators. Topology can be fun.

Two Perpendiculars

Neither Euclidean theorem is wrong. Mine is.

The mistake is the assumption that P and Q are different points. In fact, P and Q *coincide* – this follows from Euclid's two theorems, and is highly plausible if you draw an accurate picture.

May Husband and Ay ...

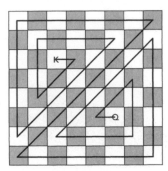

A 15-move solution.

The smallest number of moves is 15. The path shown, and its reflection about the diagonal, are the only solutions. (Remember – each square is visited *exactly once*; that is, the path cannot cross itself.)

What Day is It?

Today is Saturday. (As I told you right at the start, the conversation took place yesterday.) Darren's answers imply that the day of the conversation is exactly one of Friday, Monday or Thursday. Delia's imply that it is Saturday, Sunday or Friday. The only common day is Friday. So *when the conversation took place*, it was Friday.

Logical or Not?

The logic is wrong. If the weather is bad, then pigs don't fly. As a consequence, we don't know whether they have wings. So we don't know whether to carry an umbrella.

It may seem strange that a deduction can be illogical when – as here – the conclusion is entirely sensible. Actually, this is very common. For example:

$$2 + 2 = 22 = 2 \times 2 = 4$$

is nonsense as far as logic goes, but it gives the right answer. All mathematicians know that you can give false proofs of correct statements. What you *can't* do – if mathematics is logically consistent, as we fervently hope – is give correct proofs of false statements.

A Question of Breeding

We are told that Catt breeds pigs.

Hamster does not breed pigs, hamsters, dogs or zebras. So he breeds cats.

Now, Dogge breeds either hamsters or zebras; Pigge breeds dogs hamsters or zebras; Zebra breeds either dogs or hamsters. Since the namesake of Zebra's animals breeds hamsters, Zebra must breed dogs. Therefore Dogge breeds hamsters, so Pigge breeds zebras.

Fair Shares

Here's Steinhaus's method. Let the three people be Arthur, Belinda and Charlie.

(1) Arthur cuts the cake into three pieces (which he thinks are all fair, hence subjectively equal).

(2) Belinda must either
- *pass* (if she thinks that at least two pieces are fair) or
- *label* two pieces (which she thinks are unfair) as being 'bad'.

(3) If Belinda passed, then Charlie chooses a piece (which he thinks is fair). Then Belinda chooses a piece (which she thinks is fair). Finally, Arthur takes the last piece.

(4) If Belinda labelled two pieces as 'bad', then Charlie is offered the same options as Belinda – pass, or label two pieces 'bad'. He takes no notice of Belinda's labels.

(5) If Charlie did nothing in step 4, then the players choose pieces in the order Belinda, Charlie, Arthur (using the same strategy as in step 3.)

(6) Otherwise, both Belinda and Charlie have labelled two pieces as 'bad'. There must be at least one piece that they *both* consider 'bad'. Arthur takes that one. (He thinks that all the pieces are fair, so he can't complain.)

(7) The other two pieces are reassembled into a heap. (Charlie and Belinda both think that the result is at least two-thirds of the cake.) Now Charlie and Belinda play I-cut-you-choose on the heap, to share what's left between themselves (thereby getting what they each judge to be a fair share).

The Sixth Deadly Sin

In the early 1960s John Selfridge and John Horton Conway independently found an envy-free method of cake division for three players. It goes like this:

(1) Arthur cuts the cake into three pieces, which he considers to be 'fair' – of equal value *to him*.

(2) Belinda must either
- *pass* (if she thinks that two or more pieces are tied for largest) or
- *trim* (the largest) piece (to make the two the same). Any trimmings are called leftovers and set aside.

(3) Charlie, Belinda and Arthur, in that order, choose a piece (one they think is largest or equal largest). If Belinda did not pass in step 2, she must choose the trimmed piece unless Charlie chose it first.

At this stage, the part of the cake other than the leftovers has been divided into three pieces in an envy-free manner—a 'partial envy-free allocation'.

(4) If Belinda passed at step 2, there are no leftovers and we are

done. If not, either Belinda or Charlie took the trimmed piece. Call this person the 'non-cutter', and the other one of the two the 'cutter'. The cutter divides the leftovers into three pieces (that he/she considers equal).

Arthur has an 'irrevocable advantage' over the non-cutter, in the following sense. The non-cutter received the trimmed piece, and even if he/she gets all the leftovers, Arthur still thinks that he/she has no more than a fair share, because he thought that the original pieces were all fair. So *however* the leftovers are now divided, Arthur will not envy the non-cutter.

(5) The three pieces of leftovers are chosen by the players in the order non-cutter, Arthur, cutter. (Each chooses the largest piece, or one of the equal largest, among those available.)

The non-cutter chooses from the leftovers first, so has no reason to be envious. Arthur does not envy the non-cutter because of his irrevocable advantage; he does not envy the cutter because he chooses before he/she does. The cutter can't envy anybody since he/she was the one who divided the leftovers.

Recently, Steven Brams, Alan Taylor and others have found very complicated envy-free methods for any number of people.

When it comes to sharing cakes, avoiding the second deadly sin[*] is more tricky, in my experience.

Weird Arithmetic

The *result* is correct, though as teacher said, you should cancel 9 from the top and the bottom to simplify it to $\frac{2}{5}$. But Henry's presumed *method* leaves a lot to be desired.

For instance,

$$\frac{3}{4} \times \frac{8}{5} = \frac{38}{45}$$

is wrong.

[*]   Gluttony.

So when does his method work? An easy way to find one more solution is to turn Henry's upside down:

$$\frac{4}{1} \times \frac{5}{8} = \frac{45}{18}$$

But there are other solutions. With the stated limits on the number of digits, we are trying to solve the equation

$$\frac{a}{b} \times \frac{c}{d} = \frac{10a+c}{10b+d}$$

which boils down to

$$ac(10b+d) = bd(10a+c)$$

where a, b, c and d can each be any digit from 1 to 9 inclusive.

There are 81 trivial solutions where $a = b$ and $c = d$. Aside from these, there are 14 solutions, where $(a, b, c, d) = (1, 2, 5, 4)$, $(1, 4, 8, 5)$, $(1, 6, 4, 3)$, $(1, 6, 6, 4)$, $(1, 9, 9, 5)$, $(2, 1, 4, 5)$, $(2, 6, 6, 5)$, $(4, 1, 5, 8)$, $(4, 9, 9, 8)$, $(6, 1, 3, 4)$, $(6, 1, 4, 6)$, $(6, 2, 5, 6)$, $(9, 1, 5, 9)$ and $(9, 4, 8, 9)$. These form seven pairs (a, b, c, d) and (b, a, d, c), corresponding to turning the fractions upside down.

How Deep is the Well?

The depth of the well is

$$s = \tfrac{1}{2}gt^2 = \tfrac{1}{2}10(6)^2 = 180 \text{ metres} = 590 \text{ feet}$$

which agrees very well with what the *Time Team* measured (about 550 feet) when you take into account the difficulty of timing the fall by hand. A more accurate figure for g is 9.8 m s^{-2}, leading to a depth of 176 metres or 577 feet. Presumably the exact time was slightly less than 6 seconds.

Yes, the well really *was* that deep. How did they dig it, so long ago? The mind boggles.

McMahon's Squares

The 24 tiles can be assembled as shown. There are 17 other solutions, plus rotations and reflections.

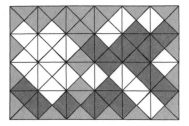

One of the 18 basically different solutions.

One feature of the tiles helps us work out how to assemble them. Choose a border colour, say grey. There are four tiles that have two blue triangles opposite each other and no other blue triangles. Their remaining triangles are grey/grey, black/black, white/white and black/white. The only way to fit these tiles in is to stack four of them together across the narrow width of the rectangle:

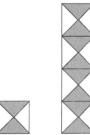

The four tiles like the left-hand one have to stack together. The white triangles can be any combination of black and white.

There are still lots of ways to proceed, but this observation helps to restrict the possibilities. There are 18 basically different solutions, which lead to 216 solutions altogether by swapping colours, rotating the picture or reflecting it. Note the stack in the third column of the sample solution above.

Archimedes, You Old Fraud!

Let's say that Archimedes can exert a force sufficient to lift his own weight, call it 100 kg. The mass of the Earth is about 6×10^{24} kg. To keep the analysis simple, suppose that the pivot is 1 metre from the Earth. Then the law of the lever tells us that distance from the pivot to Archimedes is 6×10^{22} metres, and his lever is $1 + 6 \times 10^{22}$ metres long – about 1.6 million light years, or about two-thirds of the way to the Andromeda Galaxy. If Archimedes now moves his end of the lever one metre, the Earth moves $1/(6 \times 10^{22}) = 1.66 \times 10^{-23}$ metres.

Now, a proton has a diameter of 10^{-15} metres—

Yes, but it still *moves*, dammit!

True. But suppose that instead of using this huge and improbable apparatus, Archimedes stands on the surface of the Earth and *jumps*. For every metre he goes up, the Earth moves 1.66×10^{-23} metres down (action/reaction). Jumping has exactly the same effect as his hypothetical lever. So the place to stand is on the Earth – but you don't stand *still*.

The Missing Symbol

Well, the symbols $+, -, \times, and \div$ don't work, because $4 + 5$ and 4×5 are too big, and $4 - 5$ and $4 \div 5$ are too small. Neither does the square root sign $\surd$, because $4\sqrt{5} = 8.94$ and that's too big as well.

Give up? How about the decimal point, 4.5?

Where There's a Wall, There's a Way

How to make the wall.

Rotation and reflection yield three other solutions. The component shapes are called *tetrahexes*.

Connecting Utilities

No, you can't. As stated – and without 'cooking' the puzzle by, say, working on a surface that isn't a plane, passing cables through a house, whatever – the puzzle has no solution.

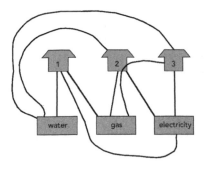

A cheat that works by passing a cable through a house.

If you experiment, you'll soon become convinced that it's impossible – but mathematicians require proof. To find one, we first connect things up without worrying about crossings, like this:

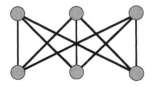

Can this be redrawn without crossings?

While we're at it, I've replaced the buildings by dots.

Now, suppose that we could redraw this picture, keeping all the connections at the dots, to eliminate the crossings. Then the lines would form a kind of map in the plane. This map would have $E = 9$ edges (the nine connections) and $V = 6$ vertices (the six dots). Euler's formula for maps (page 177) tells us that if F is the number of faces, then

$$F - E + V = 2$$

so $F - 9 + 6 = 2$ and $F = 5$. One of these faces is infinitely large and forms the outside of the whole diagram.

Now we count the edges in another way. Each face has a boundary formed by a loop of edges. You can check that the possible loops in the diagram contain either 4 or 6 distinct edges, nothing else. So there are six possibilities for the number of edges in the five faces:

```
4  4  4  4  4
4  4  4  4  6
4  4  4  6  6
4  4  6  6  6
4  6  6  6  6
6  6  6  6  6
```

which respectively total 20, 22, 24, 26, 28 and 30. But every edge forms the border between two faces, so the number of edges has to be half of one of these numbers: 10, 11, 12, 13, 14 or 15.

However, we already know that there should be 9 faces. This is a contradiction, so we can't redraw the diagram to have no crossings.

People often claim that 'You can't prove a negative.' In mathematics, you most certainly can.

Don't Get the Goat

No, there isn't. The contestant doubles their chance of success by changing their mind. But this is true *only* under the stated assumptions. For example, suppose that the host (who knows where the car is, remember) offers the contestant the opportunity to change their mind *only* when they have correctly chosen the door with the car behind it. In this extreme case, they always lose if they change their mind. At the other extreme, if he offers the contestant the opportunity to change their mind only when they have chosen a door with a goat behind it, they always win.

Fine – but what if my original assumptions are valid. The fifty–fifty argument then looks convincing, but it's wrong. The reason is that the host's procedure does not make the odds fifty–fifty.

When the contestant makes their initial choice, the probability that they have the right door is one in three. So on average and in the long run, the car is behind that particular door one time in three. Nothing that happens subsequently can change that. (Unless the television people surreptitiously move the prizes ... OK, let's assume that doesn't happen either.)

After a goat is revealed, the contestant is left with two doors. The car must be behind one of them (the host never reveals the car). One time in three that door is the one that the contestant has chosen. The other two times out of three, it must be behind the *other* door. So, if you don't change your mind, you win the car one time in three. If you do, you win it two times in three – twice the chance.

The trouble with such reasoning is that unless you've spent a lot of time learning probability theory, it's not always clear what works and what doesn't. You can experiment using dice to decide where the car goes: 1 or 2 puts it behind the first door, 3 or 4 behind the second, 5 or 6 behind the third, say. If you try this twenty or thirty times, it soon becomes clear that changing your mind really does improve the chance of success. I once got an e-mail from some people who had been arguing about this problem in the pub, until one of them got out his laptop and programmed it to simulate a million attempts. 'Don't change your mind' succeeded on roughly 333,300 occasions. 'Do change your mind' succeeded on the remaining 666,700 occasions. It's fascinating that we live in a world where you can do this simulation in a few minutes in a pub. Nearly all of which is taken up writing the computer program – the actual sums take less than a second.

Still not convinced? Sometimes people see the light when the problem is taken to extremes. Take a normal pack of 52 playing cards, held face down. Ask a friend to pull a card from the pack, without looking at it, and lay it on the table. They win if that card is the ace of spades (car) and lose otherwise (goat). So now we have one car and 51 goats, behind 52 doors (cards). But you now pick up the remaining 51 cards, holding them so that you can see their faces but your friend can't. Now you discard 50 of those cards, none being the ace of spades. One card remains in your hand; one is on the table. Is it *really* true that each of these two cards has a fifty–fifty chance of being the ace of spades? So why were you so careful to hang on to that particular card out of the 51 you started with? Clearly you have a big advantage over your friend. They got to choose one card, without seeing its face. You had a choice of 51 cards, and you *did* see their faces. They have one chance in 52 of being right; you have 51 chances. This is a *fair* game? Pull the other one!

All Triangles are Isosceles

The mistake is the innocent assertion that X is inside the triangle. If you draw the picture accurately, it's not. And it turns out that *exactly one* of the points D and E is outside the triangle, too. In this particular case, D is not between A and C. But the other point is 'inside' the triangle—well, on its edge, but not outside it. Here E lies between B and C. This diagram makes it clear what I mean:

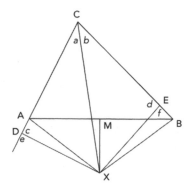

The correct picture.

Now the argument collapses. We still find that CE = CD and DA = EB (steps 5 and 9). But in step 10, CA = CD − DA, not CD + DA. However, CB is still CE + EB. So we can't conclude that lines CA and CB are equal.

Fallacies like this one explain why mathematicians are so obsessive about hidden logical assumptions in proofs.

Square Year

We are looking for squares either side of 2001. A little experiment reveals that $44^2 = 1936$, and $45^2 = 2025$. With these figures, Betty's father was born in $1936 - 44 = 1892$ (so he died in 1992), and Alfie was born in $2025 - 45 = 1980$.

To rule out any other answers: the previous possible date for Betty's father would be $43^2 - 43 = 1806$, so he would have died in 1906, making Betty well past retirement age. The next possible date for Alfie would be $46^2 - 46 = 2070$, so he wouldn't get born in 2001.

Infinite Wealth

Whatever amount you win, it will be *finite* (unless the game goes on for ever, with you always tossing tails, in which case you win an infinite amount of cash, but you have to wait infinitely long to get it). So it's silly to pay an infinite entry fee. The correct deduction is that whatever finite entry fee you pay, your expected winnings are bigger. Your chance of a big win is of course very small, but the win is so huge that it compensates you for the tiny chance of success.

But that still seems silly, and this is where the mathematicians of the time started scratching their heads (and very likely their tails too, though we don't mention such things in polite company). The main source of trouble is that the expected winnings form a *divergent series* – one with no well-defined sum – which may not make a great deal of sense.

As a practical matter, the sums involved are limited by two features that the simple mathematical model fails to take into account: the largest amount that the bank can actually pay, and the length of time available to play the game – at most one human lifetime. If the bank has only £2^{20} available, for instance, which is £1,048,576, then you are justified in risking £20. If the bank has £2^{50} available, which is £1,125,899,906,842,624 – a little over a quadrillion pounds, which exceeds the annual Gross Global Product, then you are justified in risking £50.

There is a more philosophical point: how sensible the *long-term* average winnings (expectation) actually is when the 'long term' is far longer than any player can actually play for. If you are playing against a bank with £2^{50} in its coffers, it will typically take you 2^{50} attempts to make the big win that justifies you spending £50, let alone a much larger amount. Human decisions about risk are subtler than the mindless computation of long-term expectations, and the subtleties are important exactly when the gain (or loss) is very big but its probability is very small.

A related point is the relevance of long-term averages over numerous trials, if in practice you only get to play once, or just a few times. Then you have an extraordinarily small chance of a big win, and the pragmatic decision is not to throw money at something so unlikely.

On the other hand, in cases where the expectation *converges* to a finite sum, it may make more sense. Suppose that you win £n if the first toss of a head is on the nth throw. Now the expectation is

$$1 \times \frac{1}{2} + 2 \times \frac{1}{2^2} + 3 \times \frac{1}{2^3} + 4 \times \frac{1}{2^4} + \dots$$

which converges to 2. So here you should pay £2 to break even, which seems fair enough.

What Shape is a Rainbow?

The arcs are parts of circles. For a given colour, the arc concerned is very thin. All the circles involved have the same centre–which is often below the horizon. The interesting question is – why? The answer turns out to be distinctly complicated, though very elegant. Teacher was right to direct our attention to the colours, though she did miss an opportunity to do some really neat geometry.

Consider light of a single wavelength (colour), and look at a raindrop in cross-section. Raindrops are spheres, so in section we get a circle. A ray of light from the Sun hits the front of the drop, is refracted (bent through an angle) by the water, reflects off the back of the drop, and is refracted a second time as it leaves the drop and heads back roughly the way it came.

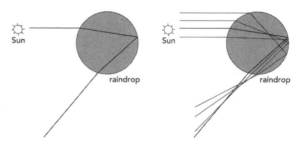

(Left) The path of one ray. (Right) Many rays.

That's what happens to one ray, but in reality there are lots of them. Rays that are very close together usually hit the same drop, but turn through slightly different angles. But there is a focusing effect, and most of the light comes back out along a single 'critical direction'. Bearing in mind the spherical geometry of the drop, the end result is that effectively each drop emits a *cone* of light of the chosen colour. The axis of the cone joins the drop to the Sun. The vertex angle of the cone is about 42°, for a raindrop, but it depends on the colour of the light.

When an observer looks at the sky, in the direction of the rain, she observes light only from those drops whose cones meet her eye. A little geometry shows that these drops themselves lie on a cone, whose tip is at her eye, and whose axis is the line joining her eye to the Sun. Again, the vertex angle is about 42°, depending on the colour of the light.

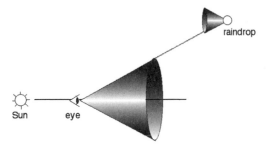

The eye receives a cone of light.

If you place a cone to your eye and sight along it, what you see is the edge of its circular base. More accurately, the *directions* of the incoming light are perceived as if the light were being emitted by the circular base. So the upshot is that the eye 'sees' a circular arc. The arc is not up there in the sky: it is an illusion, caused by the directions of the incoming light rays.

Usually, the eye sees only part of this circular arc. If the Sun is high in the sky, most of the arc is below the horizon. If the Sun is low, the eye sees almost a semicircle. From an aircraft a complete circle can sometimes be seen. If the rain is nearby, the arc may appear to be in front of other parts of the landscape. The arc is often partial – you see the returning light only when there's rain in that direction.

Because different colours of light lead to different vertex angles for the cone, each colour appears on a slightly different arc, but they all have the same centre. So we see 'parallel' arcs of colours.

Sometimes you can see a second rainbow, outside the first. This is formed in a similar way, but the light bounces more times before coming back out of the drop. The vertex angle of the cone is different, and the colours are in reverse order. The sky is brighter inside the main rainbow, very dark between this and the 'secondary' rainbow, and medium-dark outside that. Again, all this can be explained in terms of the geometry of the light rays. René Descartes did that in 1637.

A really informative website is en.wikipedia.org/wiki/Rainbow

Alien Abduction

Each alien is going after the pig that initially is nearest to it. If they chase the *other* pig, they will soon catch it.

Why? The way to catch a pig is to drive it into a corner. If the position looks like the next picture, and it is the *pig's* turn to move, it will be abducted. However, if it is the *alien's* turn to move, the pig can escape.

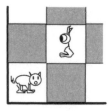

How to catch your pig
– provided the pig has to move.

Which of these happens depends on the *parity* (odd or even) of the distance (in moves) from alien to pig. If the pig is an even number of moves away – as it is if each alien goes for the pig it is initially facing – then the pig always escapes. If it is odd – as it is if the aliens switch pigs – then the pig can be driven into a corner and abducted.

Disproof of the Riemann Hypothesis

The argument is logically correct. However, it doesn't disprove the Riemann Hypothesis! The information given is contradictory: it implies that an elephant has won *Mastermind*, and also that it has not. We can now prove the Riemann Hypothesis false by contradiction:

(1) Assume, on the contrary, that the Riemann Hypothesis is true.

(2) Then an elephant has won *Mastermind*.

(3) But an elephant has not won *Mastermind*.

(4) This is a contradiction. So our assumption that the Riemann Hypothesis is true is wrong.

(5) Therefore the Riemann Hypothesis is false.

The same argument proves that the Riemann Hypothesis is true, of course.

Murder in the Park

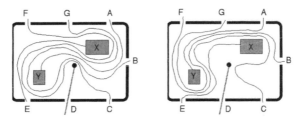

The two possible topological types of path.

Topologically speaking, there are just two cases to consider. Either the butler went to the north of Y on his way to X (left-hand diagram) or he went to the south (right-hand diagram). The gamekeeper must then have gone to the south (respectively north) of X on his way to Y.

The tracks of the youth and the grocer's wife must then be as shown, perhaps with additional wiggles. In the first case, the grocer's wife's path from C to F cuts off the youth's path from the part of the park that contains the body. In fact, only she and the butler could have approached the place where Hastings's body lay. The same is true in the second case. Since the butler has a confirmed alibi, the murderer must have been the grocer's wife.

The Cube of Cheese

The corners of the hexagon lie on various midpoints of sides of the cube, like this:

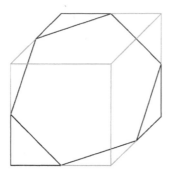

Hexagonal slice of a cube.

Drawing an Ellipse – and More?

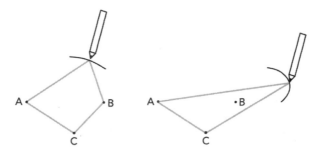

The pencil draws arcs of various ellipses.

When the pencil is in the position shown on the left, the length of string AC+CB is constant, so the pencil moves as if you had looped a shorter string round A and B. Therefore it draws an arc of an ellipse with foci A and B. When it moves to a position like the one on the right, it draws an arc of an ellipse with foci A and C. The complete curve therefore consists of six arcs of ellipses, joined together. Since basically this isn't new, mathematicians aren't (terribly) interested.

The Milk Crate Problem

The milkman is correct for 1, 4, 9, 16, 25 and 36 bottles, but wrong for 49 and any larger square number.

If you think about this the right way, it's obvious that when the number of bottles gets sufficiently large, the square lattice packing can't possibly be the best. (What *is* the best is horribly difficult to work out, and nobody knows.) The square lattice must fail for a large number of bottles, because a hexagonal lattice packs bottles more closely than a square one. When there aren't too many bottles, 'edge effects' near the walls of the crate stop you exploiting this fact to make the crate smaller, but as the numbers go up the edge effects become negligible.

It so happens that the break-even point is close to 49 bottles. And it has been proved that 49 bottles of unit diameter can fit inside a square whose side is very slightly less than 7 units. The difference is too small to be seen by the naked eye, but you can easily see big regions of hexagonally packed circles.

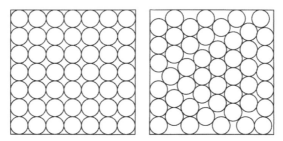

(Left) 49 unit bottles in a 7 × 7 square. (Right) How to fit the same bottles into a slightly smaller square.

Incidentally, this example shows that a *rigid* packing – one in which no single circle can move – need not be the closest packing possible. The square lattice is rigid for any square number of bottles inside a tightly fitting square crate. Or, indeed, on the infinite plane.

Road Network

The shortest road network introduces two new junctions and makes the roads meet there at *exactly* 120° to each other. The same layout rotated by 90° is the only other option. The total length here is $100(1 + \sqrt{3}) = 273$ km, roughly:

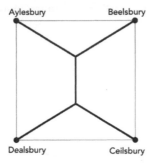

The shortest network.

Tautoverbs

- No news is no news. (Experts consider this the smallest but most perfectly formed tautoverb, a kind of tautohaiku.)
- The bigger they are, the bigger they are.
- Nothing ventured, nothing lost.
- Too many cooks cook too much.
- You cannot have your cake and eat it too, *unless you do them in that order*. What's difficult is to eat your cake and have it too.
- A watched pot never boils *over*. (Unless it's custard.) The time taken for a liquid to boil is not influenced by the presence of an observer, except in certain esoteric forms of quantum field theory. It merely seems longer for psychological reasons. Do not be deceived.
- If pigs had wings, the laws of aerodynamics would still stop them getting off the ground. I mean, let's be *sensible*. The porcithopter is not technologically feasible.

Scrabble Oddity

$$\text{TWELVE} = 1 + 4 + 1 + 1 + 4 + 1$$

Dragon Curve

Dragon curves can be made by repeatedly folding a strip of paper in half – always folding the same way – and then opening it out to make all folds into right angles.

Making dragon curves by paper-folding.

These curves determine a fractal (page 189). In fact, the infinite limit is a space-filling curve (page 83), but the region it fills has a complicated, dragon-like shape. The sequence of right (R) and left (L) turns in the curve goes like this:

Step 1 R

Step 2 R **R** L

Step 3 R R L **R** R L L

Step 4 R R L R R L L **R** R R L L R L L

In fact, there is a simple pattern: each sequence is formed from the previous one by placing an extra R at the end, followed by the reverse of the previous sequence with R's and L's swapped. I've marked the extra R in the middle in bold.

The dragon curve was discovered by John Heighway, Bruce Banks and William Harter – all physicists at NASA – and was mentioned in Martin Gardner's Mathematical Games' column in *Scientific American* in 1967. It has lots of intriguing features – see en.wikipedia.org/wiki/Dragon_curve

Counterflip

Assume that there is an odd number of black counters—so in particular, there exists at least one of them. As play progresses, counters that are removed create gaps, breaking the row of counters into connected pieces, which I'll call *chains*. We start with one chain.

I claim that any chain with an odd number of black counters can be removed. Here's a method that always works. (The sample game doesn't always follow it, so other methods work too.)

Starting from one end of the chain, find the first black counter. I claim that if you flip that counter, then there are three possibilities:

(1) The chain originally consisted of one isolated black counter, and when you flip it it is removed, with no effect on any other counters.

(2) You now have a single shorter chain having an odd number of black counters.

(3) You now have two shorter chains, each having an odd number of black counters.

If this claim is true, then you can repeat the same procedure on the shorter chains. The number of chains may grow, but they get shorter at each step. Eventually they all become so short that we reach case 1 and they can be removed entirely.

The claim is proved by seeing what happens in three cases of a single chain, which exhaust the possibilities:

(1) The chain concerned consists of a single black counter. It has no neighbours, so when it is flipped it disappears.

(2) The chain has a black counter at one end. Flipping the end counter results in a shorter chain which has an odd number of black counters.

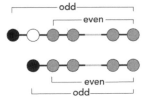

Grey counters may be either black or white. The black counter on the end disappears, and its neighbour (here shown white) changes colour. The overall change in the number of black counters is either 0 or 2, so an odd number of black counters remains.

(3) The chain has white counters at both ends. Flipping the first black counter from one end (it doesn't matter which) results in

two shorter chains. One has a single black counter (which is odd) and the other has an odd number of black counters.

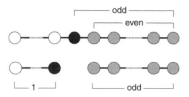

The first black counter (from the left) disappears, and its neighbours (one white, one grey – i.e. black or white) change colour. Two chains are created; one has one black counter, the other has an odd number of black counters.

It does matter which black counter you flip. For instance, if the chain has at least four counters, with three black counters next to one another and the rest white, then it is a mistake to flip the middle black counter. If you do, you get at least one chain containing no black counters at all, so this chain cannot be removed.

Oops ...

To complete the analysis, here's why the puzzle can't be solved if the initial number of black counters is even:

(1) If there are no black counters (zero is even!) you can't get started.

(2) If the initial number of black counters is even (and non-zero), then whichever black counter you remove, at least one shorter chain is created that also has an even number of black counters. Repeating this process eventually leads to a chain with no black counters but at least one white one. This chain cannot be removed since there is no place to start.

Spherical Sliced Bread

All slices have exactly the same amount of crust.

At first sight this seems unlikely, but slices near the top and bottom are more slanted than those near the middle, so they have more crust than you might think. It turns out that the slope exactly compensates for the smaller size of the slices.

In fact, the great Greek mathematician Archimedes discovered that the surface area of a slice of a sphere is equal to that of the corresponding slice of a cylinder into which the sphere fits. It is obvious that parallel slices of a cylindrical loaf, of equal thickness, all have the same amount of crust ... since they are all the same shape and size.

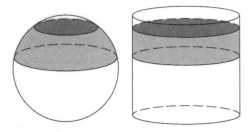

The surface area of the spherical band (pale blue) is the same as that of the corresponding band on a tightly fitting cylinder.

Mathematical Theology

I asked you to start from $2 + 2 = 5$ and prove that $1 = 1$ and also that $1 = 42$. There are lots of valid answers (infinitely many, in fact). Here are two that work:

- Since $2 + 2 = 4$, we deduce that $4 = 5$. Double both sides to get $8 = 10$. Subtract 9 from each side to get $-1 = 1$. Square both sides to get $1 = 1$.
- Since $2 + 2 = 4$, we deduce that $4 = 5$. Subtract 4 from each side to get $0 = 1$. Multiply both sides by 41 to get $0 = 41$. Add 1 to each side to get $1 = 42$.

FREUD

and

WOMEN

Lucy Freeman *and*
Herbert S. Strean, *D.S.W.*

CONTINUUM · NEW YORK

Acknowledgments

To Philip Winsor, our editor, warmest thanks for his artistry, skill, and patience in helping to shape the book. Our thanks also to Frederick Ungar, Edith Friedlander, and Ruth Selden for their many valuable suggestions. We also express our appreciation to Edward L. Bernays for granting special interviews.

1987
The Continuum Publishing Company
370 Lexington Avenue
New York, NY 10017

Printed in the United States of America

Designed by Jacqueline Schuman

The lines from "Tribute to the Angels" by H.D.
are used by permission of John Schaffner Associates, Inc.

Library of Congress Cataloging-in-Publication Data

Freeman, Lucy.
  Freud and women.

  Bibliography: p.
  Includes index.
  1. Freud, Sigmund, 1856–1939—Relations with women.
  2. Pschoanalysts—Austria—Biography.   I. Strean,
  Herbert S.   II. Title.
  BF173.F85F68   1987   150.19′52   87-8855
  ISBN 0-8264-0385-9 (pbk.)

Contents

Part Four
FREUD'S CONTRIBUTIONS TO UNDERSTANDING WOMEN

Introduction to the Paperback Edition

Recent assaults by Jeffrey Moussaleff Masson on Freud's theory of women's seduction fantasies have awakened new interest in psychoanalysis. Freud came to see these fantasies—reflecting a little girl's wish for sexual involvement with her father—as the cause of the women's later neurosis. In Masson's much-publicized book *The Assault on Truth: Freud's Supression of the Seduction Theory*, he attacked Freud's integrity and that of current analysts.

When Freud started to form his theories, his first patients were women suffering from hysteria who claimed their fathers had sexually seduced them when they were little girls. At first Freud believed they told the truth. But as he listened to this charge from the couch by more and more women, he concluded that most of them only fantasized such attacks. They were, instead, victims only of their own strong wishes for incest, based on their passionate oedipal feelings. These were natural to all little girls at this stage of life so that, as adults, they could transfer their passion to a more appropriate man.

Masson claimed that Freud changed his theory because if all the women had actually been seduced by their fathers, this meant that the entire theory of psychoanalysis and its use for treatment were invalid. These charges and Masson's conclusion have been proved false both by psychoanalysts and informed laymen. While there is no doubt that some little girls are victims of incest, it is the incestuous fantasies that are universal, part of the development of the normal sexual impulse that is first attached to the mother, then shifted to the father as the mother becomes the rival. Even if, as Masson wrongly asserted, all women who claimed they were seduced by their fathers actually were, it does not follow that psychoanalysis must fall. Freud's findings about the important role played by the unconscious part of the mind, his oral, anal, and phallic stages of development, and his later discovery of the aggressive impulse as equal in power to the sexual, have nothing to do with his seduction theory.

Masson's words apparently persuaded a number of readers to believe him at first, though his claims have since proved fraudulent. He has, however, inspired even greater numbers of readers to want to know the truth about Freud—not only his theories but his emotional life as a child, adolescent, and man.

New books about Freud and later figures on the psychoanalytic scene such as Melanie Klein, who studied babies and their emotional development, have appeared. The use of psychoanalytic concepts has become ever more common—even President Reagan during a press conference apologized for a "Freudian slip." Athletes discussing their ups and downs as heroes and villains speak of the impact of the "unconscious" on their ability to play. Psychoanalysts are even portrayed as sleuths in today's murder mysteries—in books, the movies, and television.

Freud himself wrote twenty-four volumes chiefly on the relationship between women and men, emphasizing the development of our sexual or erotic desire throughout life from infancy to death. It certainly appears pertinent to the current controversy aroused by Masson that the facts be known about Freud's relationship to the women in his own life.

Freud has been called everything from "celibate" to "drug addict," from a hater of women to an idolizer of them. As we look at all the accusations hurled at him, it becomes clear that the public is entitled to greater access to the truth about how he related to the women close to him over the years from birth to death.

The reader will obtain here as full a portrait of Freud as it is possible to draw about his sexual feelings and his aggressive impulses, as well as his acts, related to the women he loved, admired, helped and, in some cases, depended on. As these relationships are examined, the reader will come to understand that Freud, like all mortals—men *and* women—possessed prejudices and blind spots toward women. For one, he was victim of the attitudes toward women in nineteenth-century Vienna. For another, he had a strong attachment to his mother, who called him "my golden Sigi" and, as he grew up, made him feel he could conquer the world like his boyhood hero, Alexander the Great. Freud conquered the human mind.

But, as it becomes clear, Freud also felt a large amount of unconflicted love and profound and genuine respect for women. During an era when women were prohibited from entering the ranks of professional life, he encouraged them to become psychoanalysts, a very difficult profession, one of three "impossible professions," as he called them, along with parenthood and statesmanship. It is significant that though none of Freud's three sons became a psychoanalyst, his daughter Anna, until her death in 1983, was considered the foremost child analyst in the world and one of the important theorists on the functioning of the "ego" as it acted as mediator between our "id" and "superego," or conscience.

It is hoped that this book, researched with respect and admiration for the man who discovered the way to ease mental torment, will bring new awareness of the intellectual and emotional capacities in Freud that enabled him to make his monumental contribution to history. It is a contribution that has influenced every area of our society, from our courts, schools, and hospitals to our art, culture, and personal lives.

Lucy Freeman
Dr. Herbert S. Strean

1987

Introduction

Sigmund Freud is one of the most controversial figures of the twentieth century. He has been idealized as saint and condemned as archdemon. Highly esteemed for his brilliant discoveries, he has also been contemptuously demeaned for alleged prejudices. He has been concomitantly abhorred and revered for his views on sexuality. Few are silent when it comes to their feelings about Freud.

Those who regard him as a genius compare him to Darwin, Einstein, Galileo. They cite his many fundamental discoveries: the nature of the unconscious; infantile sexuality, particularly the Oedipus complex; the interpretation of dreams; the process of free association that leads to the secret world of hitherto unspoken thoughts; the importance of transference and the power of repression and resistance.

Those who revile and reject Freud see him as champion of the status quo and a male chauvinist who has exaggerated the importance of the phallus, erroneously posited a death instinct, foolishly contended that people inherit certain characteristics, and incorrectly averred that women possess less of a sense of social justice than men. They call his theories farfetched and nonsensical.

Even among psychoanalysts and psychotherapists there is little unanimity on the validity of Freud's findings. Some claim nothing has been discovered about theory or therapy since Freud made his monumental discoveries. Others insist he knew little about the oral stage of development, particularly the infant's hostility toward the mother, maternal rejection, the ego and its autonomous functions, narcissism and psychosis. Some therapists believe Freud overemphasized the importance of sexuality in the development of neuroses, while others avow he attributed too much to the role of aggression—and there are those who claim he does not attribute enough. A number of therapists say Freud was too subjective as a psychoanalyst, became at times emotionally involved with patients, while others accuse him of lack of empathy. Therapists of many persuasions regard Freud's perspective as too deterministic and his philosophy as too gloomy.

In recent years, Freud's positions on women have aroused intense debate. He has been praised for leading women to sexual and emotional liberation and attacked for his antifeminist, denigrating attitude toward women. There are those who insist Freud changed the concept of women from passive, inhibited, frightened creatures chained to children and kitchen to that of a human beings entitled to equality with men, while others accuse Freud of viewing women as second-class citizens who wish to castrate men.

Though Freud's relationships to the women in his life reflect compassion and empathy, one can also see in them his fear of women and his hostility toward them. While he could not fight for the emotional and sexual liberation of women unless he felt respect, admiration, and concern for them, some of his theoretical blind spots and therapeutic oversights seem to reflect a fear of his own feminine fantasies, an anxiety about his hostility toward women, and a failure to resolve some of his Oedipal conflicts.

As Dr. Reuben Fine points out in his recent A *History of Psychoanalysis*, there are several gaps and biases in Freud's writings, particularly in his self-analysis. He tended to stress his personal experiences too much. The death of his father precipitated his own great inner turmoil and, accordingly, he wrote that the death of a man's father is "the most important event, the most incisive loss of a man's life." Some psychoanalysts would disagree, saying it is the death of a mother, or a child, or a beloved grandparent, or perhaps even, if the parents survive, the loss of a man's home as he is forced to flee his native land.

In attempting objectively to assess Freud's views, one is compelled to note his predominant emphasis on male psychology and his frequently self-acknowledged confessions of ignorance about women. In an essay on "The Psychology of Women" written in 1932, Freud said: "Throughout the ages the problem of women has puzzled people of every kind . . . you too will have pondered over this question insofar as you are men; from the women among you that is not to be expected, for you are the riddle yourselves."

When feminists criticize Freud's emphasis on the phallus, they find a certain justification in some of his writings. In "Three Essays on the Theory of Sexuality," he wrote in 1905, "The significance of the factor of sexual overvaluation can be best studied in men, for their erotic life alone has become accessible to research. That of women . . . is still veiled in an impenetrable obscurity." Commenting on the phallic phase, Freud said in 1923, "Unfortunately we can describe this state of things only as it affects the male child; the corresponding processes in the little girl are not known to us."

Many of the facts about Freud's relationships to women have not been carefully studied. What transpired between him and his mother that shaped his feelings toward women? What role did his nurse play in his early life? What of growing up with five younger sisters? Was Freud's relationship with his wife, Martha, unambivalent, as Ernest Jones, Freud's biographer, claims? Is it true that Freud was not a very concerned parent, and this stopped him from undertaking the analysis of young children? What did Freud really feel toward female patients, and how did he behave with them? What about his relationships to women psychoanalysts?

Because Freud's views toward women are constantly discussed—in the media, in the classroom, in the bedroom—we believe it timely to take a careful look at Freud and the women in his life—his personal relationships with his mother, sisters, nurse, wife, sister-in-law, daughters, and friends, as well as women patients and colleagues.

As with any mortal, Freud's feelings toward women were at times biased, distorted, and ambivalent—containing both love and hate. But because he was also a genius, he discovered dimensions of female psychology that are startling and unsurpassed.

We write this book not to idealize Freud or repudiate him, but to show him as human—he above all others showed us that no man was a god—and to try to reveal faithfully and to evaluate objectively those attitudes toward women that affected Freud's theories about the emotional conflicts and psychosexual development of the female sex.

What I have said about femininity is certainly incomplete and fragmentary and does not always sound friendly. But do not forget that
I have only been describing women insofar as their nature is determined by their sexual functions. It is true that the influence extends
very far; but we do not overlook the fact that an individual woman
may be a human being in other respects as well. If you want to know
more about femininity, inquire from your own experiences of life, or
turn to the poets, or wait until science can give you deeper and more
coherent information.

—**Sigmund Freud**

Part One

The Women in Freud's Personal Life

The Oedipal Conquistador

The woman who had the most influence on Sigmund Freud was his mother, Amalie Nathanson Freud. "For each of us fate assumes the form of one (or several) women . . . ," he wrote his friend and colleague, Dr. Sandor Ferenczi, on July 9, 1913.

From the day Freud was born at Schlossergasse 117 in Freiberg, Moravia, (Free Mountain)—a peaceful little town about one hundred and fifty miles northeast of Vienna—an unusually strong bond existed between Freud and his mother.

Her birthplace was Brody, in northwest Galicia, then part of Poland near the Russian frontier. She was born August 18, 1835. A tombstone in town bore the name of her ancestor, Samuel Charmatz, a rabbinical scholar. She spent her adolescent years in Vienna, the city that dominated Freud's life. Her father was listed as "a business agent."

Amalie met and married Jakob Freud in Vienna on July 29, 1855, when she was twenty-one and he forty. He was in the business of selling wool, a widower with two sons, one of whom had a son Johann, a year older than Sigmund, which meant Freud was already an uncle when he was born.

As a young woman, Amalie was slender, with dark hair parted in the center, then drawn tight behind her ears in the style of the day. She had an attractive face, flashing brown eyes, and a wide, strong mouth. Her nephew, Edward L. Bernays, known as dean of public relations in America, possesses a number of photographs of the Freud family. One shows Amalie standing beside Freud, when he was about twenty, appearing so youthful that she looks like his sweetheart or sister. In later years, according to Mr. Bernays, Amalie became "rather heavy and very corseted, with a pronounced bosom."

Another photograph, taken August 18, 1905, when she was seventy-one, shows her posed against the backdrop of a city street wearing a fancy dark hat atop her white hair. She has a queenly quality; her face holds a stern expression, yet also a certain sadness. She could not have had an easy life taking care of eight children, one of whom died in infancy, a tragic loss. She also had to please an elderly husband, who lived without much money most of the time. Jakob described himself in 1852 as a widower. Sally Kanner, the mother of Freud's half-brothers, Emmanuel and Philipp, had died and, it is suggested by one of Freud's recent biographers, Ronald W. Clark, that Jakob married a woman called Rebecca between the death of his first wife and his marriage to Amalie, having considered himself divorced (the religious law in those days) from a wife who could bear no children.

Freud's oldest son, Martin, in his book *Glory Reflected*, recalls family gatherings in Amalie's apartment, located in the Cottage district of suburban Vienna, every Christmas and New Year's Eve, holidays observed by many Viennese Jews. Amalie ignored the Jewish holidays, for neither she nor Jakob were Orthodox Jews; they had been married by a rabbi of the Reformed Jewish faith.

Freud always arrived much later than anyone else because he saw patients until nine in the evening. Amalie knew this but, says Martin, "it was a reality she could never accept. Soon she would be seen running anxiously to the door and out to the landing to stare down the staircase. Was he coming? Where was he? Was it not getting very late? This running in and out might go on for an hour but it was known that any attempt to stop her would produce an outburst of anger which it was better to avoid by taking as little notice as possible. And my father always came at very much his usual time, but never at a moment when Amalie was waiting for him on the landing."

Amalie and her son seemed to have formed a mutually strong and intense symbiotic relationship which was never disrupted. Freud saw his mother almost every Sunday of his life, except when he or she were away on summer vacations. On Sunday morning, accompanied by one or two of his children, he would visit Amalie, often carrying a symbol of their fertile relationship—a flowering plant. He would also see her that evening, when she and any of his five sisters who were available, would come to his home for dinner. As soon as the meal ended, he would leave for his study to write further on his psychoanalytic discoveries. Saturday night was reserved for recreation: chess with friends until he was fifty, when he gave it up as occupying too much time, and the card game Tarok, played by four people, which he enjoyed as much as his mother did her card games.

Amalie possessed great vitality and a zest for life that found her,

Freud at sixteen with his mother. *Austrian Press and Information Service.*

when she was ninety-four, still playing cards "at an hour most old ladies would be in bed," according to Dr. Ernest Jones in his three-volume biography of Freud. Martin writes, "I saw my Grandmother Amalie often . . . She [had] had great beauty, but all traces of this had gone when I remember her first."

Amalie resented growing old. When she was ninety, she declined the gift of a beautiful shawl, saying, "It would make me look too old." At that age she was still wearing earrings. Six weeks before she died at ninety-five, a photograph of her appeared in a newspaper, and she remarked, "A bad reproduction; it makes me look a hundred."

Hanns Sachs, colleague of Freud and friend of the family, reports that on her ninetieth birthday Amalie received an ovation at Ischl, a summer resort in the Austrian Alps she had visited regularly for more than thirty years. She was serenaded by the town band, and the mayor presented her with the freedom of the town as visitors brought gifts. In the evening one of her granddaughters said, "You must be terribly tired, Granny." She replied, "Why? I haven't done a stroke of work all day."

Not long before she died, on the eve of Freud's seventieth birthday, he went to see her, hoping to spare her the trip to his home the next day, which she always took on his birthday. The following morning, according to Sachs, the first visitor to ring Freud's bell was his mother, refusing to be separated from her beloved son on the day commemorating his birth.

But nature took care of their final separation. In August 1930, when Freud was vacationing with his family in the Austrian Salzkammergut, and Amalie was in Ischl as usual, she was struck down by gangrene of the leg. Dr. Paul Federn, another colleague and friend of Freud, took her home to Vienna, where she died on September 12, at ninety-five.

Freud wrote Jones three days later: "I will not disguise the fact that my reaction to this event has because of special circumstances been a curious one. Assuredly, there is no saying what effects such an experience may produce in deeper layers, but on the surface I can detect only two things: an increase in personal freedom, since it was always a terrifying thought that she might come to hear of my death; and secondly, the satisfaction that at last she has achieved the deliverance for which she had earned a right after such a long life. No grief otherwise. . . ." He told Jones he did not attend the funeral, that Anna, his youngest child, represented him. He was, at the time, suffering from one of his many operations for cancer of the jaw.

On September 16 he wrote of his mother's death in a letter to Ferenczi, "I was not allowed to die as long as she was alive, and now I may. Somehow the values of life have notably changed in the deeper layers." Perhaps he felt that now that his mother was dead he could wish to die because of the pain from his jaw operations and because he did not want to live without her. This was a wish he found in the lives of patients as they mourned the death of parents, unconsciously wishing to rejoin them, so painful the loss, so deep the dependency, and so profound the symbiotic merger.

During her lifetime, Freud tried to spare his mother from grief. When he was at Auersperg Sanatorium after his first operation, he wrote her a rather formal letter October 17, 1923, no doubt hiding feelings of loss and anger he did not want to express because they might disturb her. He said simply, addressing her "Dear Mother": "Everyone you ask will confirm the news that on the fourth and eleventh of this month I underwent an operation on the upper jaw which, owing to the skill of the surgeon and the excellence of the nursing, has been very successful so far. It will take me some time to grow accustomed to a partial denture that I have to wear. So don't be surprised if you do not see me for a while and I hope you will be in good spirits when we meet again." He signed the letter, "Affectionately, Your Sigm."

In spite of his ambivalence toward Amalie, Freud consciously worshipped his mother, as she did him. She believed he was destined for fame from the moment he entered the world at 6:30 P.M. on Tuesday, May 6, 1856. His very name held the German word for victory, *Sieg*. An old peasant woman assisting at the birth, on seeing the caul that covered his face, is reported to have told Amalie this was a sign she had brought a great man into the world. He had such thick black hair Amalie called him "my little blackamoor." She later changed this to "my golden Sigi," in spite of his dark hair, to more aptly describe her feeling that he was her treasure. She believed in superstitions, signs, and premonitions, as did most people in that unscientific era, and Freud was to become famous by exploring these irrational regions of the mind and the powerful part they played in determining human behavior.

One of Freud's major discoveries, described in his paper "On Narcissism" written in 1914, is that parents project their lost childhood narcissism on their own children and view them as princes and princesses. Freud was probably able to theorize that parents endow their infants with strong doses of omnipotence and grandiosity because of his experience with his adoring mother.

Amalie's belief her son would have an important future was confirmed when he was eleven or twelve. He was eating out with his parents at a restaurant in the *Prater*, the famous park on the outskirts of Vienna. A poet wandered from table to table improvising verses for a small sum. Freud's parents sent him to bring the poet to their table, and the poet showed his gratitude to "the messenger" by making up a few lines about Freud. The poet, Freud later recalled, "had been inspired to declare that I should probably grow up to be a Cabinet Minister." While this referred to becoming a lawyer and enjoying a political career, the literal meaning of the English verb to "minister" did come true in Freud's life. He was to "minister to a mind diseased"—many a mind diseased—in a way that brought far more glory than becoming an Austrian Cabinet Minister.

About his personal life, Freud wrote little; he was reluctant to have it made public. He did say he inherited his passionate nature and temperament from his mother and his sense of humor, shrewd skepticism, liberalism and free thinking from his father. Freud described his father as possessing a gentle disposition, loved by his children, and an optimist, "always hopefully expecting something to turn up," which, unfortunately for the family finances, it rarely did. Freud said he was the duplicate of his father physically and "to some extent mentally." His father would often take him for walks in the dense forests of the foothills of the Carpathian mountains, half a mile from Freiberg, as Freud would later take his children for walks in the Austrian Alps. From Jakob, Freud absorbed a love of trees, plants, and flowers, of which the orchid and fragrant gardenia were his favorites.

After the death of his father on October 23, 1896, Freud wrote his close friend, Dr. Wilhelm Fliess, a nose-and-throat specialist in Berlin, "I valued him highly and understood him very well, indeed, and with his peculiar mixture of deep wisdom and imaginative light-heartedness he meant a great deal in my life."

Yet his father could be cruel. Freud recalled that at the age of seven he "urinated" in his parents' bedroom and his father said in anger, "That boy will never amount to anything." He did not understand his son had been sexually aroused by the sight of his parents in bed and the urination had been his outlet. Freud later traced his anxiety over this experience to a repressed incestuous wish to steal his mother from his father and take his father's place in bed.

Freud's father encouraged him to develop his intellectual powers. Jakob once told a friend with great pride, "My Sigmund's little toe is cleverer than my head." Freud was later to instruct the world that when a father demeans himself vis-à-vis his son, the son develops a sense of arrogance and grandiose fantasies. Usually, when a son feels like an Oedipal victor, Freud further pointed out, he has a punitive superego and feels that he, himself, weakened the father.

In many ways Jakob seemed a lesser figure than Amalie in his son's life, yet it appears Freud received deep emotional sustenance from his father. On Freud's thirty-fifth birthday, Jakob wrote in the family Bible a verse about his son's success in which he said, "You ascended and rode upon the wings of the spirit."

But it was Freud's mother who played the dominant role in his very early life. As he was later to say, the mother is a child's original love, and all future loves are modeled after this first experience of love. He acquired from Amalie a lasting sense of self-confidence that enabled him to make discoveries that "disturbed the sleep of the world," in his words. He wrote his fiancée, Martha Bernays, on April 19, 1884, "I know I am someone, without having to be told so." His self-esteem allowed him to endure the most vituperative criticisms of his theories, when they were introduced to the skeptical medical profession of Vienna.

One reason for his confidence was that he was fed at his mother's breast "in a loving way," according to Jones. Psychoanalysts have found that the way a mother holds a baby at her breast and feeds him—either with love and protectiveness or with hate and irritation—shows how she feels about the baby and how she will treat him thereafter, so that a baby senses at his mother's breast, and during the first few months of life, whether his mother loves and wants him or hates and rejects him. All mothers have ambivalent feelings; it is a matter of whether the love overpowers the hate, or the hate, the love. What was crucial for Freud as a baby, and is so for other infants, is that if they receive consistent warmth and empathy from a loving mother, they develop what Erik Erikson calls "an inner certainty."

Even though there were rival brothers and sisters, Freud knew he was his mother's favorite. When he was three, the family moved to Leipzig for a year, and then to Vienna. Here, in a six-room apartment in the Leopoldstadt, in the Jewish quarter, Jakob, Amalie, and eventually seven children, were distributed among three bedrooms. Freud, the only member of the family who rated privacy, was allotted an airless hall bedroom in which he slept and studied, until he became an intern at the Viennese General Hospital at the age of twenty-six. When he was ten and a half and his sister, Anna, eight, Amalie, who was very musical, bought a piano, wanting Anna to learn how to play and become a music teacher. Freud complained that her practicing so disturbed him he could not study, and Amalie had the piano removed at once.

There is an amusing sidelight to this story. Freud wrote Martha on November 26, 1885, as he was studying in Paris under Dr. Jean Martin Charcot, the famous neurologist, that he should call on the wife of the Freud family's physician, who was visiting in Paris. He said, "The unfortunate woman has a ten-year-old son who, after two years in the Vienna Conservatorium, won the great prize there, and is said to be highly gifted. Now instead of secretly throttling the prodigy, the wretched father, who is overworked and has a house full of children, has sent the boy with his mother to Paris to study at the Conservatoire and try for another prize. Just imagine the expense, the separation, the dispersal of the household! Little wonder that parents grow vain about their children, and even less that such children grow vain themselves. . . The prodigy is pale, plain, but looks pretty intelligent."

The "prodigy" proved to be Fritz Kreisler. Freud's unwarranted indignation was surely related to his complaint at the age of ten, when his sister, the musical prodigy, was interrupting his studies, and his mother, realizing her "golden Sigi" was upset, got rid of the piano. In Freud's statement about "the prodigy," he also is implicitly referring to his own strong vanity, nurtured by his indulgent and worshipful mother.

Clark, in his *Freud: The Man and the Cause*, said "His [Freud's] position as his mother's first born and favorite gave him a privileged place in the home which he was not reluctant to exploit . . . it was accepted that the home should be run largely for his convenience . . ."

Freud later wrote, "A man who has been the indisputable favorite of his mother keeps for life the feeling of a conqueror, that confidence of success that often induces real success." He also said the relationship of a mother to a son "provides the purest example of an unchangeable affection, unimpaired by any egoistic considerations," and described the relation of mother and son as "altogether the most perfect, the most free from ambivalence of all human relationships."

Blinded by his rivalry with his sisters, Freud could never fully appreciate the tremendous influence a mother has on a girl's psychosocial and psychosexual development. Freud's competition with his sisters prevented him from acknowledging that if a girl has been the indisputable favorite of her mother, she, too, can keep for life "the confidence of success that induces real success."

Though Freud was his mother's indisputable favorite, he had to cope with newcomers almost continually during the first ten years of life, and the arrival of these babies aroused intense jealousy. He was followed by a brother Julius, who died at eight months, when Freud was nineteen months old. Anna was born when Freud was two and a half. Then came Rosa, and shortly after, the Freuds moved to Leipzig. In Vienna, there followed Marie (Mitzi), Pauline (Paula), Adolfine (Dolfi), and Alexander, ten years younger than Freud.

In *New Introductory Lectures*, written in 1932, he gives considerable attention to "when the next baby appears in the nursery." He says, as if describing his own plight as a child, "in cases in which the two children are so close in age that lactation is prejudiced by the second pregnancy, this reproach [against the mother] acquires a real basis, and it is a remarkable fact that a child, even with an age difference of only *eleven months* [the exact age difference between Sigmund and his brother, Julius], is not too young to take notice of what is happening. But what the child grudges the unwanted intruder and rival is not only the suckling but all the other signs of maternal care. It feels that it has been dethroned, despoiled, prejudiced in its rights;

it casts a jealous hatred upon the new baby and develops a grievance against the faithless mother."

Freud never became aware of his deep rage at his "faithless mother" and the "grievances" toward her that were displaced on other women, as he consciously sustained his idealization of his mother. Though Freud wrote that "a child's demands for love are immoderate, they make exclusive claims and tolerate no sharing," he never seemed consciously to come to terms with his own resentment of Amalie. He apparently worked overtime to protect his idealized image of his mother from the intrusion of hostile feelings.

Psychologists Dr. Robert Stolorow and Dr. George Atwood, in their recent book, *Faces in a Cloud: Subjectivity in Personality Theory*, point out: "The central conflict in his [Freud's] emotional life had thus been established; namely, the conflict between an intense, possessive need for his mother's love and an equally intense, magically potent hatred. By splitting off, repressing, and displacing the enraging and disappointing qualities of the mother image, he safeguarded his relationship to her from invasion by negative affects, protected her from the overwhelming power of his anger, and saved himself from the dreaded catastrophe of losing her."

When a child represses strong feelings of anger toward a mother, believing that to express them means risking loss of the mother's love or abandonment by her, a neurosis can result. One form the neurosis can take is agoraphobia—and it was from this phobia Freud suffered all his life. As a psychoanalyst he was to write: "The common symptom of agoraphobia . . . is usually linked with the persistence of childhood dependency, fear of abandonment by the mother, and also with fears of the patient's aggression."

Freud's ambivalence toward his mother, and toward women in general, was to express itself in countless ways. The theme of the "two mothers," one the natural mother, the other, a mother substitute who brought up the child, was central to Freud's speculations about Leonardo da Vinci. When Freud was commissioned to translate into German a volume of John Stuart Mills, he disagreed sharply with the philosopher's acceptance of the equality of women. Later Freud was to tell his fiancée, Martha, "The position of woman cannot be other than what it is; to be an adored sweetheart in youth, and a beloved wife in maturity." When an American visitor asked Freud whether it would not be better if both partners in a marriage were equal, Freud answered, "That is a practical impossibility. There must be inequality, and the superiority of the man is the lesser of two evils." Not only did Freud consistently contend that woman showed less of a sense of social justice than men, but he never wavered from his position "that envy and jealousy play an even greater part in the mental life of women than of men."

Because Freud repressed so much of the rage he felt toward his mother, the role of the mother in the development of the child was neglected by him, left to a later generation of psychoanalysts to explain. In his famous case of Little Hans, a five-year-old boy who suffered from a horse phobia, Freud excluded the mother from the treatment plan, despite her active role in the cause of Hans's neurosis. Only Freud, Hans, and Hans's father participated in the first child guidance case in history.

But out of Freud's exceptional closeness to a mother who believed he could do anything in the world and who cherished him, came an extraordinary courage. It was a courage that led him to observe in his first patients, then in himself, the powerful sexual yearning for the parent of the opposite sex, accompanied by hatred and jealousy of the parent of the same sex.

He called this the *Oedipus complex,* after the Greek hero who fulfilled the prophecy of the Delphic oracle and murdered his father, King Laertes, and unknowingly married his mother, Queen Jocasta. Freud discovered that every child, at about the age of four, shows ardent feelings for the parent of the opposite sex and jealousy of the parent of the same sex. This is part of the natural psychosexual development toward maturity, in preparation for falling in love with a more appropriate member of the opposite sex later in life. The very early Oedipal feelings pave the way for marriage and the wish to have children. But if there is too much conflict for the child during the Oedipal period, that is, if parents are too harsh and cruel, either physically or psychologically or both, the child's sexual development will be thwarted or blocked.

Originally Freud described the Oedipus complex as the core of neurosis. Later, he was able to appreciate that pre-Oedipal conflicts, those occurring during the first three or four years of life and connected with feeding or toilet-training problems (the early stages of psychosexual development), were often as important as the Oedipal conflicts and the degree of their pathology frequently affected the later Oedipal situation. Freud had a tendency to underestimate the importance of pre-Oedipal factors because of his unconscious wish to overlook and deny some of the hates of his early pre-Oedipal years.

During his self-analysis in the last few years of the nineteenth century, Freud became aware of his erotic feelings for his mother. He described this growing awareness in letters to Dr. Fliess, who encouraged him to continue the inner journey.

When he began the self-analysis, Freud wrote Fliess on May 25, 1895: "A man like me cannot live without a hobby-horse, a consuming passion—in Schiller's words, a tyrant. I have found my tyrant, and in his service I know no limits. My tyrant is psychology, it has always been my distant beckoning goal and now, since I have hit on the neuroses, it has come so much the nearer."

Unconsciously, Freud was, in all likelihood, alluding to the major tyrant in his life—his mother. She was a beloved tyrant, but a tyrant nevertheless. The concept of Mother Nature attracted him to medicine. He commented that "it was after hearing Goethe's beautiful essay on Nature read aloud at a popular lecture by Professor Carl Brühl just before I left school that decided me to become a medical student." Freud's biographer, Dr. Ernest Jones, points out: "Goethe's dithyrambic essay is a romantic picture of Nature as a beautiful and bountiful mother who allows her favorite children the privilege of exploring her secrets."

This imagery attracted the youthful Freud. Goethe's statement about Nature neatly mirrored Freud's strong unconscious symbiotic and erotic attachment to his mother: "Nature! We are surrounded and embraced by her, powerless to separate ourselves from her, and powerless to penetrate beyond her. . . . We live in her midst and know her not. She is incessantly speaking to us, but betrays not her secret. We constantly act upon her and yet have no power over her . . ."

It seems more than a coincidence that in Freud's writings one of his most oft-used words is "nature." Speaking of the birth of psychoanalysis, he said, "They regard me as a monomaniac while I have the distinct feeling that I have touched on one of the greatest secrets of nature."

Freud wrote Fliess on June 12, 1895, it was too soon to make public his new theories because, "Saying anything now would be like sending a six-month female embryo to a ball . . ." This is a spectacular simile in itself, but its importance lies in the fact that he thought of his work as "female." His creation, his production, was identified with his mother.

Two years later he wrote Fliess on October 3, 1897, that his self-analysis was progressing and he now remembered an experience when traveling overnight on a train with his mother from Leipzig to Vienna. While sharing a compartment, his "libido towards *matrem* was aroused" when he "must have had the opportunity of seeing her *nudam* . . . ," he said. In this letter he also mentioned his "real infantile jealousy" of his younger brother Julius, admitting, "his death left the germ of guilt in me." Though Freud did not then make the connection, the fact that he mentioned erotic feelings at seeing his mother nude, then guilt at his brother's death, indicated a feeling of guilt at the wished-for (when a boy) death of his rival father. No feeling or thought ever disappears from the unconscious.

But he was soon to arrive at this thought. Twelve days later, on October 15, 1897, he wrote Fliess that only one idea of general value had occurred to him [but what an idea!]: "I have found love of the mother and jealousy of the father in my own case too, [he had seen it in patients] and now believe it to be a general phenomena of early childhood . . . If that is the case, the gripping power of *Oedipus Rex*, in spite of all the rational objections to the inexorable fate that the story presupposes, becomes intelligible . . . the Greek myth seizes on a compulsion which everyone recognizes because he has felt traces of it in himself. Every member of the audience was once a budding Oedipus in phantasy, and this dream-fulfillment played out in reality causes everyone to recoil in horror, with the full measure of repression which separates his infantile from his present state."

Two years later, in his classic *The Interpretation of Dreams*, where he fully described the unconscious mind and its relationship to the operation of the conscious, Freud broke through his usual reticence on personal matters and told of two boyhood dreams involving his mother. He now gave his "associations" to the first dream, in which "dumplings" appeared. This meant he expanded freely his thoughts about the various images and feelings in the dream, a procedure that leads to the underlying wishes and fears against which a dream protects awareness in the conscious mind because they are terrifying or humiliating.

Freud recalled that when he was six years old, his mother gave him his "first lessons" about the meaning of life and death, telling him each person was made of dust and must therefore return to dust. This was to him a fearful thought—death—and he expressed doubts about it. Whereupon she rubbed the palms of her hands together, "just as she did in making dumplings, except there was no dough between them," and showed him flakes of skin produced by the friction as proof humans were made of dust. He had felt astonished, he said, but bound to accept the belief he was later to hear expressed by Shakespeare, "Thou owest Nature a death." Though Freud talked about his own guilt over his repressed death wishes for his rival father and feared death at the hands of his father as the appropriate "eye for an eye" punishment, he once again overlooked his death wishes toward his mother. Behind his interest in Shakespeare's "Thou owest Nature a death," lay his disguised and repressed way of avenging himself on Amalie.

The second dream about his mother he called an "anxiety dream." In it he saw "my beloved mother, with a peculiarly peaceful, sleeping expression on her features, being carried into the room by two (or three) people with birds' beaks and laid upon the bed."

He remembered as a boy waking from this dream, tears in his eyes, and screaming so loudly he woke his parents. Now, more than thirty years later, as he analyzed the dream, he said the "birds' beaks" appearing in the dream reminded him of woodcut illustrations in Philippson's Bible portraying Egyptian gods in funeral masks resembling the heads of falcons.

He was also reminded of "an ill-mannered boy, a son of a *concierge*," with whom he played on the grass when they were children and from whom he first heard "the vulgar term for sexual intercourse," meaning "fuck." This memory, said Freud, related to his unconscious choice of the birds' heads as a symbol. The slang word in German for sexual intercourse was *vogeln*, from the word Vogel, meaning bird, Freud added.

He recalled being haunted by the expression on his mother's features in the dream, an expression he compared to his grandfather's whom he had seen a few days before the dream, lying in a coma, after which he died. As a boy, when he woke from the dream, Freud thought it was his mother who was dying and thus his tears and his scream. But when she rushed to his bed he felt calm as he saw her face, "as though I had needed to be reassured that she was not dead."

The feeling in the dream that his mother had died hid the deeper meaning of the dream, he explained. He felt "anxiety" not alone because he dreamed she was dying, but because of "an obscure and evidently sexual craving that had found appropriate expression in the visual content of the dream"—the falcons' heads. The deeper wish had been to possess his mother sexually with his little "bird's head." Every dream pictures in disguised form the fulfillment of some forbidden childhood wish over which there is conflict, as Freud discovered.

He also wrote Fliess he had discovered that hostile impulses against parents, a wish they would die, are an integral part of neuroses. Such wishes come to light consciously in the form of obsessive ideas and are repressed at times when "pity for one's parents is active, at times of their illness or death."

Freud commented, "It seems as though in sons this death-wish is directed against their father and in daughters against their mother." He gave as illustration: "A servant-girl makes a transference from this to the effect that she wishes her mistress to die so that her master can marry her." He explained this was "Lisel's dream about Martha and me." Lisel was governness to the Freud children at this time, and told Freud a dream in which she saw Martha dead.

Freud recognized his own wish for his father's death, but was never able to admit that he also felt hatred for his mother and wished her dead when she hurt him or he felt in any way rejected by her. He never dared face the murder in his heart for the beloved "tyrant." It was left to contemporary psychoanalysts to delve into the roots of a hatred that exists in the first years of life when a mother, as she nurtures, trains, and guides her baby, will inevitably hurt and frustrate him so that he will feel occasional rage. If she is unable to allow him to separate emotionally from her in gradual stages because of her own unconscious fears of separation, thus curtailing his natural drive for independence, his hatred may run deep. Amalie and Freud sustained their symbiosis, never really separating from each other emotionally. His unconscious hatred toward her prevented separation.

In a letter to Fliess on October 3, 1897, Freud spoke of his fear of traveling on trains, always expecting an accident, saying, "My anxiety over travel, you have seen yourself in full bloom." He did not realize the fear of separating from his mother lay behind his dread of railroad stations (the train would take him away from her to dangerous lands). Nor did he tie in his fear of trains with the trip he took at the age of four when he saw his mother nude, which aroused sexual feelings, anger at the thought of his rival father, and guilt. The dream he recalled of seeing his mother dead would bear out both his sexual and angry feelings connected with Oedipal and pre-Oedipal conflicts. Freud always emphasized the boy's sexual feelings toward the mother and downplayed the aggressive ones. It was his way of repressing his wish to see his idealized Amalie dead.

But Amalie lived to see her favorite child fulfill her dreams. The "little blackamoor" emerged as the world's white knight of mental healing. In the summer of 1930, just before Amalie died, Freud was awarded the annual Goethe prize for literature. It consisted of 10,000 marks and a medallion with his face engraved on it (he gave the medallion to his nephew, Edward Bernays). By this time, psychoanalysis was starting to be accepted in America, England, France, and Germany, and Freud had founded the International Psychoanalytical Association, as well as several international journals.

Why was Freud the one, in the history of mankind, to dare explore the secrets of sexuality? A comprehensive answer to this question would have to take into consideration his intellectual genius, his boundless energy, his strong drive endowment, and further knowledge of his childhood.

Freud's passion for psychoanalysis and his never-ending interest in all of its aspects (treatment, a theory of personality, research) was in many ways a recapitulation and sublimation of his intense passion for his mother. His relentless love of his mother was later shown in his jealous, possessive, and romantic interest in Martha Bernays, who became his wife.

As Freud delved into the unconscious, trying to get deeper and deeper into it, it would appear he was symbolically trying to plunge into his treasured mother. Freud loved archaeology—also plunging into the depths. As he traveled the royal road to the unconscious, he was on an incestuous journey, much like Oedipus.

Like Oedipus, he too had to be punished for the crime he committed in fantasy. Freud accepted the punishment inflicted by his superiors and colleagues in the form of their outrage at and rejection of his work. For, as he pursued psychoanalysis like an Oedipal conquistador (Hannibal, Oliver Cromwell, and Alexander the Great were his heroes), he had to be punished. Again, like the Oedipal boy who wants mother and competes with father, but has mixed feelings toward father, Freud always had a "beloved friend and a hated enemy" (split of the father image).

Research has come to light which even more strongly suggests that as a little boy Freud frequently would have been very sexually aroused by the family's living conditions. Up to now it has been taken for granted his family occupied the whole house at 117 Schlossergasse in Freiberg. But Clark, in uncovering material for his biography, showed that Jakob and Amalie lived only in one room on the second floor. The other room on this floor, says Clark, was occupied by a locksmith named Zajic, whose ancestors had lived in the house for generations, and who used the ground floor as his workshop.

Freud was born in this second-floor room, and lived in it for three and a half years, possibly with his mother and father, brother Julius while alive, and then Anna. The one room would have been the family's home all this time, unless they acquired the other room on the second floor, or the whole house, by 1859 when Jakob left Freiberg.

Such extremely intimate living in the early years of his life, during which Freud had to be aware of what he later called "the primal scene," undoubtedly aroused in him strong sexual stirrings which, along with a child's natural passionate curiosity about the mysteries of sex, had to be repressed.

His quest for the answers to the many facets of sexuality was an intellectual way of solving the most intense conflicts of his life. They revolved around a sexual desire aroused by parents who, because of poverty, had to conduct their sexual activities in the presence of what they believed to be a sleeping child, as many parents in the world are forced to do.

It just happened that their son Sigmund was a particularly brilliant, sensitive little boy and, suffering inner pain over the torments of his forbidden love for the pretty young mother who adored him, later chose, not promiscuity, not homosexuality, not celibacy, but a search for the answers to the searing questions of his life. His mother believed he could do anything; but there was one thing he could not do—possess her completely. He could, however, try to find out why he felt such strong passion for her and why he suffered so, and perhaps help other men relieve a similar agony.

He once told his friend and colleague Theodor Reik, "What this world needs are men of strong passions who have the ability to control them."

He was speaking of himself, in relation to his mother.

The Mystery of Nannie

One other woman, in addition to his mother, was important in Freud's most vulnerable years, She was a nurse he referred to simply as "Nannie," who took care of him the first two and a half years of his infancy. She was later identified as Monika Zajic, possibly related to the Zajics with whom the Freud family shared the second floor. She also worked for the wife of Emmanuel, Freud's half-brother.

She came to the house daily. A Czech and an ardent Catholic, she took Freud, as a little boy, every Sunday to church services. One wonders that his parents permitted this, though they were not religious Jews.

Freud dimly recalled Nannie, remembering he gave her all his pennies. He relied on his mother's later description of her as "old and ugly but very sharp and efficient." Freud said in a letter to Fliess on October 15, 1897, "I asked my mother whether she remembered my nurse. 'Of course,' she said, 'an elderly woman, very shrewd indeed. She was always taking you to church. When you came home you used to preach and tell us all about how God conducted his affairs.'" Jones speculates that Nannie's "terrifying influence" contributed to Freud's later dislike of Christian beliefs and ceremonies. Freud may have associated church ceremonies with the devil and guilt at "evil doings," for he always avoided ceremonies of any kind, including weddings—he attended only three in his life, other than his own.

He apparently both loved and feared Nannie. He told Fliess, "From what I can infer from my own dreams her treatment of me was not always excessive in its amiability and her words could be harsh if I failed to reach the required standard of cleanliness," then he added, "It is reasonable to suppose that the child loved the old woman who taught him these lessons, in spite of her rough treatment of him."

This phrase, "rough treatment," shows, no doubt, she was "harsh" in the way she cleaned him, perhaps rebuking him severely when he got dirt on his clothes or face while playing. In a letter to Fliess on October 3, 1897, he mentions her as being responsible for his neurosis, and yet also as a source of his early self-esteem. He writes: "My primary originator [of neurosis] was an ugly, elderly but clever woman who told me a great deal about God and hell, and gave me a high opinion of my own capabilities." But then, discussing his current feelings of anxiety, he says, "If they [the reasons for his anxiety] emerge, and I succeed in resolving my hysteria, I shall have to thank the memory of the old woman who provided me at such an early age with the means for living and surviving."

We can only speculate as to what she "provided" that enabled him to survive, but perhaps he sensed in her a love for him that he felt in her daily caring for him and her attempt to instill in him the difference between right and wrong.

He must have considered her an emotional support and thus felt devastated when she mysteriously disappeared. This occurred at the time his mother was giving birth to his sister Anna, eleven months after Julius had died. Freud imagined that his half-brother Philipp, twenty years older, was the one involved in Nannie's disappearance, and later remembered asking Philipp what had become of Nannie. Philipp had said, "She has been locked up in prison." There is no indication Freud asked why.

Many years later (Freud does not say how many), the mystery of Nannie's disappearance was cleared up when his mother explained, "She turned out to be a thief, and all the shiny Kreuzers and Zehners and toys that had been given you were found among her things. Your brother Philipp went himself to fetch the policeman, and she got ten months."

Freud theorized that his Nannie disappeared at the moment his mother lay in bed with the new baby, and several fantasies had become intertwined in his child-mind. The changes in his mother's figure before the baby was born told him, he said, the source of the baby, but not how, specifically, it had happened that a newcomer would appear on the scene with whom he would once again have to share his mother's love, just after he found himself in full possession of it as a result of the death of baby Julius.

Forty years later Freud had what he thought an apparently unintelligible memory from childhood. He saw himself standing before a "chest" and tearfully demanding something of Philipp, who was holding the chest open. He remembered Philipp's words to him as a child about Nannie: "She has been locked up in prison." Now in memory, Freud saw his mother, not pregnant but slender once again, come into the room, presumably having left the house on an errand. At first Freud supposed she had interrupted Philipp's teasing of him. But in digging deeper into memory, he uncovered thoughts of missing his mother and fearing *she* might be locked in the "chest" by Philipp. He dimly recalled begging Philipp to open the chest to reassure him his mother was not locked in it. Philipp did so, whereupon Freud, seeing the empty chest, began to cry.

Freud theorized that the "chest" symbolized the womb, and his anxious request to his brother related not merely to his mother's momentary absence, but to the more terrifying question as to whether another unwelcome little brother or sister had been put in her womb. Since Philipp was the one who had reported Nannie to the police—in Freud's boyhood mind, the one to put people in a "chest" —he had formed the fantasy that his half-brother and his mother, about the same age, had cooperated in producing his rival sister, Anna. Jones comments that Freud never liked his first sister. Jones also says it was natural, in a child's eyes, for Freud to pair off his mother and Philipp, the younger ones, and his father and Nannie, the two older "forbidding authorities."

While Freud's references to his mother during his early years are exclusively positive, always picturing her as an object of his erotic feelings and his tender love, when he describes his relationship with Nannie, his ambivalence is quite conscious, his hostility apparent, as he blames her for his neurosis. Freud needed to picture Nannie as an ogre responsible for his neurosis so he could preserve his mother as an ideal and displace the hostility he felt toward her onto the nurse.

Freud's attitudes toward Nannie included a large measure of rage. There must have been many times he wished she would "drop dead," or at least disappear. Nannie's eventual disappearance, much like his brother Julius', was probably experienced by him as a "wish come true." As all children do, because they are as yet unaware of the limits of reality and believe they are omnipotent, Freud blamed himself for these desertions.

Because he had, as a child, actually witnessed Julius and Nannie disappear, Freud had to be very careful to avoid feeling or expressing hostile wishes toward Amalie—lest she vanish too. Though she gave him reason to resent her because she brought many "intruders" into the family, Freud managed to preserve his positive image of Amalie and direct his resentments toward Nannie, his siblings, and his father.

Jones remarks that Freud chose a "curious day" on which to open his first office in the practice of "neuropathology," Easter Sunday, April 25, 1886. Everything in Vienna was closed on that holy day. Jones speculates that Easter may have had an "emotional significance" for Freud, standing for the Catholic Nannie, who once took him to church services in Freiberg. Perhaps his coming of professional age had something to do with the strict conscience his Nannie helped instill in him. Or, perhaps, it was defiance of her emphasis on Catholicism and, by extension, all organized religion, a rebellion against Nannie's religious dogma, as his father had rebelled against orthodox Jewry.

But in spite of conflicting feelings about Nannie, Freud must have felt deeply upset as a little boy when she suddenly vanished—without a goodbye. He had transferred to Nannie all the feelings he felt for his mother, and no doubt carried with him the rest of his life the belief he had been abandoned by someone he loved-hated.

The Sisters Five

Until Freud was ten years old and his brother Alexander was born, except for the eight months his brother Julius lived, he was surrounded by women—his mother, his Nannie, and his five sisters. Unconsciously, he may have set up the same situation in later life when he was again surrounded by women—still his mother and sisters, plus his wife, three daughters, a sister-in-law, patients, pupils and colleagues, friends, and faithful servants.

His first sister, Anna, was born when he was two and a half, his second, Rosa, when he was four. Martin remembered a painting that hung in his grandmother's drawing room showing the five little girls, Freud, then ten and a half, and Alexander, the baby. Martin remarked that none of his aunts seemed particularly pretty little girls in the picture, but his father, as a boy, "was not only goodlooking but even beautiful." (Mr. Bernays has a photograph of Freud before he grew the beard that covered his lower face, showing a mouth with strength and classic beauty, like his mother's.)

Martin says that of the five little girls in the painting, Rosa was Freud's favorite. When he was poor, in the years before his practice flourished, he said it hurt when, for the first time, he was unable to buy Rosa even a small birthday present. She had a happy disposition, charm, and dignity, and grew up to be pretty, and people compared her to the actress Eleanora Duse, according to Martin, who added, "As a widow, well on in her sixties, she could still command love from young men, something about which she was very proud and not in the least discreet." He recalls, "But shadows began falling around her when she lost her highly gifted children and when, to make more emphatic her loneliness, she became totally deaf."

Freud wrote his fiancée Martha on February 10, 1886, that he and Rosa were the "neurasthenic" members of the family, but that as a neurologist he was as worried by it "as a sailor by the sea." When Rosa married Heinrich Graf, a prominent lawyer in Vienna, they lived for a few years in an apartment on the same floor as Freud, and Rosa occasionally took care of Freud's children when he and Martha went on brief vacations. She lost her only son, Hermann Graf,

25

at the age of twenty when he was killed on the Italian front in the summer of 1917, the only loss the Freud family suffered in World War I.

Anna was also of cheerful temperament, "truly a Viennese," says Martin. She married Eli Bernays, Martha's brother. Freud did not attend the wedding, partly because of family friction and partly because of his dislike of formal occasions. Perhaps, also, his decision reflected his feeling that Anna was his least favorite sister.

Freud did not write much about his sisters, but he mentioned Anna in the interpretation of one of his memories. He recalled an experience from "very early youth" when his father gave Anna and him a book with colored illustrations that described a journey through Persia, telling them they could destroy it. Freud was five at the time and Anna, almost three. He recalled the image of "the two of us blissfully pulling the book to pieces (leaf by leaf, like an *artichoke*, I found myself saying)." This was almost the only memory of that particular time of his life, and he related it to fantasies of masturbation.

As Freud was later to discover, when a young boy has erotic fantasies toward a girl but finds these feelings evoke anxiety and produce guilt, he may convert the warm desire into hatred. Since Freud related "the artichoke episode" with Anna to mutual masturbation, he may have formed a phobic response to her to avoid facing forbidden sexual fantasies. This tends to be corroborated by Anna, who wrote that her brother, Sigmund, exercised a definite control over her reading. "If I had a book that seemed to him improper for a girl of my age, he would say: 'Anna, it is too early to read that book now,'" she recalled.

She also complained he not only resisted seeing her as a sexual young woman by controlling her reading, but kept his male friends away from her. Instead, her "learned brother" used his male friends exclusively "for scientific discussion." Though Jones says that Freud's absence from Anna's wedding was due to "reasons that cannot be given," we may conjecture that just as Freud did not want to permit Anna to be sexual, to read forbidden works, or to flirt, he may have also been jealous of Eli, who could do with Anna what Freud fantasied but found forbidden. Freud was jealous, too, of Eli's control of the Bernays' purse strings.

Freud on the left with his five younger sisters and their baby brother
Alexander. An anonymous oil painting. *Austrian Press and Information Service.*

Of the younger sisters, not much has been chronicled. Marie went
to Paris for a year as a lady's companion and earned money she sent
her mother. In March 1886, she married Moritz Freud, a distant relative from Romania. Pauline is the least known of the sisters; possibly
she made less impression on Freud than the others.

Martin says Dolfi, the youngest, was Freud's second favorite. In a
letter to Martha on September 9, 1883, Freud says, "Yesterday I
went on an excursion with Dolfi. . . . She waited for me while I was
seeing my patient, and then we walked back via Dornbach [a sub-
urb of Vienna]. She is the sweetest and best of my sisters, has such
a great capacity for deep feeling and alas an all-too-fine sensitiveness
. . . her instinct allows her to guess what her judgment cannot
provide."

On April 21, 1884, he wrote Martha about Dolfi, after their sister
Pauline had fallen in love: "Dolfi is the only one still unattached.

Yesterday—I had invited her to tea to get her to mend my black coat—she said: 'It must be wonderful to be the fiancée of a cultured man, but a cultured man wouldn't want me, would he?' " Here, Dolfi was probably fantasying herself as her brother's wife, though feeling quite inferior to him. She did not seem to have much confidence in herself as a woman and spent the rest of her life taking care of her mother.

Martin wrote of his aunt, "Dolfi was not clever or in any way remarkable and it might be true to say that constant attendance on Amalie had suppressed her personality into a condition of dependence from which she never recovered."

Freud wrote his brother Alexander on April 5, 1918, that his mother was refusing to leave financial arrangements to Dolfi "and [Amalie] torments her terribly every time money has to be spent . . . Dolfi hasn't much energy left." He suggested that Dolfi, unknown to her mother, be given a sum that would allow her to keep the household going "without our old lady realizing the expenses." Freud suggested that Alexander and he each contribute a thousand kronen, in addition to the usual five hundred. He also suggested they pay all the expenses for his mother's summer vacation, explaining, "You realize as well as I do that a summer will come when we won't have the chance of repeating the assistance." Since Jakob had died, the two sons supported both their mother and Dolfi.

There is some indication of how the sisters felt toward their older brother in the gifts they bought him for his thirtieth birthday, when they all were very poor. Freud wrote Martha on May 6, 1886, that in addition to the clythia (an evergreen plant) she had sent, "Paula and Dolfi brought me a beautiful little brush box, Mitzi a large photograph of herself and two Makart bouquets [consisting of dried palm branches, reed, bamboos and a peacock feather, named after the Austrian artist Hans Makart], Mother a cake, and Rosa a lovely framed blotter for my writing table."

Though Freud remained emotionally tied to his sisters all his life, he was frightened of his sexual feelings toward them and was never outwardly affectionate. When he became engaged, he told Martha he bestowed on her more kisses in the two days after their engagement than he had given his sisters in the twenty-six years of his life.

Freud often lectured his sisters and appeared to be a warm, caring, but authoritarian big brother. He helped them with their lessons and told them how to think politically. Not only did he control Anna's reading (advising her against Balzac and Dumas when she was fifteen), but supervised all his sisters' choice of books. Freud, who often fantasied himself a general and identified with the conquerors, at the age of fourteen when the Franco-Prussian war broke out, kept a large map on his desk and followed the campaign with small flags, describing the military maneuvers to his sisters.

He appeared to have had a tendency to stop his sisters from sexual pleasure. As the discoverer of the many dimensions of incest, Freud defended against his sexual feelings toward his five sisters by treating them at times as if they should be as celibate as nuns. According to Pauline, who married a man named Winternitz, Freud could be very severe when he found his sisters doing something of which he did not approve. He caught Pauline spending money for sweets when she needed it for necessities and admonished her with such grimness that fifty years later she had neither forgiven him nor forgotten the episode, telling it to Martin with much feeling. This memory probably represented in capsule form many incidents in which her brother lectured and intimidated her.

Freud wrote Rosa in July 1876, when she was staying at Bozen with their mother, against having her head turned by a slight social success after she had given a performance on the zither. Freud's letter, according to Jones, was full of advice "on how unscrupulous people are in overpraising young girls to the detriment of their later character," warning that such experiences could end in girls becoming "vain, coquettish and insufferable." Again we see Freud as big brother acting in an authoritarian manner and stopping his sisters from achieving, and from having pleasure. In many ways he was a male chauvinist with them.

But he also cared deeply for them, they were welcome at his home all their lives, as were his nieces and nephews. Once, as a student, he was invited out to lunch but found it difficult to eat roast meat, thinking how hungry his sisters were. Jakob was never successful at business; and Amalie's relatives sometimes sent money. At times, to help with the finances she rented a room to a boarder after two of her daughters left home.

The lives of the five little girls in the painting ended in tragedy, except for Anna, who had gone to live in America in 1892 with her husband, and who died in New York in her nineties. When Freud fled the Nazis in 1938, he left behind his four sisters—Mrs. Rosa Graf, Mrs. Marie Freud, Mrs. Paula Winternitz, and Dolfi Freud. He and Alexander had given them 160,000 Austrian schillings, about $22,400.

Toward the end of 1938, Marie Bonaparte, Princess George of Greece and Denmark, who helped Freud escape to London, also tried to rescue his four sisters but could not get the cooperation of the French authorities, as she had done with Freud. Dolfi died of starvation in the Jewish ghetto in Theresienstadt, and the three others were murdered by the Nazis, "probably at Auschwitz," Martin reports.

Freud never knew their fate. By that time, he too had died, victim of the cancer that had plagued him for sixteen years, spared at least the knowledge of the horror of his four sisters' last days on earth.

The Girl Friend

When Freud was sixteen, he formed a passionate attachment to a girl named Gisela, a year younger than himself. As he was to note later, when he pondered "the riddle of adolescence" and pointed out that the young person recapitulates in his teen-age years his earlier attachment to the parent of the opposite sex, Freud formed a symbiosis with Gisela that was quite similar psychologically to the earlier one with his mother.

Freud met Gisela when he went on a summer holiday to Freiberg in 1872, the first time he had returned to his birthplace. He stayed with the prosperous Flüss family, with whom his father had kept in touch. Mr. Flüss, unlike Jakob, had become a success in the textile business.

Freud was too shy to tell the Flüsses' fifteen-year-old daughter how he felt, and after a few days she went away to school. He wandered in the woods, thought how pleasant life might have been if his parents had not left Freiberg and he had grown up a country boy, like Gisela's brothers, and married her. At one point in their relationship, Freud expressed his symbiotic cravings by writing Gisela, "I really believe we shall never part, though we became friends from free choice, we are so *attached* to each other as if *nature* would have made us *blood relations*." Reflecting further on his intense tie, he also said, "I think we are so far gone that we live in one another . . ."

Though Freud adored Gisela's beauty, her aquiline nose, her long black hair, her tight-lipped mouth, and the suntanned color of her face, it was really Gisela's mother he loved. At sixteen, when young Freud would again feel his strong erotic passions for Amalie (but had to control them, lest he act them out) he displaced these revived incestuous fantasies on Frau Flüss. Similar to the way he described and experienced his own mother, he wrote a friend about Frau Flüss: "She obviously recognizes that I also need encouragement to speak or to help myself, and she never fails to give it . . . This is where her dominion over me shows; as she guides me, so I speak, so I present myself, I shall retain a beautiful memory of a good and noble human being . . . Her eyes sparkle with intelligence and fire."

Freud's romance with Gisela was later associated in his mind with the time when he was twenty-four, and he discovered that his father and half-brother Emmanuel wanted him to stop his intellectual pursuits, marry Emmanuel's daughter Pauline, and become a business-man in Manchester, England, where Emmanuel lived.

Just as Freud was unable to tolerate his resentment toward his mother, particularly the anger that emerged from threats of being displaced by competitors, he had to keep from consciousness his dis-placed hostility toward Gisela and her mother. Though for a lifetime the color yellow affected him, "because of the yellow dress she [Gisela] was wearing when we first met," he worked hard to deny he was upset when Gisela, not many years later, married another man. Rather, in his paper "Screen Memories," written twenty-seven years after his romance with Gisela, he demeaned the relationship and claimed indifference to her, referring to the romance as "calf-love."

Freud frequently used the defense mechanism of displacement when he was angry with a woman. When Gisela married and he lost her and Frau Flüss as love objects, he handled his depression and anger by becoming preoccupied with a poem about a lover whose sweetheart married another man. The betrayed lover verbally attacks his lost one and, says Freud, his "dire envy fills the lines in which the [smug?] appearance of the bride is disguised, and the groom's anticipa-tion of delight."

Freud's unconscious hatred and ambivalence toward Amalie influ-enced many of his perceptions, biases, and blind spots. Thirty-five years after Gisela was married, when the name "Gisela" was men-tioned by a patient, Freud did not feel any conscious emotion, he said, but found himself putting three exclamation marks after the name as he wrote it down. As Clark insightfully comments, "but the appar-ent absence of any reference to Frau Flüss in Freud's correspondence or reminiscences should be noted, together with his view that the theory of repression is the cornerstone on which the whole structure of psychoanalysis rests."

The "Gisela" involvement appears to be Freud's only romance with the opposite sex until the evening when, at the age of twenty-six, he found the love of his adult life, the woman to whom, as far as is known, he remained faithful until the day he died.

The "Admirable Manager"

After receiving his medical degree from the University of Vienna on March 31, 1881, Freud decided to become a teaching assistant at the Vienna Institute of Physiology, doing research, which he enjoyed. He still lived at home to save money and help support the family.

But his professional, as well as his personal life, was to change drastically as a result of an evening in April 1882. After his day's work, he arrived home intending to rush straight to his room to study, as he always did, regardless of visitors. He was arrested by the sight of "a merry maiden peeling an apple and chatting gaily at the family table," as Jones describes the scene.

Freud wrote Martha on June 19, 1885, "You know, after all, how from the moment I first saw you, I was determined—no, I was compelled—to woo you." On June 26 of that year, he wrote teasingly that he was becoming quite superstitious, "Since I learned that the first sight of a little girl sitting at a well-known long table talking so cleverly while peeling an apple with her delicate fingers, could disconcert me so lastingly."

To everyone's surprise, that night in April 1882, Freud did not go to his room but joined the family, as he did each time Martha visited during the next few weeks. Then he decided to court her straightforwardly and sent her a red rose every day, with a compliment written on his visiting card. He recalled later that the first compliment was to liken her to "the fairy princess from whose lips fell roses and pearls," with, however, the doubt "whether kindness or good sense came more often from Martha's lips." (In later letters, he addressed her as "My dearest Princess," "My dearest treasure," "My sweet Marty," and sometimes "My little woman," or, in the body of the letter, occasionally, "girl"—after years of marriage it became at times "My Beloved Old Dear.")

Martha was dining with the Freud family on June 13, 1882, two months after Freud met her, when he took her name card as souvenir, and she pressed his hand under the table. Four days later they were engaged, and they celebrated the seventeenth of every month for a number of years.

Martha Bernays was born July 26, 1861, in Hamburg, Germany, of a family distinguished in Jewish culture. Her grandfather, Isaac Bernays, had been Chief Rabbi of Hamburg during the reform movement which swept Orthodox Judaism around 1848, a liberalism he fought determinedly. Her father, Berman Bernays, was a merchant who had come to Vienna in 1869 when Martha was eight, dying ten years later. His son, Eli, then for several years took his father's place as secretary to Lorenz von Stein, famous Viennese economist.

Because Freud had no means of supporting Martha, their courtship lasted four and a quarter years, until he could open an office of his own as a specialist in nervous diseases. During three of these years, they were separated when her mother took her to Wandsbek, a town just outside Hamburg. Mrs. Emmeline Bernays, learning of Martha's engagement to an impoverished researcher in physiology, decided the young couple were better off separated; furthermore, she had been unhappy in Vienna after her husband's death and wanted to return to her former home.

On the advice of Dr. Ernest Brücke, director of the Institute of Physiology, a man Freud respected, he gave up the financially unrewarding research he had hoped to make his life's work and became an intern at the Viennese General Hospital to prepare for general practice, a more lucrative profession, so he could marry Martha.

A photograph of Martha in the possession of Edward Bernays shows her as a pretty little girl of three or four, plump and fat-cheeked, wearing a gingham dress with pantaloons, and laced, dark shoes. She stands by a chair clutching a pillow, her short dark hair parted down the middle and combed behind the ears. Another photograph pictures her at twenty-one with a soft smile, her hair worn in the same fashion but now long and pinned at the back of her neck. The resemblance to the photograph of young Amalie is remarkable, except that Martha's mouth is smaller and less strong than Freud's mother. It is interesting that Freud became engaged to Martha when she was twenty-one, the age of his mother when she gave birth to him. Martha told Freud he resembled her father.

During their courtship Freud wrote Martha 900 letters. They corresponded almost daily. Four pages was a short letter, some ran to twelve closely written pages, and there was one of twenty-two pages. Though essentially love letters, Freud's also contained descriptions of

Martha Bernays at the time of her engagement to Freud. *Austrian Press and Information Service.*

his clinical and laboratory activities ("I drug myself with work"), news about friends and superiors, vivid sketches of new friends, and discussion of their respective families.

The letters are revealing in many ways. They give a portrait of the developing mind of Freud between the ages of twenty-six and thirty, when he married. They show his brilliance, his eloquent, conversationlike literary style—one he never lost, even when describing his most complicated theory—his sensitivity, curiosity, intensity, warmth, as well as less appealing qualities—jealousy, possessiveness, rage, selfishness, and dependency.

Because he was accustomed to getting his own way from boyhood on (shades of sister Anna's lost piano), Freud could be difficult at times. "I am afraid I do have a tendency toward tyranny," he wrote Martha on August 22, 1883. He was impatient with what he thought stupidity; his caustic tongue impaled many an enemy or naive spirit. "I would rather you didn't make me out so good-natured; I can hardly contain myself for silent savagery . . ." he wrote Martha on January 16, 1884.

Freud's eternal yearning for the symbiosis to which he aspired with Amalie and Frau Flüss appeared intensely in his relationship with Martha. He asked her, above all else, to be very honest with him, to tell him all of her thoughts and her feelings (as he was later to do with patients). "Please don't be taciturn or reticent with me, rather share with me any minor or even major discontent which we can straighten out and bear together as honest friends and good pals," he wrote on August 17, 1882. So strong was this wish to merge with Martha and so hungry was he to know every detail of her life that he wrote on January 20, 1886, "I would rather think of you with false teeth in your mouth than one dishonest word." Apparently his wish to own her was stronger than his desire to have Martha look attractive to him.

Freud describes himself on June 19, 1884: "I am very stubborn and very reckless and need great challenges. I have done a number of things which any sensible person would consider very rash. For example, to take up science as a poverty-stricken man, then as a poverty-stricken man to capture a poor girl—but this must continue to be my way of life: risking a lot, hoping a lot, working a lot. To average bourgeois common sense I have been lost long ago."

On June 30 of that year he reminisces about their quarrels: "Do you remember how you often used to tell me that I had a talent for repeatedly provoking your resistance? How we were always fighting, and you would never give in to me? We were two people who diverged in every detail of life and who were yet determined to love each other, and did love each other . . . [you] so seldom took my side that no one would have realized from your behavior that you were preparing to share my life; and you admitted that I had no influence over you . . . you were hard and reserved and I had no power over you. This resistance of yours only made you the more precious to me, but at the same time I was very unhappy . . ."

What Freud was not able to fully comprehend was that his possessiveness infuriated Martha and this, together with his demanding manner "provoked [her] resistance." Though Freud's and Martha's romance has been considered idyllic by many writers, their mutual love was frequently interrupted by power struggles. Freud wanted Martha to cater to his omnipotence, grandiosity, and desire for superiority, while Martha, in many ways, wanted to be her own woman and not "give in." Freud's demands for one hundred percent loyalty conflicted with Martha's desires for some autonomy. Naturally, Martha's resistance to his attempts to control her attracted him, for he fantasied himself as a conquistador who enjoyed battling for what he wanted.

In a letter to Martha, also written in June 1884, Freud reveals that once, after a lover's quarrel, his hopes sank and he walked away from her "like a soldier who knows he is defending a lost position." When they had a reconciliation, he felt he had "become another person, many wounds that went deeper than you knew have been closed . . ."

On August 17, 1884, he describes himself as "a human being who is still young and yet has never felt young," an apt description of what he was later to call a neurotic depression. He promises, on April 29, 1885, after asking for a letter, not the mere postcard she sent, "once we have got through this terrible period of waiting, you won't have to touch a pen for years."

Freud's humor threads through his letters to Martha. He writes on June 19, 1882, two days after their engagement, "As yet we don't see humor in the same thing" (one wonders if they ever did). On January 21, 1882, when she did not send a letter for a few days, then explained she could not write because the room was too cold, he wrote, "Darling, is it possible that you can be affectionate only in summer and that in winter you freeze up? Now sit down and answer me at once so that I will still have time to get myself a winter girl." Six months before their marriage, on March 19, 1886, he wrote, as they tried to get enough money together to buy furniture, "Oh, my little darling, you have but one minor fault; you never win the lottery."

Jones says Freud's attitude toward Martha, as seen in the letters, was "a veritable *grande passion*," that Freud "was to experience in his own person the full force of the terrible power of love with all its raptures, fears, and torments." The four-year courtship "aroused all the passions of which his intense nature was capable. If ever a fiery apprenticeship qualified a man to discourse authoritatively on love, that man was Freud."

Martha's letters, according to Jones, the only one the Freud family allowed access to all of them, show she "truly and deeply loved him" in spite of his possessive jealousy. At one point he demanded she give up Fritz Wahle, a young man who had been her music teacher, now engaged to a friend of hers, but also paying her attention. Sensing Fritz was a rival, Freud tried, at first, to intellectually understand the relationship between Martha and her former teacher. He told himself Martha had been Fritz's pupil so that what looked like weakness of character on Fritz's part (courting someone else's loved one) was merely a peculiarity of teachers. He wrote Martha: "One has to consider the past, since without understanding it one cannot enjoy the present; nor can one understand the present without knowing the past." Thus years before his great discoveries about the influence of earlier trauma on later behavior, Freud could put into one sentence the purpose of the psychoanalytic treatment he was to evolve.

But Freud was a vulnerable and jealous creative genius, and his great intellectual understanding of Fritz did not prevent him from feeling intense rage. Freud wrote Martha, "When the memory of your letter to Fritz comes back to me I lose all control of myself, and had I the power to destroy the whole world, I would do so without hesitation."

Freud's rage, reminiscent of his anger when Gisela married another man, appears to reflect the revival of his hatred toward his mother for

Martha Bernays and Freud in Wandsbek in 1885. *Austrian Press and Information Service.*

betraying him in favor of his earlier rivals, Julius and Anna. He displayed similar fury when Martha failed to take his side in the antagonisms which developed between him and Martha's brother and mother. He acknowledged a "talent for interpreting" between the lines of Martha's letters and finding evidence of her disloyalty to him. According to Jones, Freud required from Martha ". . . complete identification with himself, his opinions, his feelings and his intentions. She was not really his unless he could perceive his 'stamp' on her . . . The demand that gave rise to the most trouble was that she should not simply be able to critisize her mother and brother . . . but she had also to withdraw all affection from them—this on the grounds that they were his enemies, so that she should share his hatred of them. If she did not do this, she did not really love him." But she refused to forsake her mother and brother.

Stolorow and Atwood believe that Freud's need to mold Martha into an idealized image of perfection, a mother-surrogate who loved him exclusively, stemmed from his unconscious desire to prevent a repetition of the traumatic betrayals he had experienced with his mother. Most importantly, "he sought to ward off the dreaded emergence of the image of the hated mother and of his own repressed, omnipotently destructive rage at her. His jealous outbursts were more often than not followed by self-reproaches, which can be understood as serving his need to restore the intactness of Martha's [i.e. his mother's] perfect image."

At the start of their relationship, on August 14, 1882, Freud wrote Martha of his wish to possess her: "From now on you are but a guest in your family, like a jewel that I have pawned and that I am going to redeem as soon as I am rich . . . no matter how much they love you, I will not leave you to anyone, and no one deserves you; no one else's love compares with mine."

Martha did give up Fritz, but was steadfast when it came to her family. She was deeply attached to her mother and much influenced by her, but here too she had her own sense of independence, for she did not listen when her mother opposed the marriage, both at the beginning of courtship and just before the wedding ceremony.

Freud's obsessive love he was later to describe psychoanalytically as similar to "madness." He told Martha his troubles vanished "as with a stroke of magic" as soon as he was in her company, which was not often because he could not afford to travel to Wandsbek. He wrote in August 1882, that he had been ill with a sore throat, which for several days prevented his swallowing or speaking, and on recovering, he was seized with "a gigantic hunger like an animal waking from a winter sleep," accompanied by an intense longing for her, "a frightful yearning . . . uncanny, monstrous, ghastly, gigantic; in short, an indescribable longing for you."

He spoke of the pain in parting from her: "Why don't we make a friend of everyone? Because the loss of him or any misfortune happening to him would bitterly affect us. Thus our striving is more concerned with avoiding pain than with creating enjoyment." This was to be one of the fundamental tenets of psychoanalytic theory. Freud distinguished between the "pleasure principle" and the "reality principle." There exists in us the wish to feel pleasure—we want what we want when we want it. This conflicts with reality, restrictions imposed by parents and society. We live in conflict between the wishes from the "id" that demand gratification and the conscience that forbids it.

That Freud wanted Martha in the total sense, as a child wants his mother, is shown in one letter: "I miss you so much . . . I hardly live like a decent human being. I miss you in every way, because I have taken you to myself in every respect, as sweetheart, as wife, as comrade, as working companion, and I have to live in the most painful privation." Another time, as though to deny these feelings, he described himself as a "cheerful pessimist," saying, "I am a virtuoso in finding the good side of things."

That he was aware of the repetition of the incestuous theme in his life is apparent in the letter he wrote Martha on July 13, 1883. He said that in a visit to Breuer's home one evening, he and Breuer referred to the women they loved, Breuer, his wife, and Freud, his fiancée, as "Cordelia," though in a different way. Freud said he told Breuer that Martha was "in reality a sweet Cordelia [loyal daughter], and we are already on terms of the closest intimacy and can say anything to each other," whereas Breuer always called his wife Cordelia "because she is incapable of displaying affection to others, even including her own father."

Freud could behave like a dejected and pouty little boy with Martha, angry at her for not being an omnipotent mother-figure. He felt he had missed fame by a hairsbreadth because he had left Vienna to see her. In *An Autobiographical Study*, written in 1925, he says he wishes to "go back a little and explain how it was the fault of my fiancée that I was not already famous at an earlier age." He describes how, in 1884, he was in the middle of a study of the anesthetic properties of cocaine, when "an opportunity arose for making a journey to visit my fiancée, from whom I had been parted for two years." He hastily wound up his investigation of cocaine, prophesying in an article about it that further uses would soon be found. When he returned from his holiday with Martha, he discovered that one of his friends, Dr. Carl Koller, to whom he had spoken about the anesthetic qualities of cocaine, had made experiments on the eyes of animals and had reported the experiments at the Ophthalmological Congress at Heidelberg. Koller was, therefore, regarded as the discoverer of local anesthesia by cocaine, which became for a while important in minor surgery (including an operation on the eyes of Freud's father). Freud concludes, in telling this story, "but I bore my fiancée no grudge for her interruption of my work."

The unnecessary initial and concluding remarks about Martha suggest someone ought to be blamed, and there is plenty of evidence that it was himself Freud really blamed, says Jones. In another context Freud does admit his own responsibility, saying, "I had hinted in my essay [on cocaine] that the alkaloid might be employed as an anesthetic, but I was not thorough enough to pursue the matter further." Jones also says that Freud, in conversation, once ascribed the omission to his "laziness."

Several engagements occurred in the Freud and Bernays families at about the same time. Anna became engaged to Eli, Martha's brother, and they were married on October 14, 1883, though Martha and Freud's engagement in June of 1882 had preceded theirs at Christmas 1882. No doubt Freud envied Anna for being able to marry at once. Minna, Martha's younger sister, then sixteen, was already engaged to a friend of Freud's, Ignaz Schönberg, a Sanskrit scholar. Freud looked forward to the time when Martha and he and Minna and Ignaz would become a happy marital quartet. He did not include in this fantasy Anna, his least favorite sister, and Eli, with whom he quarreled at times over how the Bernays money should be invested.

Freud wrote Martha that two of the quartet were "good people," Martha and Ignaz, and two were "wild, passionate people," Minna and himself. (He had once written Martha, "I still have something wild within me, which as yet has not found any proper expression." Here, he was referring to the "id" that he was to discover later.) He explained further that two of the foursome were adaptable and two wanted their own way, saying, "That is why we get on better in a criss-cross arrangement; why two similar [strong-willed] people like Minna and myself don't suit each other specially; why the two good-natured ones don't attract each other."

He described the kind of woman who would attract him: "A robust woman who in case of need can single-handedly throw her husband and servants out of doors was never my ideal, however much there is to be said for the value of a woman being in perfect health. What I have always found attractive is someone delicate whom I could take care of." Martha was fairly delicate during the days of her engagement when she suffered mild anemia and had to drink wine. In the early years of her marriage she seemed quite healthy, except just before she gave birth, when Freud would mention to Fliess she seemed fatigued. In March 1919, she fell ill with pneumonia, from which she took months to recover. In later life, she appeared in good health.

Freud's relationship to his future mother-in-law was quite hostile from the start. He saw her as a competitor for Martha and wanted her out of the way. Edward Bernays recalls that she was known as "the small grandmother because she was petite." He adds, "She was, however, in spite of her size, or perhaps in compensation, determined, strong, sometimes edgy—a powerful woman."

Martin describes his grandmother, Emmeline, as "a much less vital person than Amalie, but to us she was a character too . . ." While his maternal grandmother looked mild, soft and "angelically sweet," she was always determined to have her own way. He gave as an example the day the family were out on an errand and were caught in a torrential rain. Following the motto "Old People and Children First," he says, his parents put Emmeline and her grandchildren into the only available carriage. She insisted the windows of the carriage be kept closed all the way home, in spite of the children's strong protests they were suffocating.

Occasionally she stayed with the Freud family, and on Saturdays they would hear her singing Jewish prayers in a soft but firm, melodious voice. Martin says: "All of this, strangely enough in a Jewish family, seemed alien to us children who had been brought up without any instruction in Jewish ritual." Freud did not believe in God and wrote that the need to have the illusion there was a God related to the childhood fantasy of the father as god.

Mrs. Bernays met her match in Freud. He wrote a frank letter to Minna on February 21, 1883, telling her that her fiancé, Ignaz, had informed Emmeline she was "selfish" and in bitterness broke all relations with Emmeline. Freud asked Minna not to "side with Mama in your letters nor to believe all the complaints that you hear about us from her." He said he saw his future mother-in-law as "a person of great mental and moral power standing in our midst, capable of high accomplishments, without a trace of the absurd weaknesses of old women [at this time she was fifty-three] but there is no denying that she is taking a line against us all, like an old man." He said she should be content "to know that her three children were fairly happy and ought to sacrifice her wishes to their needs . . ."

Freud opened his first office at Rathausstrasse 17 in April of 1886. He and Martha planned to marry in September. Early in the spring he searched for an apartment. He heard that few people wanted to live in the Sühnhaus, called the House of Atonement. It was built by Emperor Franz Josef at Maria Theresienstrasse 5, the site of the former Ringtheater, which had burned to the ground in the fall of 1881 killing 600 persons during a performance of "The Tales of Hoffmann." Freud wrote Martha of the building's morbid associations, asked if she felt superstitious about living there. She telegraphed she was willing to take the apartment.

Just before his marriage he endured what he described in a letter to Breuer on September 1, 1886, as "the crazy four weeks," a month's Army maneuvers for which he had been called up, something he had not expected until the following year. He told Breuer he was "lying here on a short leash in this filthy hole . . . painting flagpoles black and yellow . . . So far I haven't been locked up." His stint in the Army meant the loss of a month's earnings as a doctor in nervous diseases, money which he and Martha had counted on for furnishings. Freud decided not to let this crisis interfere with their plans to marry in September, but Mrs. Bernays, never one to mince words, wrote him: "That you could for a *single* moment in the present circumstances, when you have to interrupt your practice for almost two months, think of marrying in September is in my opinion an abysmally irresponsible piece of *recklessness*. *Another* word would be more fitting [perhaps she was thinking of *idiocy*], but I will not use it. I shall *not* give my consent to such an idea . . . to run a household without the means for it is a *curse* . . . wait quietly until you have a settled means of existence . . . At the moment you are like a spoilt *child* who can't get his own way and cries, in the belief that in that way he can get everything." At least she signed the letter, "Your faithful Mother."

Having so many times had his own way with his mother, Freud was not one to give in to anyone else's mother. After serving his month in Olmütz, from August 9 to September 10, 1886, he returned to Vienna on September 11 to change from uniform to formal wedding outfit, including frock coat and silk hat. He left for Wandsbek the next day on money borrowed from Minna for the fare, having spent his last cent on a gold watch as wedding present for Martha. He was tanned from the outdoor life in the Army, a handsome man, slender but sturdy, with intense, expressive dark eyes, standing five feet, seven inches and weighing 126 pounds.

The civil marriage took place on September 13, 1886, in the Town Hall of Wandsbek. Besides the immediate family, eight relatives were invited. The bride and groom wrote a joint letter of alternating sentences to Emmeline from the first stop on their honeymoon, Lübeck. Freud's concluding thought: "Given at our present residence at Lübeck on the first day of what we hope will prove a Thirty Years' War between Sigmund and Martha." He was being sarcastic about Emmeline's jaundiced attitude toward the marriage, but also may unconsciously have been alluding to his own latent resentment regarding his marriage. The "War" may have referred to Sigmund and Martha's tempestuous power struggles that were part of their courtship. Freud did write his friend and colleague, Ferenczi, on October 2, 1910, when the latter became engaged, "Warm greetings for today to you and Frau Gisela. Soon your idyll will be over, too."

The thirty years became fifty-three and, according to Jones, the only sign of overt war ever recorded was a temporary difference of opinion as to whether mushrooms should be cooked with or without their stalks.

After continuing their honeymoon in Holstein on the Baltic, and Berlin, Dresden and Brünn, Martha and Freud returned to Vienna on October 1 to settle down to marriage and raising a family.

Jones describes Martha as "an excellent wife and mother . . . an admirable manager . . . never the kind of *Hausfrau* who puts things before people." He says that Freud's comfort and convenience always ranked first with her. She took care of him almost as a mother would, buying all the groceries, ordering his clothes—light colored suits for vacation made from British cloth, and dark wools and black ties for winter. She never interfered with his work by making demands on his time, though she was left alone evenings as he studied and worked until past midnight.

Edward Bernays describes his aunt as "much more than just a housewife." He says, "She was quiet and gentle. She gave complete tranquility to wherever she was—quiet pervaded the corner where she sat knitting or sewing. I'm sure Freud loved her, among other reasons, because he was very disturbed at times by attacks on his work and contention among his disciples and to come home and feel this serenity from his wife must have been a most calming influence."

Frau Professor, as she was called, never became an easygoing, fun-loving Viennese, but kept her rather conventional Germanic attitudes and precise Hamburg speech. However, Freud did wean her from the Jewish orthodoxy in which she had been brought up, so religion played little part in the lives of their children. One of Martha's cousins wrote, "I remember very well Martha telling me how not being allowed to light the Sabbath lights on the first Friday night after her marriage was one of the more upsetting experiences of her life."

According to Sachs, "She never wavered in her admiration—I could almost say adoration—of her husband. I do not know how much she understood of the importance of his work (her intelligence and education were certainly sufficient for it); but I am sure that he was a great man to her before a word of his books was written as well as afterwards . . ."

Though Martha overtly admired, idealized, and catered to her husband, she also muffled her anger toward him and tended at times to be servile, submissive, and somewhat self-demeaning. When Freud felt a deep desire to move into 19 Berggasse in 1891, he went to the house, inspected it, decided it suited the family's needs, and without further ado signed the lease. He returned home, told Martha he had found their ideal quarters and that evening showed it to her. Though Martha was dissatisfied, thought the neighborhood a poor one, the stone stairs dark, steep, and dangerous and the accommodations insufficient for their growing family, she did not protest. Freud had signed the lease, liked the house, and she swallowed her anger and lived at 19 Berggasse almost the rest of her life without complaining. She organized the house entirely for Freud's convenience, managing servants with humaneness and generosity, perhaps treating them with the kind of empathy she wished from her husband.

Jones says that in the early years Freud discussed his cases occasionally with Martha, but later "it was not to be expected that she should follow the roaming flights of his imagination any more than most of the world could." Theodor Reik recalled, "She really never understood why her husband became famous. Nor did she understand his work. She was not particularly interested in his research. She had little awareness of the important role of the unconscious. Once she said to me about a hysterical woman, 'She'll get over it if she'll use her will.'" Like many people in those days (and even today), it was difficult for Martha to appreciate the role of the unconscious in determining human behavior.

Martha dutifully concealed her disbelief in psychoanalysis though one time her true feeling emerged as she remarked to a visitor, "I must admit that if I did not realize how seriously my husband takes his treatments, I should think that psychoanalysis is a form of pornography."

Disposed to self-sacrifice, she was the obedient wife. Once when Freud left the house in the rain, she ran after him to give him his galoshes. Though she loved the opera, she always arrived late at the performances because she wanted to be at home when Freud ate dinner, and he always ate around 9:00 P.M.

Martin pictures his mother as "ruling her household with great kindness and with an equally great firmness." She believed in punctuality, something then unknown in leisurely Vienna, he says. At the stroke of one, everyone in the household was seated at the long dining-room table and, at the moment one door to the room was opened by the maid entering with the soup, a second door opened and Freud walked in from his study to take his place at the head of the table, facing Martha at the other end.

This meal was the principal one of the day, consisting of soup, meat, vegetables, and dessert. Freud was not fussy over food, except for cauliflower, which he would not eat, and he was not fond of chicken. He would say, "One should not kill chickens, let them stay alive and lay eggs." His favorite dish was *Rindfleisch*, boiled beef, served three or four times a week with different sauces. Martha must have "shared a Viennese secret, how to make the *Rindfleisch* so juicy and tasty," Martin comments.

As long as Martin could remember, the household had a cook, who worked only in the kitchen; a maid who waited on the table and admitted Freud's patients (for thirty years this was the faithful Paula Fichtl, who moved with such speed that Martin said it seemed she would rather run than walk); a governess for the elder children, and a nannie for the younger; and a charwoman, who came each day to do the heavy cleaning.

Sachs says of Martha that "instead of being regarded by her servants as a curse, as these model housewives usually are, she was extremely popular with them." She "never tolerated the idea that the life of a human being shall be subordinated to the welfare of the furniture." She bought Christmas presents, choosing them carefully, for each servant and their relatives, causing Minna to say, "We draw a line at the niece of our milkman."

Both sisters had a reputation for the beauty and precision of their needlework, and during World War I, Sachs reveals, they gave some of their "wonderful creations" as a bribe to the woman who kept the tobacco shop where Freud bought his twenty cigars a day, so she would provide the extra ration he needed. He wrote Stefan Zweig on February 7, 1931, ". . . I ascribe to the cigar the greatest share of my self-control and tenacity in work."

Martin says his mother kept busy from morning until night: "I cannot remember her enjoying a quiet moment to sit down and to relax with a good book: and she loved to read a good book." Few afternoons passed without her receiving at least one visitor, often several, as Freud saw patients. While she entertained her friends, the children were left in charge of the governess.

Freud with his mother and wife 1905. *Austrian Press and Information Service.*

Martin saw his mother as having her own special courage. In 1938, when she was seventy-seven, Nazi Storm Troopers invaded the apartment. Martin, who hurried there as soon as he learned of the invasion, said his mother's "inner strength" did much to keep the situation calm, as the Nazis confiscated the five thousand shillings in his father's safe. He describes the scene: "It was no small thing to a woman of mother's housewifely efficiency to see her beautifully run home invaded by a pack of irregulars. Yet she treated them as ordinary visitors, inviting them to put their rifles in the sections of the hall-stand reserved for umbrellas and even to sit down . . . her courtesy and courage had a good effect. Father, too, retained his invincible poise, leaving his sofa where he had been resting [recovering from still another jaw operation] to join mother in the living-room where he had sat calmly in his arm-chair throughout the raid." During the hour's raid, all passports were confiscated and formal receipt given for the confiscated money. When Martha told Freud how much money had been taken from his safe, he remarked dryly, "I have never received so much for a single visit."

Because Martha gave so generously of her energy and time to Freud, there is some indication she did not fully relate emotionally to her children. While she could chase after her husband in the rain to give him his galoshes, her maternal indifference to her son is reflected in a story Martin tells. One day, as a boy, while swinging on a trapeze that hung between two rooms, he lost his grip and fell against a sharp corner of a piece of furniture instead of on the mattress he had placed on the floor to cushion falls. He gashed his forehead and blood started to flow from a cut several inches long. His mother, who had been sewing in one of the rooms, did not drop a stitch, but asked the governess to telephone the doctor who lived a few doors away (telephones were rare, but doctors were the first to have them, though Freud hated using his because, Martin theorized, he wanted to see the face of the person to whom he was talking). Martin scrambled to his feet without anyone's help, horrified at the blood that covered his face and quite surprised, he says, that the accident produced no panic in his mother, "not even one outcry of horror."

As psychoanalysts have discovered, when a wife devotes her whole self to her authoritarian husband, repressing resentment, she may displace some of the hostility on her children. Perhaps Martha, who mothered her husband day in and day out with almost no complaints, may have unconsciously fantasied hurting him in revenge, and Martin's bloody face may have reflected what she secretly wanted to happen to Freud.

In the marriage of Martha and Freud, there was mutual respect and admiration. But a modern-day psychoanalyst would have to characterize some of their behavior as sadomasochistic. Freud's first love was his work—Martha had to play second fiddle to his dominant passion, psychoanalysis. Her resentment was repressed, but came out in occasional barbs at psychoanalysis, and possibly in some indifference toward the children. Freud seemed to have a loving wife in Martha but burned the midnight oil, working while she slept. Apparently he got more sexual satisfaction from psychoanalysis than he did from his marriage and gave more sexual stimulation to the world than to his wife. [There are hints in some of his letters that she was not a very sexually warm woman but then women of that era were not encouraged to be sexually free.]

Just as most of Freud's biographers have portrayed his courtship of Martha as idyllic, when it contained many power struggles, their marriage has also been overidealized. Recently, it has come to light that when Freud was in his early fifties, he received a letter from Emma Jung, wife of Carl Gustav Jung, who recalled "the conversation on the first morning after your arrival, when you told me about your family. You said then that your marriage had long been 'amortized,' now there was nothing more to do except—die . . . you said you didn't have time to analyze your children's dreams because you had to earn money so that they could go on dreaming. Do you think this attitude is right?"

Dr. Max Schur, Freud's personal physician for many years, wrote that, as the marriage continued, "there developed a small cloud in their relations which they did their best to conceal from the world." Dr. Schur notes what appeared to be Freud's forgiveness of the pedantic attitude sometimes shown by his wife as, like Martha with him, he tried to control his irritation. Perhaps, Schur was referring to her inability to understand Freud's work.

Freud never resolved his strong incestuous tie to his mother. As he later wrote, it is difficult for a man to experience a wife as a sexual woman if he is psychologically still a little boy yearning for, but feeling guilty about his sexual wishes toward an attractive mother. At the age of forty-one, he wrote Fliess, ". . . sexual excitation is of no more use to a person like me." He made a telling statement in a letter to Dr. James Jackson Putnam, an American psychiatrist, saying, "I stand for a much freer sexual life. However, I have made little use of such freedom . . ." Another time Freud wrote that he showed a remarkable understanding "from personal experience" of how discouraging it was to be married to a woman not interested in sex.

Just as he had to repress his incestuous wishes toward his five sisters by keeping them at arm's length, so he had to try to make Martha an inhibited and somewhat asexual woman. He wrote fourteen years after marriage, "I was very annoyed because I thought my wife was not being sufficiently reserved toward some people sitting near us."

Freud's fear of his own sexual feelings probably account for his superior attitude toward women, which he used to ward off intimacy and lust. Only the fear of seeing a woman as a sexual being could make him say, "A thinking man is his own legislator and confessor and obtains his own absolution but the woman, let alone the girl, does not have the measure of ethics in herself. She can only act if she keeps within the limits of morality, following what society has established as fitting. She is never forgiven if she has revolted against morality, possibly rightly so."

Because Freud was so sexually stimulated as a boy sleeping in his parents' bedroom and because his father put limited controls on him, he lived a life of sublimated sexuality, always dreading he would emerge as a turned-on Oedipus overwhelmed by a sexy Amalie. When he writes about women having little sense of justice or morality, this is the statement of a man who must defend himself against the possibility that women will become temptresses.

One of Freud's great discoveries was the "inverted Oedipus complex." He pointed out that when a boy has strong incestuous feelings toward his mother but fears to compete with his father, he may demean women and prize men. Freud, in many ways, demeaned women and elevated men to the point of idolization. He said, "There must be inequality and the superiority of the man is the lesser of two evils." Of the scientists Hermann Ludwig Ferdinand von Helmholtz, Freud said, "He is one of my idols." Of the physician, Theodore Meynert, he reflected, "in whose footsteps I followed with such veneration." Of his collaborator, Dr. Josef Breuer, he said emotionally, "He radiates light and warmth." Though Freud was in many ways an Oedipal conquistador, he was also an inverted Oedipal boy who, like many men, possessed vulnerabilities and sexual anxieties.

Though the marriage held satisfactions and dissatisfactions, pleasures and frustrations, both partners consistently defended themselves from any overt expression of animosity. After Freud died, Martha wrote to his friend and colleague, Dr. Ludwig Binswanger, " . . . How terribly difficult it is to have to do without so much kindness and wisdom beside one! It is small comfort to me to know

Martha and Sigmund Freud in the garden of their London home in 1939.
Austrian Press and Information Service.

that in the fifty-three years of our married life not one angry word fell between us, and that I always sought as much as possible to remove from his path the misery of everyday life. Now my life has lost all content and meaning."

Edward Bernays visited his Aunt Martha in London in 1956. He reports she was living with Anna in a house Martha's son Ernst, an architect, had reconstructed for them in 1939, the one at 20 Maresfield Gardens. Anna also had quarters in the house where she psychoanalyzed children. Freud's desk and his artifacts were left just as he had arranged them.

Mr. Bernays says Martha lived quietly, visited by her children, Martin, Mathilde and Ernst, who also lived in London. She died in 1952 at the age of ninety.

She had given her husband six children and an atmosphere in which he could live with as much serenity as possible with a large family. Just after their golden wedding anniversary celebration at 19 Berggasse on September 13, 1936, with four of the surviving children present (Oliver was absent), Freud wrote Marie Bonaparte, "It was really not a bad solution of the marriage problem and she [Martha] is still today tender, healthy and active."

Thus his philosophic, heartfelt comment on his long marriage to the woman who tried her best to provide for his comfort so he could work in peace, even though she did not grasp how much he contributed to the understanding of the human mind.

Friend and Companion

The most controversial woman in Freud's life was his sister-in-law, Minna, who lived as part of the household for forty-two years. After the six children were born, Minna became a permanent member of the Freud ménage, joining the family late in 1896.

Martha and Minna, who was six years younger, were very close. Perhaps Martha felt she had to take care of her younger sister. They were "marked contrasts," says Sachs. "In exterior, Frau Professor was small and lithe, very mobile. . . . Tante Minna was tall and not slender, rather statuesque and self-reliant, of few but not uncertain words to which she liked to give an epigrammatical turn."

Martin described his aunt as "an extraordinarily well-read person with a great gift for discriminating, and sometimes sharp, criticism." Jones also mentioned her "pungent tongue," which, he said, Freud admired. Minna and Freud always "got along well," Jones says, in that Freud found her a stimulating and amusing intellectual companion. She gave support to him in his work. Jones comments, "She certainly knew more about Freud's work than did her sister, and he [Freud] remarked once that in the early nineties Fliess and she were the only people in the world who sympathized with it." When Freud was poor and thinking of leaving Vienna, Minna suggested he stay in Austria until his fame reached America, then American patients would flock to him in such numbers that he would be saved the trouble of emigrating, a prediction that came true, though it took thirty years.

Minna appeared to play an important part in both her sister's and Freud's lives. Martha was left alone most evenings of the week as Freud worked and must have enjoyed her sister's company at the opera or as a companion. The children were also fond of Tante Minna —it is often easier to love an aunt than a mother, who has to do all the disciplining.

Freud accepted Minna from the beginning, when he was twenty-six and she, the sixteen-year-old younger sister. He wrote her a long letter on August 28, 1884, after she asked if she could have a photograph of him, saying he would send it. He told her he had received "several urgent letters from your beloved," his friend Ignaz. Freud also said he himself felt weary and apathetic and "want to see no other friendly faces but yours and that of my Martha." He asked Minna to please help "scare away" intruders on their time, "and in return we will often take you with us and be very nice to you," a promise that came true. He ended the letter, "Looking forward to a happy meeting next month, my dear little sister," and urged, "Pull yourself together and write to me at once what I can bring you . . . ," signing the letter, "Your brother, Sigmund," though he was not as yet married to Martha.

The following year Ignaz fell ill with tuberculosis and called off his marriage to Minna. Freud wrote Martha on June 23, 1885, concerning Ignaz, "He cannot marry her [Minna] now, this is clear for every possible reason; he will not be able to marry her if he dies from his illness, and he ought not to see her as someone else's wife if he remains alive . . . As for Minna herself, she won't want to do anything but stand by Schönberg, as long as there is such a person . . . poor Minna . . . will have a lot to forget. So for the moment do let her cling to the shred of hope that remains."

That August 14, Freud wrote Martha that he had been to Baden where Ignaz was being taken care of by his brother and sister-in-law, and said, "I don't think the end will come for some time." He reported that Ignaz "hoped her [Minna's] vivacious temperament would soon help her to forget him and that she will behave as an unattached girl; I promised to do everything to make things easy for him . . . I leave it to your judgment, which in Schönberg's case has always turned out to be so correct, how much of this you will feel like passing on to Minna."

Then, on December 18 of that year, he wrote, "Tell Minna from me that whenever we entertain friends, there will always be a place laid for her." He was feeling empathy and sympathy for her at this time of loss.

After his friend Ignaz died early in February 1896, he wrote a long letter to Minna in which he said: "Your sad romance has come to an end, and when I think it over carefully I can only consider it fortunate that the news of Schönberg's death should reach you after such a long time of estrangement and cooling off. Let us give him his due by admitting that he himself tried and succeeded in sparing you the pain of losing your lover, even though it was less his high-minded

Minna Bernays, Freud's sister-in-law. *Austrian Press and Information Service.*

intention than the moral weakness of his last years that prompted him to do so . . ."

This is an unusually judgmental phrase for Freud to use. We will never know the basis for it. Possibly his jealousy of the man who almost married the sister-in-law of whom he was so fond colored his views.

Freud also said, "You haven't been made as unhappy as you so easily could have been; fate has grazed you only lightly; but you have suffered a lot, you have had little joy and a lot of worry, and in the end a great deal of pain from this relationship. You were hardly out of your teens when you took upon you tasks normally faced in life only by adults . . . try to regain something of the lost youth during which one is meant to do nothing but grow and develop, give your emotions a long rest and live for a while quietly with the two of us who are closest to you . . . " Thus very early in Minna's life Freud assumed the role of older brother and probably father, since Minna's father had died.

Freud also borrowed money from Minna during the first months of private practice, when he saw few patients. He wrote her in Wandsbek that he had earned only forty-five dollars that month and needed one hundred twenty dollars on which to live. He explained he had pawned the gold watch given him by his half-brother, Emmanuel, and now Martha's gold watch would have to be sold, unless Minna could help out, "which, of course, she did," Jones reports. In the next month Freud's financial tide began to turn as he saw more patients, and he may have repaid Minna and also bought her the coral necklace he wanted to give her at the time of his wedding but could not afford.

Minna never seems to have taken another man seriously after her fiancé died. For a while she served as a "lady's companion," an occupation she did not enjoy—as a girl she had done housework with a duster in one hand, a book in the other. Her interests were primarily literary. She also had a special talent for choosing the right gift; Jones heard her say if the moment ever came when she could not think of the proper gift, it would be "time to die."

Since Martha did not enjoy traveling, especially at Freud's fast pace, and Minna liked to travel, Minna often accompanied Freud, who hated to travel alone. He was a restless traveler, eager to crowd in as many experiences as possible in a short time. Minna once said his ideal was to sleep in a different place every night. When he was away, Freud sent Martha a postcard or telegram practically every day, and a long letter every few days. It was important for him to keep in constant touch with her, to let her know what he was experiencing.

Freud and his sister-in-law had much affection for each other, exchanging photographs in the early years. When Freud in his letter to Minna referred to Ignaz as a man with "moral weakness," he sounded more like a jealous competitor than a benign brother-in-law. On telling Minna she should be relieved her romance was over, Freud may have been relieved to have Minna, as well as Martha, all to himself.

With her love of literature, Minna may have wanted some sort of career involving books, perhaps as a teacher, and because of her frustration may have received some fulfillment by vicariously participating in Freud's scholarly world. Whatever role she played in his life and whatever role he played in her fantasies remains unknown. That they were fond companions seems certain. That they supported and enhanced each other's lives is clear. Writing to Max Halberstadt, a photographer who became engaged to his daughter Sophie, Freud said on July 7, 1912, ". . . the two mothers—my wife and sister-in-law—know you, your mother, your family, and according to them, have always considered you a member of the intimate circle of our relations." He implies that in addition to her other roles, he saw Minna in fantasy as a mother to his children.

After the children grew up and left 19 Berggasse (all except Anna) Minna slept in a corner of the apartment overlooking the garden, in a bedroom adjacent to Freud's and Martha's bedroom, which could be reached only by going through their room. She also had to pass through their bedroom to use the bathroom. Her "sitting room" was in another corner of the apartment.

Following her romance with the doomed Ignaz, Minna was never seriously involved with another man, except Freud. We know his dominant passion, psychoanalysis, gave him much sexual stimulation, and it was psychoanalysis he shared with Minna more than with his wife. Minna supported his work, talked with him about it, and often traveled with him on vacations, when Martha preferred to stay home.

She emerges as a woman with an unresolved Oedipal conflict—similar to the many women with whom Freud spent time in his consultation room, having professional tête-à-têtes. She was participant in a triangle, with Freud as the loving father and herself in the position of the daughter competing with the mother, Martha. Perhaps, like an Oedipal girl who feels guilty because of her wish to usurp her mother's role, Minna was capable only of a part-time relationship with a man, as is true of many conflicted women involved in affairs with married men.

Since Freud made Martha his mother-surrogate, and, therefore, eventually had to avoid her sexually lest his incestuous wishes emerge and terrify him, it is possible to surmise that, like many men involved in an extra-marital affair, he could only derive sexual satisfaction, partial or complete, from a woman not his "wife-mother."

Virtually all his biographers have sought to repudiate the idea that Minna and he were sexually involved. Says Jones, "It is an entirely untrue legend that she [Minna] displaced his wife in his affections. There was no sexual attraction on either side . . . to say that she in any way replaced her sister in his affections is sheer nonsense . . . Freud was quite peculiarly monogamous. Of few men can it be said that they go through the whole of life without being erotically moved in any serious fashion by a woman beyond the one and only. Yet this really seems to have been true of Freud . . ."

Theodor Reik also described Freud as a highly moral man, but one who was "always very gallant to women." Reik said, "He might have felt very much attracted to certain women but it was never more than that . . . He was very stimulated by women, especially by Princess Marie Bonaparte, a very beautiful woman. But it never went far."

Edward L. Bernays says of his Aunt Minna and Uncle Sigmund: Terrible lies have been published about them. My uncle no more had sexual relations with her than Caesar did with Pocahontas. She was very intellectual and he liked to talk to her about his ideas."

Helen Puner in *Freud: His Life and His Mind* claims that Freud became disillusioned with Martha sexually but decided, because of the strict moral code under which he was brought up, not to sexually desire another woman, even though he might dream of it.

Clark mentions in his biography that Jung reportedly was approached by both Martha and Minna, each of whom spoke separately of Freud's passion for his sister-in-law. But, Clark protests, "The story sounds decidedly unlikely, as does the even less substantiated suggestion that Freud's emphasis on sex in his work was a reflection of his own voracious longings. Maybe he was an unsublimated randy old man, but there is not a tatter of evidence for this, and the known context of his life makes it distinctly improbable. Were it not for the secrecy with which his papers have long been shrouded—almost certainly a reflex action to the attacks made on his theories during his lifetime—the idea would long ago have died a natural death."

Freud confessed he was a Victorian to Dr. Georg Groddeck, a pioneer in psychosomatic medicine, when Groddeck, a rather eccentric man, decided to marry a woman with whom he had lived for years. Freud told Groddeck he never felt at ease with a man who came to psychoanalytic conventions with his mistress, as Groddeck had done.

Dr. Martin Grotjahn, a psychoanalyst and an outstanding interpreter of Freud, calls Freud's "late sexual maturity and his not very sensual personality—smoking was his only passion—" the "trademark of genius," and comments: "He stated freely that he was proud to proclaim freedom from repression and freedom of sexual expression but that he did so without participating in the privileges this freedom would offer."

It would seem that, in addition to his biographers, everyone close to Freud believed it impossible that he ever was sexually intimate with his sister-in-law, though he cared very much for her. In the summer of 1900, when she had tuberculosis, he took her to a sanitorium in Merano, Italy. In 1936 she was operated on for glaucoma in both eyes and had to give up her beloved embroidering. When the Freud family left for London in 1938, Minna preceded them by a month. She left the sanitorium where she was staying on May 5, 1938, and when the Freuds arrived in their new London home she was already ensconced on the second floor. Freud wrote Max Eitingon on June 6, 1938, "Under the impact of the illness on the floor above me (I haven't been allowed to see her yet) the pain in my heart turns into an unmistakable depression." He said of Martha, "My wife . . . has remained healthy and undaunted." Minna lived with the Freuds for several months before being taken to a nursing home, where she died, almost completely blind.

The Daughters Three

The Freuds' first child was born in the Sühnhaus a year after they were married. It was a girl, Mathilde. Emperor Franz Josef sent an aide-de-camp with a letter of congratulations for the new life arising from the spot where six hundred had been lost. Two sons, Martin and Oliver, were also born in that apartment.

Five years later, in August 1891, the Freud family moved to 19 Berggasse, a larger place. This street consisted of massive eighteenth-century houses in which there were a few shops, including a butcher's store, on part of the ground floor of number 19. The entrance to the main house was wide enough to let a horse and carriage into the stable beyond. On entering the house, half a dozen stone steps on the right led to a professional apartment of three rooms in which, from 1891 to 1908, Freud saw patients. A longer flight of stone steps led to the next level, called the mezzanine, where Freud and his family lived. In 1908 he gave up the ground floor professional apartment and took over for his study and consulting room what had been his sister Rosa's apartment. The entire mezzanine now served as both his living quarters and professional apartment. He lived and worked at 19 Berggasse until the day he left Vienna.

Mathilde was born at 7:45 P.M. on October 16, 1887. Just after midnight, Freud wrote Emmeline and Minna: ". . . You will have heard already by telegram that we have a little daughter. She weighs nearly seven pounds, which is quite respectable, looks terribly ugly, has been sucking at her right hand from the first moment, seems otherwise to be very good-tempered and behaves as though she really feels at home here . . . lies snugly in her magnificent carriage and doesn't give any impression of being upset by her great adventure."

He said of Martha, "She has been so good, so brave and sweet all the way through. Not a sign of impatience or bad temper, and when she had to scream she apologized each time to the doctor and the midwife . . . I have now lived with her for thirteen months and I have never ceased to congratulate myself on having been so bold as to propose to her before I really knew her; ever since then I have treasured the priceless possession I acquired in her, but I have never seen her so magnificent in her simplicity and goodness as on this critical occasion, which after all doesn't permit any pretenses." Freud was not only expressing his love and admiration for Martha but probably identifying Martha, now a mother, with his own esteemed mother.

The baby was named after Mathilde Breuer, the wife of Dr. Josef Breuer, a woman Freud admired and loved. When he was an impoverished student and came to her house, sometimes to borrow money, she always made him feel welcome. He wrote Martha on May 29, 1884, of the Breuers, "they are both dear, good and understanding friends . . . She always insists on my taking a small apartment before long, and hanging out a sign, just a sign, a beautiful sign . . ." She had designed the nameplates for his first office, both the one on the street and another for the door.

After Mathilde came Jean Martin (Martin), born December 6, 1889, named for Dr. Jean Martin Charcot, Freud's teacher in neurology who, Freud said, had changed his life, convincing him he wanted to study the causes of hysteria. Oliver, the second son, born February 19, 1891, was named after Oliver Cromwell, one of Freud's boyhood heroes. According to Martin, Oliver, or Oli, as Freud called him, was "the handsome son, an unusually pretty boy with big black eyes" who "always drew admiring glances from strangers." He became an engineer and Martin, a lawyer and publisher of the *Psychoanalytischer Verlag*, a firm his father established a few doors away from their home.

Ernst, the third son, born at 19 Berggasse on April 6, 1892, became first an architect, then an engineer. He was named after Dr. Ernst Brücke, Freud's physiology instructor. None of Freud's sons became a doctor. "Medicine as a profession for any of his sons was strictly banned by father," Martin says. Freud had little use for the anti-Semitic physicians of Vienna.

The fifth child was a second girl, Sophie, born April 12, 1893. The last child, another daughter, Anna, was born December 8, 1895. Both these daughters were named in honor of Professor Paul Hammerschlag, who taught Freud the Scriptures and Hebrew in school. Sophie was named after the professor's niece, Sophie Schwab, and Anna, after the professor's daughter, Anna Hammerschlag, who was also a favorite patient of Freud's. Freud said of his former professor, "He has been touchingly fond of me for years; there is such a secret sympathy between us that we can talk intimately together. He always regards me as his son." Freud also had the highest regard for Mrs.

Hammerschlag: "I do not know any better or more humane people, or so free from any ignoble motives."

Two significant observations can be made about the naming of Freud's children. First, Freud, always in the driver's seat of the marriage, was the parent to select the names, and Martha merely acquiesced, as she did in the selection of the Freud house.

Secondly, all the names were connected with *the men* in Freud's life. It was out of his admiration for Breuer that Freud chose Mrs. Breuer's first name for his first daughter, Mathilde. Martin and Ernst had as their namesakes two admired male mentors of their father, while Oliver was named after one of Freud's favorite military heroes. It was because of his intimate relationship with Professor Paul Hammerschlag that Freud chose the name of the professor's daughter for Anna, and the professor's niece, Sophie Schwab, for a second daughter. Freud always yearned for a strong father and in his identification with his children gave them the names of persons who were strong fathers or associated with strong fathers. Each time he called a child by name, he could thus resurrect the fantasy of a powerful, benign father.

The six children were born within eight years. Freud, whose first love was his work, was not openly affectionate to them. Just as he feared affectionate displays between his five younger sisters and himself, he did not often hug or kiss his children. In a letter to Jones on January 1, 1929, he wrote, "It is not in my nature to give expression to my feelings of affection, with the result that I often appear indifferent but my family knows better."

But he agonized if any of his children fell ill. When Mathilde was almost six, she nearly died of diptheria. During this crisis he asked what she would like best in the world and she said, "A strawberry." Strawberries were out of season but a shop managed to produce a few. When Mathilde tried to swallow one, she burst into a fit of coughing that completely removed the obstruction in her throat. The next day she was on her way to recovery, "her life saved by a strawberry—and a loving father," says Jones.

When she turned eight, Freud wrote Fliess on October 16, 1895, "At the moment I have on hand a noisy children's party of twenty for Mathilde's birthday." During her tenth year, Freud wrote Fliess on May 16, 1897: "Mathilde now has a passion for mythology, and

Freud at eighty-two taking a walk with his oldest daughter Mathilde Hollitscher at his place of refuge, Hampstead, England, *Pictorial Parade*.

recently wept bitter tears because the Greeks, who used to be such heroes, suffered such heavy blows at the hands of the Turks."

When she was fourteen, Mathilde later recalled, her father invited her to walk on his right side during their strolls. A friend who saw them criticized Freud, saying a father should always walk on the right side of a daughter. Mathilde said proudly, "That is not so with my father. With him I am always the lady."

Martin recalled the day Mathilde was on vacation at Koenigsee, bronzed almost to blackness from the sun, and how startled she was when "a Prussian lady" emerged from a crowd of excursionists, licked her finger and pressed it firmly on the back of Mathilde's neck. Her husband, a tall monocled man with a Kaiser Wilhelm mustache, shouted at her, "What on earth are you doing, Theodora?" She replied, "It's all right, Justinian, I was only trying to find out whether the color would come off."

Sachs describes the atmosphere of the Freud home as "peaceful and temperate friendliness." Freud was "der Papa" to the children and "Sigi" to Martha and Minna. He took his daughters and sons to art galleries in Vienna and taught them Tarok. During vacations he showed them the different wildflowers, instructed them in the art of finding mushrooms, which they collected and took back to their mother, who cleaned and skinned them before cooking. Martha also removed burrs from her daughters' skirts, saying jokingly that to follow them on their rambles she would "need antlers like a deer," according to Martin.

A sign of how Freud set priorities was the way he handled summer vacations. He would send his daughters, sons, and wife away every June and work alone into July. Just as he burned the midnight oil while Martha was in bed alone, so he used a good part of vacation to be alone with his labor of love. But he joined his family in mid-July, returning to his practice mid-September. Martin says "something like a tribe" took vacation together—mother, father, six children, Minna, and sometimes Freud's younger brother, Alexander, who married late in life. Martin recalls his mother was "never perturbed and also never overlooked a detail, exchanging her normal role of an ordinary practical housewife for the cold and calculating organizing genius of a senior officer of the Prussian General Staff." This quality had attracted Freud and she may have unconsciously assumed the role of organizer and executive on vacation, particularly in her husband's absence.

The children were each given a traveling satchel that could be slung over their shoulders and, days before the journey, would pack and unpack the satchels in anticipatory glee. Once Sophie asked her father, several days prior to their leaving, "Can one already *see* the journey?" This line became a classic, drawing laughter in the family whenever it was repeated. On May 12, 1938, when Freud was waiting permission from the Nazis to leave Vienna, he wrote his son Ernst in London, "One can already see the journey." This held particular poignancy because Sophie had died eighteen years before.

Martin said his sisters were more selective than he and his brothers in their friendships during summer vacations, recalling, "Once when they were asked why they did not play with the daughter of a ranger who lived nearby, a sweet and pretty little girl, they replied demurely that they had made the shocking discovery when playing with her on the swings that she did not wear drawers." This is an interesting statement from the daughters of Sigmund Freud, who made childhood sexuality more acceptable and understandable in terms of its development into adult sexuality.

Though Freud was the scholar, the intellect, and the educator, his obsession with work prevented him from relating responsibly to his children's education, and he left problems of schooling entirely in Martha's hands. "But," Martin says of his father, "he always came down from his Olympian heights to help rescue us when anything happened to a child that assumed the proportions of a tragedy to the child." Freud enjoyed the role of rescuer, and Martin recalls that when he felt discouraged during his first year as a law student and wondered whether he would succeed in law, "I went to my father for consolation, something he was always willing to offer, no matter how deeply he might be preoccupied with his own work. We had a long talk in his study."

The life of the family revolved around Freud, as his life revolved around his work. Sachs observed, "Friends of the family frequently made fun of the solemn way they spoke of everything concerning their father." He was viewed as the emperor or god he unconsciously aspired to be. Sachs further recalls, "For instance, it was said that when one of the children had been absent for some time and was met by another, the first word to the newcomer was, 'Father now drinks his tea from the green cup instead of from the blue.' "

His letters on his travels through Italy "show his humanity and his deep love for his children," says Grotjahn. "Some of these letters or notes to his daughters are the most moving human documents in this collection. They show Freud as a father in his shy tenderness and love, his patience, and his great compassion," belying his reference to them as "brats" in letters to Martha (a somewhat dubious affectionate appellation of the day).

When Mathilde was twenty-one, she fell physically ill, suffering stomach attacks, and was sent to Merano in South Tyrol to recuperate, staying with the family of a Dr. Raab. Mathilde wrote her father how ill and depressed she felt (the death of her Aunt Rosa's husband, Dr. Heinrich Graf, may have had something to do with it). Freud answered her letter on March 19, 1908, noting it was the first time she had asked him for help, and said, "You don't make it difficult for me . . . for it is easy to see that you are very much overrating your trouble."

He wrote he was not going "to offer you any illusions, neither now nor at any other time; I consider them harmful." He said he thought her stomach pains might be caused by a floating kidney and assured her that "by the time the question of marriage arises in your life, you will be completely free of it."

In fatherly frankness he suggested she probably associated her pains "with an old worry about which I should very much like to talk to you for once." He said he had guessed for a long time that in spite of all her common sense, "you fret because you think you are not good-looking enough and therefore might not attract a man." He assured her he had "watched this with a smile, first of all because you seem quite attractive enough to me, and secondly because I know that in reality it is no longer physical beauty which decides the fate of a girl, but the impression of her whole personality."

He added: "Your mirror will inform you that there is nothing common or repellent in your features, and your memory will confirm the fact that you have managed to inspire respect and sympathy in any circle of human beings. . . . The more intelligent among young men are sure to know what to look for in a wife—gentleness, cheerfulness, and the talent to make their life easier and more beautiful. . . ."

Mathilde soon recovered, fell in love with, and married Robert Hollitscher, and Freud was pleased. But when Sophie, without first consulting her parents, became engaged four years later to Max Halberstadt, a photographer who lived in Hamburg, Freud was upset. After meeting Max, Freud wrote Sophie in Karlsbad, where she was living, saying he approved of the engagement but rebuked her for ignoring her family in the choice of fiancé, and then, on a visit home, acting very anxious. He said, "the degree of your remorse may be judged by the fact that you even succeeded in upsetting your aunt, normally so imperturbable," referring to Minna. Apparently Freud's pride was punctured. The Emperor had to be consulted. He added reassuringly, "Anyhow, everything is all right."

To Max, Freud wrote on July 24, 1912, that he approved of him but it seemed nothing he could write Sophie "seemed detailed or loving enough for her. . . . It is very strange to watch one's little daughter suddenly turn into a loving woman." Just as Freud wanted to be the one and only with Amalie, Gisela, Martha, and other women in his life, he found it difficult to release Sophie to another man. But three days later he wrote Max another letter (perhaps in answer to one Max had sent him, for Freud never let a letter go unanswered) in which he said, reminiscing about his marriage, with its long engagement years before, "You are quite right in not wanting to follow our example. I have really got along very well with my wife. I am thankful to her above all for her many noble qualities, for the children who have turned out so well, and for the fact that she has neither been very abnormal nor very often ill. I hope that you will be equally fortunate in your marriage, and that the little shrew will make a good wife." Evidently Freud was still harboring a grudge against Sophie because she had not consulted him before announcing her engagement and also, perhaps, because she was getting married. But this letter showed his capacity to be a warm, understanding and benign father—something he always wished to be—for he signed the letter, "Your new and old father-in-law, Freud."

After the wedding the Halberstadts moved to Hamburg and two years later Sophie gave birth to a son. When a second son was born, Freud told all his friends that on January 1, 1919, little Heinz Rudolf Halberstadt had come into the world. He visited Sophie on his vacation in September to see his new little grandson, who seemed special to him.

One year later, he learned that Sophie, his most beautiful daughter, suddenly had become seriously ill. She was struck with "influenzal pneumonia," of which there was an epidemic that year throughout Europe. He and Martha could not go to Hamburg at once, but Oliver and Ernst, then living in Berlin, immediately set out for Hamburg. They did not reach her in time.

Freud got a telegram from them on January 25, saying she had died two days before. She was twenty-six years old and had been in perfect health up until then. Her first son was six, little Heinz, thirteen months.

The next day Freud wrote his mother: "I have some sad news for you today . . . our dear lovely Sophie died from galloping influenza and pneumonia . . . Martha is too upset; one couldn't let her undertake the journey to see Sophie before her death and in any case she wouldn't have found Sophie alive . . . She is the first of our children we have to outlive. What Max will do, what will happen to the children, we, of course, don't know as yet . . . I hope you will take it calmly, tragedy after all has to be accepted. But to mourn this splendid, vital girl who was so happy with her husband and children is of course permissible." He signed it, "I greet you fondly. Your Sigm."

That day he also wrote Jones, whose father was very ill, "Sorry to hear your father is on the list now, but we all must be and I wonder when my turn will come. Yesterday I lived through an experience which makes me wish it should not last a long time." A day later he wrote his friend, the pastor Oscar Pfister, who believed in psychoanalytic theories, that Sophie "was blown away as if she had never been," and "Tomorrow she is to be cremated, our poor Sunday child!" To his friend Ferenczi, he wrote, "Deep down, I sense a bitter, irreparable narcissistic injury."

Freud paid considerable attention to the phenomenon of death and was frequently concerned about his own death as well as the deaths of others. He became so preoccupied, he even postulated a death instinct that he felt existed in everyone. In his brilliant paper, "Mourning and Melancholia," he pointed out that when a loved one dies there are almost always reactions of sadness, emptiness, feelings of loss and abandonment and loneliness. This is the "mourning syndrome." But when the relationship with the deceased has been ambivalent and there has been excessive repressed hate, melancholia ensues—agitation, self-debasement, guilt and sometimes inability to

Sophie Freud Halberstadt, Freud's middle daughter. *Austrian Press and Information Service.*

eat or sleep. Freud probably felt melancholic after his father's death because of his great ambivalence. But he mourned Sophie, whom he especially loved.

Ten years later he wrote Ludwig Binswanger on April 11, 1929, "My daughter who died would have been thirty-six years old today. Although we know that after such a loss the acute state of mourning will subside, we also know we shall remain inconsolable and will never find a substitute. No matter what may fill the gap, even if it be filled completely, it nevertheless remains something else. And actually this is how it should be. It is the only way of perpetuating that love which we do not want to relinquish." In the unconscious, as he emphasized throughout his life, no one who is loved really dies, the images and feelings always remain.

Of Freud's children, Anna was the only one who did not marry, just as in his original family the youngest daughter did not marry but lived all her adult life with her mother, taking care of her. Anna always seemed special to Freud. Sachs reports that when Freud completed his morning work and walked across the hall for lunch in the family quarters, he would express his relief that the first part of the day was past "by making a funny sound, something between a growl and a grunt, usually aimed at his youngest daughter."

He wrote Anna just before her seventeenth birthday, on November 28, 1912, when she was in Merano, a popular Alpine resort, staying with Mathilde's sister-in-law, Frau Marie Rischawy, that he hoped she was feeling better. Her back had been hurting and she had lost weight. He reported he had refurnished his room and that before she returned, "we will do your room too; writing table and carpet are in any case assured."

On December 13 he wrote that she did not have to return home until she felt better: "Your plans for school can easily wait till you have learned to take your duties less seriously. They won't run away from you. It can only do you good to be a little happy-go-lucky and enjoy having such lovely sun in the middle of winter . . . The time of toil and trouble will come for you too but you are still quite young."

Early in life, she became his companion. He wrote Ferenczi, on July 9, 1913, he was going to spend several weeks in Marienbad "free of analysis," and "My closest companion will be my little daughter [Anna was then eighteen], who is developing very well at the moment (you will long ago have guessed the subjective condition for the 'Choice of the Three Caskets.')" He was referring to his frame of mind when he wrote the article, "The Theme of the Three Caskets," in which he discussed King Lear and his three daughters.

He told Ferenczi his interest in the theme must have ben connected with thoughts of his three daughters, particularly Anna, "and she it was who, a quarter of a century later, was by her loving care to reconcile him to the inevitable close of his life," Jones comments. "The Theme of the Three Caskets," which Jones calls "a little essay," compares Bassanio's choice of the leaden casket in the scene of the three caskets in *The Merchant of Venice* with King Lear's demand for love from his three daughters, "the muteness of the lead being equated with Cordelia's silence," Jones says. The themes of love and death are the content of Freud's paper. Sophie's engagement had just been announced, Mathilde was already married so only Anna, "the leaden one," was left to him.

When Martin and Ernst left to join the Army in World War I, Freud wrote Lou Andreas-Salomé, a psychoanalyst who became a friend, "My little daughter [Anna was then twenty years old] whom you may remember and who is staying with her eighty-year-old grandmother in Ischl, has written to us in concern: 'How am I going to take the place of six children all by myself next year?'"

Though Anna had no doubt felt this a burden all her life, like her father, who wanted to be the one and only in his family, in all probability this was also Anna's wish.

She has become one of the world's authorities on child psychoanalysis and ego psychology. She founded the first training center for child psychoanalysts, Hampstead Child Therapy Clinic in London, where she still works. Freud's last home at 20 Maresfield Gardens in London is her residence now, as well as museum for his archaeological collection. She has contributed many new theories to psychoanalysis, including her ideas on ego psychology, summarized in *The Ego and the Mechanisms of Defense.*

Freud sometimes referred to Anna as "Antigone." In a letter to his friend Arnold Zweig, German writer, on May 2, 1935, he explains he is too ill to enjoy spring with him on Mount Carmel, saying, "Even supported by my faithful Anna-Antigone I could not embark on a journey . . ." It is interesting he thought of Anna as the daughter of Oedipus and Jocasta, which speaks of his fantasy of Martha and himself. He also called Anna "my favorite son." In tracing the reasons why he may have thought of her as a son, it may be significant she was born in 1895, a time he felt very close to Fliess, the only one encouraging him in his psychoanalytic discoveries.

He wrote Fliess on October 20, 1895, when Martha was pregnant with Anna, "If it's a boy, you will have no objections to my calling my next son Wilhelm! If *he* turns out to be a daughter, *she* will be called Anna." On the day Anna was born Freud wrote Fliess, "If it had been a son I should have sent you the news by telegram, as he would have been named after you. But as it is a little girl by the name of Anna, you get the news later. She arrived today at 3:15 during my consulting hours, and seems to be a nice, complete little woman."

Perhaps because he wished so strongly for a boy he could name after his beloved friend, Freud thought of Anna as a son in his unconscious. (When Fliess's wife gave birth to a son a month later, Fliess did not name the son Sigmund but Robert Wilhelm.)

All of Freud's children left home to lead lives of their own except Anna, who remained by his side, becoming more and more indispensable to her father in his work and caring for his health. When he received the Goethe Prize, he was supposed to journey to Frankfurt, where Goethe was born, and make a speech. He wrote Ferenczi on August 1, 1930, "Needless to say, I cannot go, but Anna will represent me and read what I have to say about Goethe in relation to psychoanalysis and its right to use him as subject for investigation." It was appropriate that he thus honor the man who, he said, was responsible for getting him interested in medicine because of an essay on nature.

Anna's closest friend over the years was Mrs. Dorothy Burlingham, who died November 19, 1979, in London. She was the youngest daughter of Louis Comfort Tiffany, the famous creator of the lamps, vases, and other glass-blown objects that bear his name. A wealthy, young American heiress with four children, she went to Vienna to be analyzed when her marriage to Dr. Robert Burlingham, a New York neurosurgeon, broke up. She became, in Jones' words, "almost one of the Freud family." She took an apartment above the Freuds at 19 Berggasse and worked closely with Anna on the psychoanalytic theory of child development.

Anna Freud with her father, about 1912, when she was seventeen.
Press and Information Service.

Freud wrote Arnold Zweig on May 8, 1932, "Where am I writing from? From a farm cottage on the side of a hill, forty-five minutes by car from the Berggasse, which my daughter and her American friend (she owns the car) have acquired and furnished for their weekends. We had expected that your return home from Palestine would take you through Vienna, and then we would have insisted on your seeing it." The little farm was called "Hochroterd."

Mrs. Burlingham helped the Freuds escape to London, then financed the founding of the Hampstead Child Therapy Clinic, of which she became co-director with Anna. She also collaborated on articles and books with Anna.

Freud wrote Lou Andreas-Salomé on May 16, 1935, that she should not expect to hear anything intelligent from him, that he doubted he could still produce anything, because he was in ill health and relied more and more on Anna's care, just as Mephistopheles once remarked:

> In the end we depend
> On the creatures we made.

He added, "In any case, it was very wise to have made her."

He wrote Ernst on January 17, 1938, half a year before he left Vienna, "Mama is bearing up magnificently, your aunt is about to undergo an operation for cataract which we hope will turn out well. Anna is splendid, in spirits, achievement, and in all human relationships. It is amazing how clear and independent her scientific work has become. If she had more ambition . . . but perhaps it is better like this for her later life." Only he knew what he meant by this statement.

Dr. Helene Deutsch, who was analyzed by Freud and then worked with Anna at the Vienna Psychoanalytic Institute, said she first remembered seeing Anna at the Wagner-Jauregg Clinic, where Dr. Deutsch was a staff member, when Anna attended her father's lectures there, entering the auditorium on his arm, wearing a green suit.

"From the beginning I felt a kind of tenderness toward her that never left," said Dr. Deutsch in *Confrontations With Myself*. She added, "Anna Freud was loved and admired by all around her."

She recalled that when Anna was accepted into the Vienna Psychoanalytic Society for her paper, "Beating Phantasies and Day

Dreams," her talent was evident even then. When Dr. Deutsch told Freud how much she liked Anna's paper, "he broke into the typical smile of a proud father." Anna spoke from memory every time she lectured, just as her father did.

Dr. Deutsch described Anna as a devoted teacher of young children, whose theories in child analysis were based on her teaching experience with disturbed children, developing techniques for treating them that became fundamental in child analysis. Dr. Deutsch quoted Freud as saying about the process of identification in depression, "The shadow of the object falls on the psychic life of an individual," and Dr. Deutsch added, "About Anna Freud one can say that the light of her father fell on her professional work and changed a talented teacher into a creative child analyst."

Helen Beck, who studied with Anna Freud in Vienna, and for whose book *Don't Push Me, I'm No Computer*, Anna Freud wrote the preface, says, "She has something very few people possess. I have heard her address an audience composed of both experienced psychiatrists and teen-agers and neither group has the feeling she talks down. She has deep respect for all people. She is a simple, straight-forward person."

When Jones once asked her what she considered her father's most distinctive characteristic, Anna answered with two words, "his simplicity." Jones comments that Freud disliked anything that complicated life. He adds that when he thinks of Freud "in memory," what comes to mind is predominantly his constant cheerfulness, his tolerant attitude and his easygoing manner, all of which Anna appears to have psychologically absorbed from her father.

When she traveled to Harvard University in June 1980, to receive the university's first honorary Ph.D. in psychoanalysis, she visited her cousin, Edward L. Bernays, in Cambridge. She confessed that when he had arrived in 1913 at the Freuds' summer home in Karlsbad at the age of twenty-two, she had fallen in love with him and thought of marrying him. Then, nodding at his wife, Doris Fleischman Bernays, feminist and counsel on public relations in partnership with her husband, Anna said with a smile, "You did better with her."

To which Mr. Bernays replied, also with a smile, "If I had married you, it would have been double incest." He was referring to the fact they were related both through her mother (Martha, his aunt) and her father (Sigmund, his uncle). He recalled Anna as a very young, unassuming girl, four years his junior, "feminine in manner and appearance but so gentle and self-effacing that it was a great contrast to meet her at my home sixty-seven years later and encounter a strong personality who told of her love for me."

When Freud started to suffer from the awkward prosthesis in his mouth, a magnified denture designed to shut off the mouth from the nasal cavity, "a horror," in Jones's words (Freud called it "the monster"), Anna helped alleviate the pain. It was difficult to take out or replace the prosthesis because Freud could not open his mouth very wide. One time, says Jones, the combined efforts of Anna and her father failed to insert the prosthesis after both had struggled for half an hour, and the surgeon had to be called. For the prosthesis to fulfill its purpose of shutting off the yawning cavity between mouth and nose, and so make speaking and eating possible, it had to fit fairly tightly. This produced constant irritation in Freud's mouth and left sores so that its presence was often unbearable. But if it were left out for more than a few hours, the tissues would shrink and the denture could no longer be replaced without being altered.

It also made Freud's speech defective, though this varied from time to time according to the fit of the denture. Sometimes his speech was nasal and thick, other times natural and clear. Eating was a trial, and he seldom would eat in company. The damage done to the Eustachian tube, together with constant infection in his mouth, also greatly impaired his hearing on the right side, until he became almost entirely deaf on that side.

He went through sixteen years of discomfort, distress, and pain with this prosthesis. From the onset of this illness in 1923 to the end of his life in 1939, Freud refused to have any nurse other than Anna. He made a pact with her at the start that no sentiment was to be displayed, that everything was to be carried out in a cool, matter-of-fact manner with no emotion. Her courage allowed her to adhere to this pact, Jones says, even though she must have felt agonized at her father's suffering.

Freud wrote Arnold Zweig on May 18, 1936, ". . . I have been exceptionally happy in my home, with my wife and children and in particular with one daughter who to a rare extent satisfies all the expectations of a father." He had written Andreas-Salomé on January

Anna Freud with her father at the 6th International Psycho-Analytical Congress in the Hague. *Austrian Press and Information Service.*

6, 1935, "My one source of satisfaction is Anna. It is remarkable how much influence and authority she has gained among the general run of analysts . . .

"It is surprising, too, how sharp, clear and unflinching she is in her mastery of the subject [psychoanalysis]. Moreover, she is truly independent of me; at the most I serve as a catalyst . . . Of course there are certain worries; she takes things too seriously. What will she do when she has lost me? Will she lead a life of ascetic austerity?"

Freud's question about the kind of life Anna would lead seemed a self-fulfilling prophecy. For much of her life appeared to hold "ascetic austerity." How did this come about?

Just as Freud wanted to be Amalie's "everything," he seems to have made Anna his one and only, forming with her the intimacy for which he yearned with Amalie. At various times Freud referred to Anna as "closest companion," "favorite son," "complete little woman," and said he was glad he had "made her." Anna became his professional colleague, selected to carry his torch. In the many photographs taken of Freud, he usually wore a serious expression, but not when he posed with Anna, then he smiled.

Anna was his son and daughter, nurse, secretary, colleague, and probably the object of some of his unconscious sexual fantasies. At 19 Berggasse, prominently featured on one wall, hung Brouillet's famous painting, showing Charcot demonstrating a case of "grand hysteria" to a group of physicians. Anna, as a girl, would ask her father what was wrong with the woman, who appeared to be fainting, and he would always say, "The lady was too tightly laced," implying it was her own fault. This repetitive, playful scene between father and daughter suggests a bond of intimacy, but more important, the picture shows a woman "falling for" a man, the doctor.

Paul Roazen in *Freud and His Followers* says that Anna Freud's life is a testimony to her father's statement that "a girl's first affection is for her father." The mutual love affair between psychoanalysis' Antigone and Oedipus took a form that defies all the rules of psychoanalytic procedure—Freud analyzed his own daughter Anna. According to Roazen, the analysis stretched over a number of years. One of Freud's sons, Oliver, remembered his sister going into Freud's study for her analysis in the spring of 1921. In a 1935 letter to Dr. Eduardo Weiss, who had asked Freud's advice about analyzing his own son, Freud stated that with his own daughter the analysis had gone well but that with a son it might be different. Said Freud, "Concerning the analysis of your hopeful son, that is certainly a ticklish business. With a younger, promising brother it might be done more easily. With one's own daughter, I succeeded well. There are special difficulties and doubts with a son."

By analyzing Anna, Freud was acknowledging that he would not turn her over to another professional. Only he was privy to the contents of her personal psyche!

Jones says, "There had grown up in these years a quite peculiarly intimate relationship between father and daughter. Both were very averse to anything at all resembling sentimentality and were equally undemonstrative in matters of affection. It was a deep silent understanding and sympathy that reigned between them. The mutual understanding must have been something extraordinary, a silent communication almost telepathic in quality where the deepest thoughts and feelings could be conveyed by a faint gesture. The daughter's devo-

tion was as absolute as the father's appreciation of it and the gratitude it evoked." Jones does not mention the psychological toll this
must have taken of Anna.

Freud's possessiveness undoubtedly gave Anna much gratification.
Though being both favorite son and daughter, as well as close companion to her Oedipus, was not without deep conflicts. Anna was
jealous of the women who were significant in Freud's life. She later
reported her memories of jealous feelings were a means of gauging
the importance of a woman in her father's life. Anna did not need
to compete with her mother, since Martha was already pretty much
excluded, but she did compete with the women psychoanalysts, including Ruth Mack Brunswick and Dr. Helene Deutsch.

Freud seemed aware of his reciprocal position with Anna, and his
guilt comes through in some of his letters. In 1921 he wrote a relative, "Anna is in splendid health and would be a faultless blessing,
were it not that she had lived through her 26h birthday yesterday
while still at home." Four years later he wrote his son-in-law Max,
that Anna, "of whom we may well be proud has become a pedagogic
analyst, is treating naughty American children, earning a lot of money
of which she disposes in a very generous way, helping various poor
people. She is a member of the International Psychoanalytic Association, has won a good name by literary work, and demands the respect of her co-workers. Yet she has just passed her 30th birthday,
does not seem inclined to get married, and who can say if her
momentary interests will render her happy in years to come when
she has to face life without her father?"

Though Freud consciously hoped Anna would marry, *he* controlled
her life with men. All her suitors came through her father. Anna is
reported to have been in love at various periods with at least three
men in Freud's circle—Siegfried Bernfeld, Hans Lampl, and Max
Eitingon—but her strong symbiotic tie to her father blocked the way to
marriage. Like her father, Anna may be regarded as an Oedipal victor,
who felt too guilty to have an ongoing sexual relationship with a
member of the opposite sex, for it would be experienced as forbidden
incest. "Ascetic austerity" was easier.

In some ways, Freud and Anna became like husband and wife.
After Freud contracted cancer of the jaw, Anna cared for his physical
condition, helped him adequately rinse his mouth and insert the

prosthesis. He spent less and less time with Martha, more and more with Anna. Because Anna unconsciously wished to usurp Martha's role as wife, she may have tried to convince herself that her mother neglected Freud. From all we know of Martha, she was hardly a neglectful wife—perhaps a neglectful mother at times. Martha herself may have become jealous of Anna, with subtle antagonism developing between mother and daughter, one Freud may have unconsciously encouraged, enjoying his two women fighting over him.

A year before his death, when a number of people, including Jones, were trying to persuade Freud to leave Vienna because of the danger of Fascism, he finally agreed to forsake the city which, he said, he both loved and hated. He told Jones, after the move to London, "The advantage the emigration promises Anna justifies all our little sacrifices. For us old people [referring to Minna, Martha, and himself] emigrating wouldn't have been worthwhile."

In 1939, when Freud's suffering overcame his capacity to endure pain, he and Dr. Schur agreed "enough was enough" and euthanasia might be a way out. Freud instructed Dr. Schur to "tell Anna about our talk," not his wife. Martha once said, "Anna will never marry unless she finds a man exactly like her father." She never did.

Roazen stated, "As Freud was 'Professor' with all the magic and capacity . . . that being head of a movement entails, so Anna Freud is now 'Miss Freud.' She reigns in his place; for some 'Miss Freud' communicates exactly the same aura that the 'Professor' did."

In the room of her London home where she treated children, Anna had the furniture scaled to the size of a child, including the bathroom facilities. Mr. Bernays asked why she did this and she said she did not want the children to feel intimidated by the surroundings. One wonders if, in addition to her wish to make the children comfortable, this did not also reflect her feelings about a childhood in which she was intimidated by the rest of the family, all of whom loomed "larger" than she.

As her father feared, it would seem as though Anna Freud's life was one of "ascetic austerity," an austerity she unconsciously chose and a decision with which her father unconsciously concurred. An austerity that enabled her to remain close to him, carrying on his work, continually learning from him and loving him—indeed his Antigone.

Part Two

Women Colleagues
and Friends

"His Best and Dearest Friend"

Freud's theories, highly revolutionary for their time, started to attract attention in other countries, and women, as well as men, journeyed to Vienna to study under him for a few months or be psychoanalyzed by him. Some wished to be his patient and ease their inner torment. Others sought to train as psychoanalysts.

In the early days of psychoanalysis, training was far different than today. Now, at least three or four years at a psychoanalytic institute, as well as a personal psychoanalysis and supervision on first cases, are required. But when Freud started to practice, there were no institutes. The training of the few men and women interested in becoming psychoanalysts consisted of listening to Freud's Saturday night lectures at the Vienna Psychiatric Clinic and, at his invitation, attending the Wednesday night meetings of the Vienna Psychoanalytic Society, at his home. A short personal psychoanalysis was also advised, called a "didactic analysis."

This personal psychoanalysis was usually a matter of a few months on the couch, five or six times a week. As psychoanalysis developed into a more complex theory of human behavior and treatment, the period of analysis lengthened. Today, four or five years, with sessions three or four times a week, is considered average.

No certificates or diplomas were given during the early years of training with Freud. After the personal analysis, and attendance at his lectures, the potential psychoanalyst went out into the world with only Freud's blessing. Whatever he did, his treatment was apt to be of some help to the troubled person who, up to then, depended on drugs, rest cures, hypnosis, electric shocks, or, most often, no help at all.

Several women were among the pioneer patients and potential psychoanalysts Freud accepted into his ever-growing circle of colleagues and friends. There was often some intellectual woman—patient, student or psychoanalyst—in Freud's life, whose friendship he enjoyed. Jones, who has consistently protested against the implication that Freud felt erotic desires for any woman other than Martha, points out that Freud was always interested in the intellectual "and perhaps masculine" type woman, though he enjoyed only her mind.

After these women left Vienna, he corresponded with them, for he was, in the words of Dr. Martin Grotjahn, "a passionate, untiring correspondent, whose letters fill volumes." Ernst Freud, editor of one collection, says, "My father answered every letter he received no matter from whom." Freud wrote in longhand during spare minutes between analyses, saving his evenings for articles and books.

He undoubtedly found his women colleagues less difficult and less competitive than men. They were like a long line of adopted daughters. He did not actively seek adulation from them. "By and large, he passively accepted women as members of the inner circle around him," the existence of "what resembled a royal court did not shock him," says Roazen. "In his last years these women formed what some called a 'camarilla' around him. They shielded him from visitors, arranged his vacations, and watched over his health." As an older man, Freud was surrounded by idolatrous women, a recapitulation of his childhood and adolescence when he was catered to and adored by his mother and five sisters.

Among the women who studied with him was Marie Bonaparte, Princess George of Greece and Denmark, whom Freud met rather late in his life, in 1925 when he was sixty-nine. She was married to Prince George of Greece, brother of the King, and they had a daughter, Princess Eugénie, and a son, Prince Peter. Martin Freud called Marie Bonaparte "the best and dearest friend of father's last years." He describes her as having "most of father's chief characteristics—his courage, his sincerity, his essential goodness and kindliness and his inflexible devotion to scientific truth. In this sense the similarity of character was almost startling . . . they were in every sense of the word congenial."

Marie Bonaparte was a warm-hearted, sensitive, brilliant woman, with a glowing head of red-gold hair. She was born July 2, 1882, at Saint Cloud, near Paris. Her mother, Marie-Félix, was the daughter of François Blanc, founder of the Casino at Monte Carlo, who left his daughter a fortune. Marie-Félix married Prince Roland Bonaparte, son of Pierre Bonaparte and grandson of Lucien Bonaparte, brother of Napoleon I. She died at the age of twenty-two of an embolism, a month after Marie's birth.

Marie Bonaparte, Freud's close friend and colleague. *Austrian Press and Information Service.*

Marie grew up in the care of her paternal grandmother, whom Prince Roland summoned to his mansion to care for his motherless child. The grandmother was a strict woman, who worried Marie would fall ill and die like her mother, perhaps an unconscious wish on the old lady's part, for then she would have her son all to herself. As a girl, Marie found escape from the deep depression that stemmed from so early a loss of her mother, by writing short stories and keeping a diary. She wrote in English, having learned that language and German at four on order of her father, whom she adored.

She later wrote: "All my life I was to care about the opinions, approval and love of only a few fathers, selected ever higher and of whom the last was to be my grand master Freud."

At her father's deathbed in 1924, she read Freud's *Introduction to Psychoanalysis*. After the loss of her beloved father she felt, she said, a desire to find "a second father," and went to Vienna the following year to be analyzed by Freud.

During one of her first sessions she told Freud a dream and he suggested it revealed that as a child she had witnessed scenes of sexual intercourse. She strongly protested his interpretation. But back in Paris she met by chance one of her father's former grooms, who had been an illegitimate son of her grandfather. He admitted what Freud had guessed, that when Marie was less than a year old, he and her wet nurse had sexual relations in her presence several times. Whereupon she wrote Freud his conjecture had been correct.

Delighted at such confirmation, Freud replied that now she understood how both opposition and acclamation could be a matter of indifference if one possessed "a real certainty." He added, "That was my case," no doubt referring to the time in his infancy when he slept in the room with his parents and saw or heard them while they were having sexual intercourse.

Psychoanalysis was late in coming to France, and Marie Bonaparte played an important part in establishing it in her country, translating a number of Freud's works. She became a founder in 1926 of the *Societé Psychoanalytique de Paris,* of which she was honorary president shortly before her death in 1962. She was also vice-president of the International Psychoanalytical Association.

She was a woman with many causes besides psychoanalysis. She fought for political and social justice. By paying a ransom demanded by the Nazis, she saved the lives of two hundred Jews. When she learned Caryl Chessman was to be executed in America for rape and murder, she joined the movement to save his life, flew to San Francisco and talked to the condemned man, then tried to intercede on his behalf with the governor.

As a result of her work with Freud, she showed a deep interest in the causes of murder. Her article, "The Case of Madame Lefebvre," was one of the first psychoanalytic studies of a murderer. It centered on a crime committed in 1925 in France by a woman who killed her pregnant daughter-in-law. Marie Bonaparte interviewed the convicted murderess in prison, then interpreted her many unconscious motives for the crime, primarily her close attachment to her son and her pathological jealousy of her daughter-in-law. Madame Lefebvre had been unable to accept the loss of her son and the pregnancy of his wife was the last straw, an event which promised to remove him even further from his mother's life.

Deeply interested in female sexuality, she made many contributions to psychoanalytic theory about women. In her book, *Female Sexuality*, she said she believed, as Freud did, in bisexuality, since neither "pure masculinity" nor "pure femininity" was found in anyone. She also wrote a three-volume biography, *The Life and Works of Edgar Allan Poe, A Psycho-Analytic Interpretation*. She undoubtedly identified with Poe, who also lost his mother before he was two. She related the themes of many of his poems and short stories to his early traumas, using Freud's method of interpreting fantasies and dreams.

Freud respected her opinions highly. He wrote on April 7, 1926, "I have once more admired your judgment [about the trauma of weaning]. Naturally what is concealed behind it is the denial of the castration complex." On April 16 of that year he wrote: "I am not surprised that the phrase 'the eternal suckling' has somewhat impressed you. There is indeed something in it. Oral eroticism is the first erotic manifestation, just as the nipple is the first sexual object; the libido is led along definite paths from these first positions into the new organizations . . . Generally symptoms also come about through regression to this phase . . ."

He admired and followed her judgment in areas other than intellectual. She helped him select valuable items for his collection of Greek antiquities. She was the one to suggest Dr. Schur, an internist who was psychoanalytically trained, to be his regular doctor, a wise choice, for Freud seemed content with Dr. Schur through the many tortured years of his cancer.

In answer to a letter in which she apparently raised the question of treatment for a psychotic patient, Freud wrote on January 15, 1930: "You know that with psychoses of that kind we can do nothing through analysis. Above all, a normal ego is wanting with which one can enter into contact. We know that the mechanisms of the psychoses are in essence no different from those of the neuroses, but we do not have at our disposal the quantitative stimulation necessary for changing them." Some of today's psychoanalysts have shown that the ego of a psychotic can be strengthened through long-term psychoanalytically-oriented treatment so the person can eventually become aware of his conflicts and lose the bizarre symptoms he has adopted as defenses.

Six years later Freud wrote her on December 6, 1936, that he had just received her card from Athens and her manuscript of "the Topsy book." This book, *Topsy, the Chow with the Golden Hair*, was published in Paris the following year. It described Marie Bonaparte's changing attitude, her compassion and growing tenderness for one of her chow dogs that had suffered from cancer of the mouth and had been rescued from death by a successful operation. The book obviously expressed her feelings about Freud's illness. She had given him a golden chow who had two puppies, Lum and Chow, and they became Freud's inseparable companions. He would tell patients, "They are psychoanalytic dogs."

He wrote her his reactions to her manuscript: "I love it; it is so movingly genuine and true. It is not an analytical work, of course, but the analyst's thirst for truth and knowledge can be perceived behind this production, too. It really explains why one can love an animal like Topsy or Jo-fi [his chow at the time] with such extraordinary intensity: affection without ambivalence, the simplicity of a life free from the almost unbearable conflicts of civilization. . . ." Two years later, while the Freuds were waiting for word from the Nazis that they could leave for London, Anna and he passed the time finishing the translation into German of Marie's book, which Anna had started eighteen months before.

Freud wrote to Marie on August 13, 1937, that he had just finished work on his book about Moses, which became the last book he ever wrote, *Moses and Monotheism*, published in 1939. He said, "To the writer immortality evidently means being loved by any number of anonymous people. Well, I know I won't mourn your death, for you will survive me by years, and over mine I hope you will quickly console yourself and let me live in your friendly memory—the only form of limited immortality I recognize."

His pessimism was further evident in the thought that the moment a man questions the meaning and value of life, "he is sick, since objectively neither has any existence; by asking this question one is merely admitting to a store of unsatisfied libido to which something else must have happened, a kind of fermentation leading to sadness and depression." He then made one of his typically ironic remarks, saying "an advertisement" was "floating about in my head which I consider the boldest and most successful piece of American publicity: 'Why live, if you can be buried for ten dollars!' " This was his macabre whimsy.

Marie Bonaparte not only was Freud's disciple but his complete champion. She tried for several years to persuade the Nobel Prize Committee in Stockholm to give him the Nobel Prize. He reproved her for wasting time on what he called a hopeless endeavor. And she was his financial savior, loaning him money during the winter of 1929, when his publishing house was passing through one of its periodic crises.

Most importantly, she managed to preserve his letters to Fliess in which he described his early psychoanalytic discoveries. Freud had destroyed all the letters Fliess had written him, but Fliess kept Freud's correspondence. After Fliess died in 1928, his widow sold 284 letters from Freud, along with the scientific notes and manuscripts Freud sent Fliess from time to time, to Reinhold Stahl, a bookseller in Berlin. The widow stipulated the material was not to pass to Freud, knowing he would destroy it. Stahl fled to France during the Nazi occupation and offered the Fliess correspondence to Marie Bonaparte for 100 pounds. She took the collection to Vienna, where she was still seeing Freud for occasional analytic sessions, and told him of her purchase. He was indignant and insisted she destroy the material at once. She had the courage to defy him—her analyst, teacher, and friend. In the winter of 1938, she deposited everything in the Vienna Rothschild Bank, intending to study it on her return the following summer.

But with Hitler's invasion of Austria in March, she left Paris for Vienna at once, before the bank could be rifled. She was allowed to withdraw the contents of her safe deposit box in the presence of the Gestapo. They did not detect Freud's letters to Fliess or they would have destroyed them as they had the books in his publishing firm. When she left Paris for Greece in February 1941, she deposited the correspondence with the Danish Legation in Paris. At the end of the war, "After surviving all those perils, the letters braved the fifth and final one of the mines in the English Channel and so reached London in safety; they had been wrapped in waterproof and buoyant material to give them a chance of survival in the event of disaster to the ship," Jones recounts.

Marie Bonaparte probably saved Freud's life. She was the one to persuade him to leave Vienna after Jones's pleas failed. She arrived at his home in March 1938, and with Anna went over his notes and correspondence, burning everything they considered of secondary importance. When the Nazis demanded a monetary ransom from Freud, amounting to $4,824, before permitting him to leave, Marie Bonaparte paid it, for his bank account had been confiscated by the Nazis. He repaid her after arriving in England, and she used the money to reprint his *Collected Works*, which were destroyed by the Nazis.

Freud took his last look at Vienna on June 3, 1938. He traveled by train to Paris, where he was met at the station by Marie Bonaparte and newspaper reporters and photographers. He spent from 10:00 A.M. until 10:00 P.M. at her home resting, and "she surpassed herself in tender care and attention," he wrote. Then it was on to the rented house at 29 Elsworthy Road in London, for his first new home in forty-seven years.

He wrote Marie Bonaparte on October 4, 1938: "Everything here is rather strange, difficult, and often bewildering, but all the same it is the only country we can live in, France being impossible because of the language." On November 12, he spoke of his feelings about her consistent support of psychoanalysis: "I am always prepared to acknowledge, in addition to your indefatigable diligence, the self-effacement with which you give your energy to the introduction and popular expositions of psychoanalysis."

She visited him in early February 1939, when he was now living more comfortably in a house with a large garden at 20 Maresfield Gardens. She saw him again in the middle of March. Freud wrote, after this visit, "I want to say again how sorry I am not to have been able to give you more of myself when you stayed with us. Perhaps things will be easier next time you come—if there is no war—for my pain has been better of late . . ."

She saw him again at the end of March. This visit was followed by a less cheerful letter from him on April 28, 1939: ". . . I am not well; my illness and the aftermath of the treatment are responsible for this condition, but in what proportion I don't know. People are trying to lull me into an atmosphere of optimism by saying that the carcinoma is shrinking and the symptoms of reacting to the treatment are temporary. I don't believe it and don't like being deceived.

"You know that Anna won't be coming to the meeting in Paris because she cannot leave me; I am growing increasingly incapable of looking after myself and more dependent on her. Some kind of intervention that would cut short this cruel process would be very welcome."

Marie Bonaparte went to London for Freud's eighty-third birthday on May 6, 1939, after which he wrote how much pleasure her visit had given him. He said, "We all specially enjoyed your visit, and the prospect of seeing you again soon is splendid, even if you don't bring anything from S." (Segredakei sold Greek antiquities in Paris.)

She received a letter on June 16 saying he would look forward to her next visit, when they would again discuss their work and theories. On June 29, she visited him in London, and also during the first week in August.

When she went to London on September 26, it was to attend his funeral service. He was cremated at Golders Green, his ashes placed in one of his favorite Grecian urns, and kept at the crematorium.

From 1925 on, Princess Marie Bonaparte was in analysis on and off with Freud. He would fit her in for a few sessions whenever she came to Vienna. She remained student, analysand, colleague, and friend. Freud affectionately called her "Marie" or "Prinzessin."

Sigmund Freud and Marie Bonaparte had much in common psychologically. It was the type of relationship psychoanalysts call "a narcissistic attraction," in that they saw in each other themselves. Each was exceptionally bright and enthusiastic, demanded a lot of attention, was stimulated by strong rescue fantasies, and gave and loaned money to those in need. Each was a creative, prolific writer with a strong interest in—a passion for—uncovering the inner workings of mental life.

Both were involved in marriages in which the spouse did not share their devotion to psychoanalysis. Martha, it will be recalled, referred to psychoanalysis as "pornography," and Prince George, Marie's husband, treated her involvement in psychoanalysis as if it were a toy to amuse her. Marie and her husband had a distant, though fond, relationship, not too dissimilar from Freud's relationship with Martha, though Prince George and Marie often lived apart.

Freud encouraged the transference Marie set up toward him and, as he did in his relationship with his daughter Anna, acted out his own countertransference. "Transference" is the term Freud used to describe the patient's investment in the analyst of all the feelings he originally felt for those persons paramount in his early life. The term "countertransference" applies to the feelings aroused in the analyst by the patient, feelings of which the analyst has to be aware in order that he not react in a frightened or hostile way and impede the treatment. Freud shared many thoughts, feelings, and conflicts with Marie. He kept her photograph on the desk in his study. Jones was later staggered when he learned the many intimate details Freud had divulged about himself to Marie.

In terms of today's knowledge of psychoanalytic technique, no such intimacy with a patient would be permitted but Freud, as the discoverer of psychoanalysis, was feeling his way along uncharted psychic territory. This included, at times, friendship with a patient both during and after treatment. In those days a number of patients went on to practice psychoanalysis so that they became colleagues and it was permissible to be lifelong friends.

Thirty-seven years of Marie Bonaparte's life were spent in furthering Freud's philosophy of mental healing, her belief in her "grand master" never shaken. He gave her a noble cause, friendship, trust, and affection. She repaid him by her faithfulness and affection, her generosity with time and money for publishing and republishing his works, and helping to save his life from what would have been almost certain slaughter by the Nazis, for not only was Freud a Jew but he believed all men and women entitled to emotional liberation.

"Dearest Lou"

Another very striking, unusual woman came into Freud's life before Marie Bonaparte. Though this woman became a psychoanalyst, she was never Freud's patient, but he grew very fond of her, admiring her courage, her intellect and what he called her "incurable optimism."

Her name was Lou Andreas-Salomé. Ernst Pfeiffer, her close friend, suggests Lou, novelist and biographer, first met Freud on a visit to Vienna in the 1890s when Arthur Schnitzler introduced them. She met Freud a second time through Poul Bjerre, a psychiatrist from Sweden who was her lover for two years. Bjerre suggested she accompany him to the Weimar Congress of the International Psychoanalytical Association in September, 1911, where Freud presided. She accepted eagerly, for she had become passionately interested in psychoanalysis.

She wrote Freud the following year on September 27: "Since I was permitted to attend the Weimar Congress last autumn I have not been able to leave psychoanalysis alone, and the deeper I penetrate into it the more it fascinates me. I have only one wish: to come to Vienna for a few months. May I then address myself to you, attend your lectures and obtain permission to join your Wednesday-evening seminars? My only reason for coming to Vienna is my wish to devote myself entirely to this matter."

Freud replied promptly, on October 1: "When you come to Vienna we shall all do our best to introduce you to what little of psychoanalysis can be demonstrated and shared." She had been so enthusiastic about his theories when he met her at Weimar, he had asked with a twinkle in his eyes if perhaps she had mistaken him for Santa Claus, and subsequently she referred to psychoanalysis as his "Christmas present."

She was a tall woman with a haunting face—Freud remarked on her "beautiful, wise eyes," which were a clear blue. Her hair was silverblond; though in her youth it had been tinged red. Her nose was small, her mouth soft, a sensuous lower lip. She had a quiet way of speaking and the gift of inspiring confidence in men.

She was known for her feminist stand, for seeking both sexual and intellectual freedom, and as a strong-willed, daring woman. Before meeting Freud, she had written numerous essays and books, studied religion and philosophy, and lived with a number of men, though she remained married to one man all her life.

Lou Andreas-Salomé was born on February 12, 1861, in St. Petersburg, Russia, and named Louise at birth after her mother, but soon called Lou. Her father's ancestors were Huguenots from the south of France, who settled in the Baltic provinces and intermarried German families.

She was the sixth child, the only girl, and the darling of her father, a fifty-seven-year-old Russian general, Gustav von Salomé. Her mother, nineteen years younger, was the daughter of a wealthy sugar manufacturer of North German and 'Danish descent, and was, like Lou, slender, blond and blue-eyed. Lou was born and grew up in a mansion with an army of servants, her father's official residence in the huge, crescent-shaped building of the General Staff, opposite the Winter Palace in the heart of official Russia.

The general died when Lou was nineteen and shortly after she became very ill, complaining of fatigue and coughing up blood. Her mother was advised to take her to a warmer climate, and they left for Rome in 1882, with Lou determined to study philosophy.

In Rome she met a young philosopher and adventurer, Paul Rée, who fell in love with her and wanted to marry her, but she told him she was not ready for marriage. He introduced her to his friend, Friedrich Nietzsche, one afternoon at St. Peter's Cathedral, and Nietzsche, then thirty-six, also fell in love with her. She played a "brief but poignant part" in his life according to H. F. Peters, biographer both of Nietzsche and Lou Andreas-Salomé. Captivated by her beauty and mind, Nietzsche asked her to marry him. For a number of reasons, perhaps sensing Nietzsche's emotional instability and the overt pathological jealousy of his sister Elizabeth, who called Lou a "low, sensuous, cruel, and dirty creature," Lou left Rome and Nietzsche to study in Paris.

When she was twenty-six, she married another Friedrich—Friedrich Carl .Andreas, a short, stocky man with an outwardly gentle, scholastic manner. He wore a thick black beard and had bright, dark-brown eyes. He was born in Batavia, on the island of Java, in the Dutch East Indies, where his maternal grandfather, a North German

Lou Andreas-Salomé, whose photograph hung in Freud's study *Austrian Press and Information Service.*

physician, had married a beautiful Malayan girl. Andreas's mother had married the scion of an ancient Persian family of royal blood, the Bagratuni. Andreas spent the first six years of his life in Java, then the family returned to Germany, settling in Hamburg. He was a professor in Persian and Turkish languages at the Institute for Oriental Languages in Berlin.

He sought out Lou when he was forty-one, hearing of this beautiful, brilliant young woman whose unorthodox way of life had appealed to so many men. Like Nietzsche before him, and dozens of men after, Andreas was determined to win her. The more she resisted, the more ardently he wooed her. In a desperate move, he threatened to kill himself if she would not marry him. He plunged a pocket knife into his breast, and Lou ran for a doctor. Though Andreas had inflicted a deep, triangular wound, his life was saved. Lou consented to marry him, but remained his wife in name only for the forty-three years of their marriage, resisting all his attempts to consummate the marriage, according to Peters. The reason why she could not have sex with her husband, Peters says, was partly the shock she suffered from his violent attempts to win her, and partly because she saw in him not so much a husband as a father. "All her life she remained her father's child," Peters adds. In her diaries, Lou often referred to her husband, fifteen years older, as the "little old man." He was reported to have had a lifelong affair with the housekeeper, Marie, by whom he had two illegitimate children. Marie took over Lou's duties when she traveled.

In 1903, Andreas was offered the chair of West Asiatic Languages at the University of Göttingen, where they bought a home and named it "Loufried," a combination of their names. It stood on the outskirts of town, perched on a steep slope of the Hainberg with a magnificent view of the valley and the town below, a ridge of hills on the horizon. Lou moved into the upper story where she had her bedroom and study, while Andreas and Marie lived downstairs. Their surviving daughter was to inherit the house after Lou died.

Andreas continued to teach almost until his death in 1930 at the age of eighty-four. Lou lived a life of her own, traveling as she willed, always returning home. According to entries in her diaries, it took a long time before she could have sexual intimacy with a man, but once she started, it seemed there was little to hold her back. She had withstood the advances of Nietzsche, Rée, and her own husband. among others, and the first man to win her sexually appears to have been her physician, Dr. Friedrich Pineles, a bachelor, seven years younger than she, whom she met in Vienna in the spring of 1895, two years before she met Rilke, with whom she had a two-year affair. Pineles fell in love at once, as did Lou, and in the eyes of his sister and mother, Lou was practically his wife for almost twelve years. At one point Lou became pregnant, wanted the baby, but had an abortion because she feared Andreas would never give her a divorce.

She met Rainer Maria Rilke in Munich in 1897, and he pursued her passionately, pouring out his heart to her in poetry. He won her love and, in spite of Andreas' protests, she spent the winter of 1897 in Berlin with Rilke, who at age twenty-two could almost have been her son, for she was thirty-six. When she broke off with Rilke, she returned to Pineles. Rilke and she remained friends until his death in 1926 in a Swiss sanitorium when, half unconscious with pain, he implored his doctors to "ask Lou what is wrong with me. She is the only one who knows."

This was the dramatic life of Lou Andreas-Salomé before she met Freud. When she went to Vienna to study with him, she was fifty and quite plump. Gone the fragility of face and figure, but she still had a stately beauty, charm, and warmth, combined with an unusual intellect and intuition. She had become "ceaselessly concerned with psychoanalysis," with ever-growing admiration for Freud's "ruthless consistency." This was an item in her *Freud Journal*, a red, loose-leaf notebook that had escaped the eyes of the Nazis when, after her death in 1937, they purged her library, according to Stanley A. Leavy, who translated and wrote the introduction for *The Freud Journal of Lou Andreas-Salomé*, her notes on studying with Freud in Vienna.

She attended Freud's winter lectures on Saturday evening, "Some Chapters from the Theory of Psychoanalysis," at the Vienna Psychiatric Clinic. Once when she was absent from a lecture he intended to present to her (for each lecture, it was his practice to select one person in the audience to address), he told her he missed her. She was the first woman to attend—the first woman invited to attend— the weekly Wednesday night sessions held at his home. Two years before, in April 1910, a vote had been taken as to whether to include women in these psychoanalytic discussions, and three members of the eleven present had opposed the acceptance of women. Freud had criticized the three at the time, believing women had an equal right to learn if they wished to become psychoanalysts. (Freud began and ended his life surrounded by women, and this is probably one of the main reasons why women have had an easier time becoming psychoanalysts than in entering any other profession.)

Lou became friend and lover of Dr. Victor Tausk, one of Freud's disciples, a man, however, whom Freud did not trust because he thought him too emotionally disturbed. Tausk wrote one famous paper, "On the Origin of the 'Influencing Machine' in Schizophrenia," in which he asserted that both in reality and illusion, machines are man's unconscious projection of his body image into the outer world. Tausk committed suicide in 1919 at the age of forty-two, just before he was to remarry (he had two sons by a former marriage). His brief affair with Lou had nothing to do with the suicide.

When her studies with Freud ended, and she planned to return home, Lou was invited by him on April 2, 1913, to a "tea party" at 19 Berggasse. When she arrived, he handed her a dozen red roses. In the *Freud Journal* she tells of the "little farewell party," where they talked over tea of the difference between psychosis and neurosis. Freud commented, "how seldom it is possible to remove perversions." After tea, in his study they discussed whether an analyst should primarily do therapy or research and Freud supported research. She concluded, "As I set home with his roses, I rejoiced that I had met with him on my journey and was permitted to experience him—as the turning point of my life."

On her return to Göttingen she started a career as psychoanalyst and from 1912 to 1937, when she died on February 5; for twenty-five years she and Freud exchanged letters. In their correspondence Freud began by addressing her as "Dear Frau Andreas." Gradually, his salutation changes to "Dear Lou" or "My dear Lou," and, in the last years, "Dearest Lou" or "My Dearest Lou." Once, writing from Rome, he calls her "Dearest Lady."

Through the letters, Freud found in her, according to Jones, "A perceptive and understanding sounding board for his ideas," while she sought in him advice on problems with patients. Though the main theme of the letters is the theory and technique of psychoanalysis, there emerges in the letters a strong emotional attachment on the part of both. In *Sigmund Freud and Lou Andreas-Salomé Letters*, edited by Pfeiffer, Lou describes her gratitude for Freud's friendship and his importance to her, referring to him as "the father-face which has presided over my life."

She wrote Freud her ideas on "narcissism," in which they were both interested—the primary state of the "undifferentiated self" of the baby before the struggle between instincts and ego starts. She treated several regressed patients, sensed in them, she said, the deep conflict between the wish to return to that early state and the wish to conform to society's demands.

She also asserted on November 7, 1915: "When I remember how even in this immeasurably terrible year so much good has come my way through your psycho-analysis, I feel I should like to be of more service to it then I am. Whenever your insights fall like flashes of happiness upon my mind, because they do not merely advance my theoretic knowledge, but help me in the widest human sense, I am filled with profoundest gratitude. And then I feel keenly that coming generations will be distinguished by the degree and the extent to which they enter into this thought-experience; this will distinguish them from earlier generations more than even the most revolutionary events could do."

He wrote two days later: "You know how to cheer and encourage. I would not have believed, least of all in my present isolation, that psycho-analysis could mean so much to someone else or that anyone would be able to read so much in my words. Naturally, I believe it of you. And at the same time you have a subtle way of indicating where gaps become visible and where further argument is needed."

Discussing her article "Anal and Sexual," which she asked him to submit for publication in *Imago*, one of his journals, he said he had sent it to the editors, adding, "In my opinion it is the best thing you have given me to date. Both your incredible subtlety of understanding and your impressive gift of making a synthesis out of the material you have sifted in your investigation find admirable expression in this work." When published within a few months, Freud wrote that it was a paper "which has given us a very much deeper understanding of the significance of anal erotism."

Leavy says of her, "She was not a major theoretician, not a systematic thinker in psychoanalysis. . . . She recognized in psychoanalysis a unique contribution to human experience—a path to the world which has the stuff of poetry in it—and an infinity of surprise." This unusual woman, whom Freud called "the poet of psychoanalysis," also was an outstanding proselytizer of psychoanalysis.

The first postwar meeting of Freud and Lou occurred at the International Psychoanalytical Congress at The Hague in 1920, where she introduced him to Rilke and he introduced her to his daughter Anna, who had long wanted to meet her, according to Freud.

In Lou's *Journal*, she writes of her visit to the Freuds in the fall of 1921. She reached Vienna on November 9, and was met at the station by Anna, then both rushed to the regular Wednesday night meeting of the Vienna Psychoanalytic Society. Anna, Freud, and Lou "returned home late: I took up my quarters in my magnificent room with its verandah and its rosewood bed." The only free moments Freud had for her were late at night, when they took a short walk, or in the mornings when she would go into Anna's room, and Anna "would wrap me in a wonderful rug and squat down herself by the stove," and Freud "would come in after every analytic hour for a few minutes and talk with us and share in our work."

She said she "always found him in a serene mood, never grumpy," that "fundamentally he remains, no doubt, a man of pessimistic tendency, as I have always known him, but it is not merely self-control that strikes one in him, but his very positive attitude towards life, his cheerfulness and his kindness."

After her visit, Freud wrote his son Ernst and daughter-in-law, Lucie, on December 20, 1921: "Today Frau Lou has left . . . She was a charming guest and she is altogether an outstanding woman. Anna worked with her analytically; they visited many interesting people together, and Anna very much enjoyed her company. Mama looked after her most charmingly; with my nine hours of analysis every day I did not have much time to be with her, but she was discreet and modest throughout."

In one letter to Lou, Freud thanks her for what she has given Anna, without saying what it is. This may have related to one of the established legends among Freud's pupils that Lou had been, for a time, Anna's analyst. There has been a great deal of uncertainty about this issue. But one psychoanalyst, Beata Rank, was sure that Lou analyzed Anna for a while. Since Freud, in his letter to Dr. Eduardo Weiss, had gone on record about his analysis of his daughter, Lou's would have been a secondary, perhaps briefer one.

Both Anna and Martha stayed at Lou's home on August 3 and 4, 1922. In the lecture given May 31, 1922, which earned Anna membership in the Vienna Psychoanalytic Society, she paid tribute to Lou, announcing the lecture "arose from a series of conversations with Frau Lou Andreas-Salomé, to which I am very much indebted for her interest and help." Lou and Anna spent hours discussing emotional disturbances in children. When Lou wrote a novel *Rodinka,* a reminiscence of her life in Russia, she dedicated it, "To Anna Freud, to tell her of that which I have loved most deeply."

After one of her visits to *Loufried,* Anna told her father that Lou was impoverished. She had lived on income from her family's estate in Russia until the 1917 Revolution. But now she and her husband were physically ill. She saw patients, but they could not pay much because of the high rate of inflation in Germany.

Rescuer Freud immediately came to her aid. He insisted on sending her a monthly check, writing, on October 20, 1921, "My practice has made me relatively rich in good foreign currency (American, English, Swiss). And I should like to get some pleasure out of my new wealth."

She replied that she was "positively in floods of tears" at his kindness. His financial help continued until she died in 1937. One wonders whether Martha knew Freud gave Lou money, but no doubt he considered this his decision.

Lou and Anna had become such close friends that at times in her letters, Lou, who had no children, referred to Anna as "Daughter-Anna," as though in fantasy Anna were her daughter, with Freud the fantasy father.

When Lou became a member of the Vienna Psychoanalytic Society in 1922, she wrote Freud on June 26: "Anna's night letter has just reached me with the news that I have really and truly become a full member of the Vienna Psychoanalytical Society; as in a dream, so to speak, and as otherwise only happens in childhood, when one suddenly finds lying on one's bed the present one has wished for in one's dream. For what Daughter-Anna succeeded in reality in doing, i.e., in giving the lecture required for membership [one established member had to propose a new member, giving the reasons why the person was qualified] I should never have achieved successfully. And so I feel myself grateful and quite unashamed, to have been transported by her to the place where I have from the beginning felt more at home than anywhere else."

He answered on July 3, apologizing for the delay, due to "overwork," and said, "The Vienna Society was sensible enough to congratulate itself on your admission. Shortly beforehand your shadow was to be seen flitting across the stage in the shape of Anna's really excellent lecture. I cannot find words to express how pleased I am that you take so affectionate an interest in her."

Freud continued his deep concern for Lou's personal welfare. On August 4, 1923, he wrote he had been "horrified" to learn she was analyzing patients ten hours a day, saying, "Naturally, I regard it as a badly concealed attempt at suicide." He asked her to "put a stop to this by raising your patients' fees by a half or a quarter to correspond with the cascading collapse of the mark." He added, "The fairy godmother who stood at your cradle apparently omitted to bestow on you the gift of reckoning. Please do not wave aside my warning." It did not matter that he himself was seeing ten or more patients a day: she was not well and he was concerned about her health.

The following year, on November 25, 1924, she wrote: "The slow but positive progress I made with my patients is the thing that made the winter really rewarding for me. In this connection it is not so much the natural joy in the curative process itself as the refreshing, gratifying, rejuvenating effect of the deep and continuous preoccupation with this work. For me, that goes on increasing, with every case so to speak. It always seems to me that giving and taking are nowhere so identical as in psycho-analysis. So when I am feeling well and in good shape I do not find even many hours of analysis too exhausting a task."

Her good cheer was offset by his pessimism. On July 28, 1929, he wrote, "My worst attributes, among them a certain indifference toward the world, have assuredly had the same share in the end result as my good ones, e. g., a defiant courage about truth. But in the depths of my being I remain convinced that my dear fellow-creatures are—with individual exceptions—good for nothing."

He wrote her on her seventieth birthday he was "glad of the chance of saying to you on your birthday how greatly I esteem and love you." He signed the letter, "Hearty Greetings from what is left of your Freud." She wrote an essay, "My Thanks to Freud," the following month for his seventy-fifth birthday on May 6.

Freud reported to her on March 13, 1933, that Anna had "set off for Berlin and Hamburg" the day before, adding, "I have long felt sorry for her for still being at home with us old folks . . . but on the other hand, if she really were to go away, I should feel myself as deprived as I do now, and as I should do if I had to give up smoking! As long as we are all together, one doesn't realize it clearly, or at least we do not."

He added pessimistically, "And therefore in view of all these insoluble conflicts [growing old, being in constant pain] it is good that life comes to an end some time or other." We might speculate he was feeling guilty at not having enabled Anna to become free enough of her emotional dependence on him so she, like his other two daughters, could have married and had children of her own.

When Lou died on February 5, 1937, at her home in Göttingen, Freud wrote the obituary for the journal he had founded, *Internationale Zeitschrift für Psychoanalyse*. He stated: "I am not saying too much if I acknowledge that we all felt it as an honour when she joined the ranks of our collaborators and comrades in arms, and at the same time as a fresh guarantee of the truth of the theories of analysis.

"Her modesty and discretion were more than ordinary. She never spoke of her own poetical and literary works. She clearly knew where the true values in life are to be looked for. Those who were closer to her had the strongest impression of the genuineness and harmony of her nature, and could discover with astonishment that all feminine frailties, and perhaps most human frailties, were foreign to her or had been conquered by her in the course of her life.

"My daughter, who was her close friend, once heard her regret that she had not known psycho-analysis in her youth. But, after all, in those days there was no such thing."

There seems little doubt that Freud and Lou Andreas-Salomé had a strong attachment to each other, similar to the ones Freud had with many women during the course of his life. Lou fit into the category of beautiful, brilliant, and narcissistic women who seem to have fascinated him.

While there is no evidence they ever were involved in a sexual relationship, they seemed to have held deep affection for each other. Almost like a lover, Freud presented Lou with the bouquet of red roses when they parted temporarily. As time went by, he addressed her in letters as "My Dearest Lou." In her mind, he was the "sacred" father who provided "flashes of happiness," the psychic equivalent of physical orgasms. She remained indebted to him for helping her find a profession which gave her great satisfaction, as well as a newfound identity. She became the foster mother of Freud's daughter Anna, and therefore, unconsciously, his foster wife. Freud once told a colleague in describing her: "There are some people who have an intrinsic superiority. They have an inborn nobility. She [Lou] is just one of those people."

Freud involved her in the battles with his male colleagues. He referred to an exchange of letters between Lou and Alfred Adler "that shows her insight and clarity in an excellent light and does the same for Adler's venom and meanness." Always protective of her loved father-figure, Lou commented, "Freud and Adler differ with regard to the therapeutic method as the knife differs from the salve."

In his wish to be an Oedipal conqueror, Freud seems to have competed unconsciously with Lou's lovers. When Tausk died, Freud wrote the official obituary, praising Tausk's many contributions to psychoanalysis. But in a letter to Lou, Freud wrote, "I confess I do not really miss him; I had long taken him to be useless, indeed a threat to the future." Here Freud expressed sentiments almost identical to the ones he voiced when his sister Anna had temporarily decided not to marry Eli Bernays.

In the late 1920s when Freud suffered severe pain from cancer, he and Lou walked in the parks stretching along the Tegelsee and he picked her a bunch of red, blue, and purple pansies. They reminisced for hours about their experiences since they first met. "And then," Lou wrote, "something happened that I did not understand myself, but no power in the world could have stopped me; my trembling lips in revolt against his fate and martyrdom, burst out: 'You have done what I, in my youthful enthusiasm, only raved about.' " Startled by her own words, Lou burst into tears. Freud did not answer. "I only felt his arms around me," she concluded.

Freud gave the last part of the life of this sensitive woman a new, purposeful direction as she turned to helping others with the skills she had acquired in Vienna. He was, as she put it, her "turning point."

The Dedicated Doctor

Unlike Marie Bonaparte and Lou Andreas-Salomé, who did not have medical degrees, Dr. Helene Deutsch was a physician when she learned of the world of psychoanalysis. She was born in 1884 in Przemys, a town in Galicia, a section of Poland that became part of the Austro-Hungarian Empire in 1772. Daughter of a lawyer, Wilhelm Rosenbach, she left home for Vienna and medical school.

Wanting at first to become a lawyer like her father, Hala, the name she was called by her close friends and family, the Polish diminutive of Helene, considered herself a leader in female emancipation. In her autobiography *Confrontations with Myself*, Dr. Deutsch says that as a girl she had to struggle and sacrifice to achieve what is now taken for granted, adding, "The embattled gates to equal rights have indeed opened up for modern women."

She also says, "But I sometimes think to myself: 'That is not what I meant by freedom—it is only social progress.'" To find the "freedom" she meant, she eventually sought Freud's couch. But first she trained as a physician, after getting the special permission required of all women at that time, at the School of Medicine at the University of Vienna. She had admired Dr. Käthe Kollwitz, later a famous artist, and wanted to follow in her footsteps as a pediatrician.

It was unusual in the 1910s for a woman to be a physician, much less a psychiatrist, which Dr. Deutsch determined to become after an unhappy love affair in Vienna. She transferred to the University of Munich to get psychiatric training, studying under Dr. Emil Kraepelin. She fell in love with Dr. Felix Deutsch, an internist, later a psychoanalyst, and they were married in 1912, the year she received her medical degree. At that time, as Roazen has pointed out, women had less to lose professionally than their male colleagues by going into psychoanalysis. The career of a woman training to be a psychiatrist was unlikely to get very far, whereas in a new field, like the practice of psychoanalysis, a woman faced none of the barriers of established medicine.

The Deutsches decided to move to the psychoanalytic atmosphere of Vienna during World War I. She found a job as full-time assistant at the Psychiatric Clinic of the University of Vienna. The clinic was headed by Dr. Wagner von Jauregg, a leading medical figure who believed the body the only concern of medical science, and who headed the opposition to Freud in Vienna. However, Freud was allowed to hold his Saturday evening course in the clinic's lecture hall.

Just before she started work at the clinic, Dr. Deutsch gave birth to a son, and though she had not yet officially moved into Freud's psychoanalytic circle, she named her son Martin after Freud's eldest son. Her husband belonged to a Zionist organization where Martin was a member and he became a friend.

Immediately after her son's birth, Dr. Deutsch hired a nurse, Paula, and installed her and infant Martin in a vacant room next to her office so she could breast-feed her baby. She recalls she spent "a large part of the day in the excruciating business of providing the amount of mother's milk needed."

The baby thrived, but she felt physically and psychically drained. Her husband found a solution. In spite of war shortages, he managed to acquire two goats from a farmer in a nearby rural area, trading a piano for them (he owned two) and personally escorting the valuable goats to Vienna, part of the way on foot. The goats "grazed and fought" in the courtyard of the Wagner von Jauregg Clinic during the day, providing little Martin with milk.

The goats became connected in a personal way to her experience with Freud. She had read his works and "as I absorbed Freud's teachings on the unconscious mind and began to believe in infantile sexuality and to understand fully the role of these forces in the formation of neurosis, I gradually became a devoted disciple," she says. Her position at the clinic gave her access to his weekly lectures there, and she attended them "as a kind of hostess in a white lab coat, though I did not take any part in the proceedings." She realized, she says, "At this point I knew I was witnessing 'greatness.' I was not yet able to define it as Freud's genius."

Her relationship with Freud began in August 1918, when she learned from a woman colleague that Freud had accepted this colleague for analysis. Dr. Deutsch decided to make an appointment to see Freud. She was admitted to 19 Berggasse by Paula Fichtl, the faithful maid (both Helene Deutsch and Freud thus had a "Paula"). In talking to Paula while waiting to see Freud, Helene discovered that

Martha Freud had pneumonia and was suffering because of the severe milk shortage. Says Helene, "Here the two blessed goats entered the march of events; from then on, their milk was shared between my son and the Professor's wife."

By this time the war had ended, and one of Freud's patients, an Englishman, asked Freud to recommend an internist for officers of the English occupation forces. Freud gave him the name of Helene's husband, for whose medical competence he had the highest respect. This new position meant the Deutsch family received the canned rations distributed to English officers and their employees, so Freud quickly repaid the debt of the goats' milk.

During her first interview Freud asked, "What would you do if I sent you to someone else?" She replied, "I would not go."

He accepted her for a training analysis but warned it would bring her into conflict with the Wagner von Jauregg Clinic because of the director's notorious opposition to psychoanalysis. Whereupon she gave up her job at the clinic. From then on, she says, "my personal destiny was closely bound up with Freud and the psychoanalytic movement . . . Freud became the center of my intellectual sphere."

She calls psychoanalysis "my last and most deeply experienced revolution" [she had previously supported the Socialist party], and adds that Freud, "who was rightly considered a conservative on social and political issues became for me the greatest revolutionary of the century."

There were three "distinct upheavals" in her life, she says, "liberation from the tyranny of my mother, the revelation of socialism . . . and my release from the chains of the unconscious through psychoanalysis." In each of these revolutions she was "inspired and aided by a man": first her father, then Hermann Liebermann, active in the Socialist movement, and lastly, Freud. Her husband, she said, "had his own unique place in my heart and my existence."

She started the analysis "by a remarkable coincidence" at the very time her father, who had been living in Vienna, returned to Poland at the end of the war. This of course was no coincidence; she was seeking replacement for her lost father.

She describes her analysis with Freud as "not dramatic." It began with "a certain personal relationship to Freud," the donation of the goats' milk, an act that would not have been allowed in later years after "transference" was clearly defined and all personal relationships between analyst and patient were forbidden. She admitted this experience "not only acted as an *agent provocateur* in the first stage of analysis but also influenced many situations during and after analysis."

Only isolated experiences from the analysis survived in her memory. She recalled a dream "in which I have a masculine and a feminine organ." Freud interpreted this as meaning her desire to be both boy and girl. It was only after her analysis that it became clear to her, she says, how much her whole personality had been determined by the childish wish "to be simultaneously my father's prettiest daughter and cleverest son."

The Oedipus complex and the feminine castration complex "typically dominated" her analysis. One day she left her session in a state of emotional turmoil and remained standing for a long time in front of the window of a man's clothing store on the Berggasse. "To this day I can remember each item . . . especially the green shirts," she recalls. "I was weeping bitterly as I thought, 'What will the Professor's poor wife do now?' for I was convinced that my analyst was in love with me and was about to leave his wife, and which I didn't want at all. It was a primitive Oedipal fantasy of the kind that usually develops after a certain period of analysis, when material connected with transference emerges into consciousness." She had known intellectually about the Oedipal fantasy but now felt its emotional impact.

Her analysis ended abruptly after a year, when Freud assured her, "You do not need any more; you are not neurotic." He also said, "with absolute candor," she recalled, that "he needed my hour for the 'Wolf-Man,' who, after an interruption, had returned to continue his analysis." The name "Wolf-Man" was bestowed on a patient haunted by a recurring nightmare of seven white wolves perched in a tree staring at him, a dream that became the central theme of Freud's famous case, "From the History of An Infantile Neurosis," published in 1918.

Helene knew the "Wolf-Man" was the source of important discoveries for psychoanalysis and reassured herself she was "mature enough" to react to the situation objectively. "Certainly it would have been irrational for me to expect Freud to give up for my sake the time he needed for his creative work," she says. Then she admits, "Nevertheless, perhaps from a feeling of rejection, I reacted by having the first depression of my life." This we may doubt, for adult depressions are always rooted in childhood depressions, Freud was to discover. She adds philosophically, "It was a good lesson for a future analyst. Until then it had been difficult for me to comprehend the depressions of patients who had no 'real causes' for such feelings."

But she gained solace from Freud's statement that she had no neurosis, and he provided her with added self-esteem by telling her, "You will now be my assistant." He was referring to the Vienna Psychoanalytic Society, where she sat at the same table with him at the weekly Wednesday night meetings. She took part in the sixth Congress of the International Psychoanalytical Association at The Hague and met many well-known analysts from other countries. She described this period as "a stimulating, joyous time for me."

The training of analysts was not as yet organized at psychoanalytic institutes so the control or supervisory analysis of a trainee's first cases was conducted under the supervision of an older analyst. Her control analyst was Freud. Usually the discussions were prefaced, she reports, with his modest remark, "You know more about the patient than I do, because you see him every day; I can't tell you very much—but then you don't need it anyway." When she asked for specific advice, he always gave it.

He also honored her by sending her, as her first analytic case, a member of his family (she does not say whom) and entrusted to her the complicated case of "the very disturbed Victor Tausk, as well as other patients of special interest."

Helene became active in organizing training procedures and new institutes after she became a member of the Vienna Psychoanalytic Society. In later years, she underwent a second analysis with Dr. Karl Abraham in Berlin, "and a warm friendship formed a bond between us until his death." She recently found in her home in Boston, Massachusetts, a letter in which Abraham wrote that he had just visited Freud, who spoke of her "with great warmth and affection."

She gives Freud credit for inspiring much of her creative work: "It often happened that a fresh interest of mine, seemingly independent of Freud in its origins, actually came to life at his instigation, however indirect the stimulus may have been." She mentions specifically her interest in mythology, especially the story of Dionysus, the son who saves his mother, as contrasted with Apollo, the son who killed his mother.

She was impressed by the fact that Freud always considered the personality of the potential psychoanalyst more important than formal qualifications. She comments, "His dismissal of the medical degree as a condition of acceptance into the training program stems from this emphasis on character."

Best known for her two-volume book, *The Psychology of Women*, where she explains in detail the sexual development of women from infancy through motherhood, Helene maintains that three women physicians, Ruth Mack-Brunswick, Jeanne Lampl-de Groot, and herself, were, in Freud's phrase, "pioneers in feminine psychology." She also wrote papers on the conflicts of women, explaining, "My intense interest in women stemmed from various sources: first, from my own narcissism, a wish to know myself; second, from the fact that research until then had been chiefly concerned with men. Later a third motive was added: Freud's interest in feminine psychology, which made me want to answer his questions in this field through my own investigations and so to reverse his dictum, 'Woman does not betray her secret.' "

Freud's relationship with Helene Deutsch, physician and psychiatrist, was quite different from those he had with the nonmedical women analysts, Lou Andreas-Salomé and Marie Bonaparte. Freud seemed quite ambivalent toward Helene, at times even rejecting of her.

In contrast to the warm receptivity he showed Lou and Marie, he had asked Helene if he could send her to another analyst even at the time of the first interview. In her formal analysis, Freud fell asleep during at least two of her sessions. When this occurs, it is often expression of a latent hostile countertransference. To add insult to injury, Freud did something almost unforgivable—he asked Helene to give up her personal analysis so he could see another patient.

It would appear that one of the reasons Freud resisted getting involved with Helene as a patient was because he was afraid of his own aggression toward and competition with her.

Frequently, when analysts do not want to become aware of their hostility toward patients, they do not offer their patients sufficient opportunities to express their hostility. In one analytic hour, Helene Deutsch told Freud she never produced unpleasant stories about him in her free associations, that is, she was afraid of her negative feelings toward him. Freud did not help her resolve her fear of expressing her hostility. Instead, he dismissed her fear and said, "That is because you are too decent." What an excellent way of blocking further expression of angry feelings! Perhaps by referring to Helene as "decent," Freud could feel more "decent" with her, burying his own hostility. When he later told her prematurely that she was free from neurosis, he may have been trying to give himself reassurance about his own neurosis.

Another factor that may account for Freud's animosity toward Helene Deutsch had to do with her husband. At the time Freud first contracted cancer in 1923, Felix Deutsch, then his personal physician, chose to conceal from Freud the nature of the malignancy. When Freud found out the truth, he was so angry that he dropped Dr. Deutsch as his physician, feeling he could not be trusted. Helene Deutsch is reported to have felt some anger toward Freud for this act, and anger too at her husband for causing distance to develop between Freud and herself.

It is clear that she handled her relationship with Freud by reacting masochistically. When Freud wanted to refer her to another therapist, she did not show irritation or feelings of rejection but insisted on staying with him, despite his lack of receptivity. When he fell asleep during analytic hours, she was quick to forgive him. When he could not help her resolve her negative transference, she meekly accepted his weak and incorrect interpretation that she was "too decent." Finally, when he dismissed her so he could see another patient—rejection of rejections—she obeyed him, repressed her hurt and anger and became depressed.

Her responses to Freud are similar to those of a young child who feels she needs her father so desperately that she must swallow her aggression, become depressed, and continue consciously to worship him—lest she be unloved or abandoned. This is precisely how Freud handled his relationship with his mother. Just as he had a tendency to displace his aggression toward his mother to other women, Dr. Deutsch seemed to displace some of her aggression toward Freud on her husband.

When analysts have had unsatisfactory personal analyses, they have a tendency to repeat their negative experiences with their patients; if they have been well analyzed, they usually are empathic, sensitive, and therapeutically responsive. There is the implication in Helene's writings that the woman should cling to and be dependent on her husband. Her statement that people should "give up the illusion of the equivalence [same experience] of the sexual act for the two sexes" has understandably irked feminists. To many, her work seems to be a championing of the social position of women of the past. She is seen as depreciating women when she says, "Many intellectual women are actually fugitives with impoverished emotions . . . as a rule such women are intellectualizing rather than intellectual."

Her prescriptions for married women sound much like the role she adopted with Freud when she writes: ". . . To the woman falls the larger share of the work of adjustment: she leaves the initiative to the man. . . . They [women] are always willing to renounce their own achievements without feeling they are sacrificing anything. . . ."

Helene's second analyst, Dr. Karl Abraham, seemed more understanding of her. When she told him she was depressed over her relationship with Freud, he helped her get in touch with her own aggression toward Freud. No doubt Abraham helped Helene understand more of her relationship with her mother. Many of Helene's outstanding contributions to the psychoanalytic literature center on the dynamics of the mother-daughter relationship. For example, she showed that frequently the conflicts of female homosexuals stem from the pre-Oedipal tie to the mother. Freud tended to regard female homosexuality as exclusively caused by a woman's unconscious identification with her father.

In contrast to Freud's relationships with men, whom he tended to idealize or dismiss, he accepted ambivalent relationships with women. He held on to such relationships in the same way he held on to his mother despite his unconscious rage toward her. After he ended Helene's personal analysis, they became friendly colleagues. Freud referred patients to her and read her papers with interest.

In 1934, Helene was invited to Boston, where she still lives, to do research in psychosomatic medicine. Living in America, she could see more clearly the threat of the Nazis and in 1935 persuaded her husband to join her. Before finally deciding to leave Vienna, she consulted with Freud. He did not want her to go to America. But he would not tell her he needed her—which she wanted him to. Instead, he said the psychoanalytic community in Vienna would suffer from her loss. As Helene reported to Roazen in an interview, it seemed Freud ordered her not to go to America but, because he did not say it the way she wanted to hear it (still looking for the master's unconditional love), she left his office hurt and more determined than ever to emigrate.

While the relationship for each was highly ambivalent, Helene and Freud both profited from it. He paved the way for her to become a leading psychoanalyst. She proved to be one of his most ardent exponents at a time when he most needed advocates.

The Poet Patient

One of Freud's women patients, an American, had no desire to become a psychoanalyst but wanted only to continue life as poet and author. Hilda Doolittle, who wrote under the pen name H. D., was author of a small volume about her analysis, called *Tribute to Freud*, published in 1944 in London. It was dedicated to "Sigmund Freud, blameless physician."

The original title was *Writing on the Wall*, which, as she explains in the preface, referred to "the picture writing, the hieroglyph of the dream," which is "the common property of the whole race; in the dream, man, as at the beginning of time, spoke a universal language, and man, meeting in the universal understanding of the unconscious or the subconscious, would forego barriers of time and space; and man, understanding man, would save mankind."

When the book appeared, Jones reviewed it in *The International Journal of Psychoanalysis*. He described it as "surely the most delightful and precious appreciation of Freud's personality that is ever likely to be written. Only a fine creative artist could have written it. It is like a lovely flower, and the crude pen of a scientist hesitates to profane it by attempting to describe it . . . it will live as the most enchanting ornament of all the Freudian biographical literature."

In the book H. D. touches on her childhood, the breakup of her marriage, the birth of her daughter, the death of her brother in service in France, and the consequent death of her father from shock. She also mentions her literary circle in London which included D. H. Lawrence and Ezra Pound.

She speaks of starting to read Freud's work in 1932, when "there was talk of my possibly going to Freud himself in Vienna." The one to whom she talked was Hanns Sachs, whom she had seen professionally in Berlin. A year earlier, she had begun an analysis with Mary Chadwick in London, with whom she had twenty-four sessions when the collapse of a friend threatened her emotional health. Still earlier, she recalls informal conversations with Havelock Ellis in Brixton at the end of World War I, remembering him chiefly, she says, in terms of Norman Douglas' *bon mot,* "He is a man with one eye in the country of the blind."

In his work with her Freud showed his deep understanding of her feeling of rejection by her mother who, she felt, far preferred her brother. H. D. relates that once when she "painfully unravelled" the story of two unhappy friendships, Freud said, "But why did you worry about all this? Why did you think you had to tell me? *Those two didn't count.* But you felt you wanted to tell your mother." H. D. thought of the many times she had wanted to speak to her mother of her troubled feelings but did not, fearing to cause her mother worry and pain.

She recalled of one dream, "My mother, my mother, I cry. I sob violently, tears, tears, tears." Her mother, Helen Wolle Doolittle, was an artist and musician. She had married an older man, a widower, and H. D. was one of their children. He was a former New Englander who "mapped the stars at night and napped until noon," professor of mathematics and astronomy at Lehigh University and, from 1895 to 1912, Professor of Astronomy at the University of Pennsylvania. He also wrote books on astronomy.

She learned from Freud, she said, that "the child in me has gone. The child has vanished and yet it is not dead." During one session she mentioned vacationing in Corfu when she saw during a sunset, as if projected on the wall of her hotel bedroom, "picture-writing," a series of shapes Freud interpreted as representing a desire for union with her dead mother.

H. D. felt she was her father's favorite child, even though eventually she was a disappointment to him, while her older brother was her mother's favorite. She thought of herself as "an odd duckling" in her mother's eyes. In her poem, "Tribute to the Angels" she writes:

> what is this mother-father
> to tear at our entrails,
> what is this unsatisfied duality
> which you can not satisfy?

In her second year at Bryn Mawr she became engaged to Ezra Pound. When her parents objected to him as a son-in-law, she left college and moved to New York and from there to London. There she married the writer Richard Aldington, whom she divorced after a stormy marriage, telling the story in her book *Bid Me to Live,* while Aldington gave his version in *Death of a Hero.*

Both Pound and Lawrence disapproved of her going into psychoanalysis, but she went to Freud anyhow "for my own immediate benefit and also to fortify me for the future." She later wrote, ". . . there is no evil in Sigmund Freud . . . I am salvaged, saved; ship-wrecked like the Mariner, I have sensed bell-notes from the hermit's chapel."

She started a series of sessions with Freud in March 1933, which lasted between three and four months. He was seventy-seven, she, forty-seven. She returned for a second series the end of October 1934, and saw him for five weeks, five days a week.

Recalling the first time she saw Freud, she told how she walked down the Berggasse, turned in at No. 19, Wien IX. "The stone staircase was curved," she wrote. "There were two doors on the landing. The one to the right was the Professor's professional door; the one to the left, the Freud family door . . . there was the Professor who belonged to us, there was the Professor who belonged to the family."

She described the room where she lay on the couch: "The wall with the exit door is behind my head, and seated against that wall, tucked into the corner, in the three-sided niche made by the two walls and the back of the couch, is the Professor. He will sit there quietly, like an old owl in a tree. He will say nothing at all or he will lean forward and talk about something that is apparently unrelated to the progression or unfolding of our actual dream-content or thought association. He will shoot out an arm, sometimes somewhat alarmingly, to stress a point. Or he will, always making an 'occasion' of it, get up and say, 'Ah—now—we must celebrate *this*,' and proceed to the elaborate ritual—selecting, lighting—until finally he seats himself again, while from the niche rises the smoke of burnt incense, the smoldering of his mellow, fragrant cigar."

One day he invited her to look at the antiquities in his study, "the room beyond." He placed in her hands a small object, saying, "*This* is my favorite." It was a small bronze statue, helmeted and clothed in a carved robe "with upper incised chiton or peplum," Pallas Athene, Winged Victory. Freud commented, "She is perfect, only she has lost her spear."

H.D. adds, "He might have been talking Greek. The beautiful tone of his voice had a way of taking an English phrase or sentence out of its context . . . so that, although he was speaking English without a perceptible trace of accent, yet he was speaking a foreign language."

She describes him: "His beautiful mouth seemed always slightly smiling, though his eyes, set deep and slightly asymmetrical under the domed forehead (with those furrows cut by a master chisel), were unrevealing. His eyes did not speak to me. I cannot even say that they were sad eyes . . ."

Once he told her that if he lived another fifty years, he "would still be fascinated and curious about the vagaries and variations of the human mind." He called it "striking oil" when he made an important interpretation. When she said one day that time went too quickly, "he struck a semi-comic attitude, he threw his arm forward as if ironically addressing an invisible presence or an imaginary audience and said, 'Time gallops.' " She wondered if he knew he was quoting Rosalind's quip "Time gallops withal," from *As You Like It*.

Norman Holmes Pearson who wrote the foreword for *Tribute to Freud*, included an appendix containing nine letters Freud's heirs had selected from those he wrote her. One, dated May 24, 1936, in answer to a letter from her, said: "I had imagined I had become insensitive to praise and blame. Reading your kind lines and getting aware of how I enjoyed them I first thought I had been mistaken about my firmness. Yet on second thought I concluded I was not. What you gave me, was not praise, was affection and I need not be ashamed of my satisfaction. . . . Life at my age is not easy, but spring is beautiful and so is love." He signed it, "Yours affectionately, Freud."

He wrote in September 1936 to congratulate her on her fiftieth birthday, saying, "Belated, though sincere, congratulations . . . from an eighty-year-old friend."

In one of his first letters to her, dated December 18, 1932, he had mentioned he had read her book *Palimpsest*, and in one of his last letters, on February 26, 1937, he said he had just read her play *Ion* and was "deeply moved" by it "and no less by your comments . . . where you extol the victory of reason over passions."

She had learned from Freud. In her tribute she gives an eloquent picture of moments on the couch that hold the essence of psycho-analytic treatment as sharply as any writer yet has sketched. She had been silent for years before the analysis, she says, terrorized by inner torment, but after analysis she was able to write a war trilogy, several novels and short stories. Freud had helped free her to create again.

In 1961, near the end of her life as she lay in a hospital with a broken hip, H. D. wrote: "Of course, as the Professor said, 'there is always something more to find out.' I felt that he was speaking for himself (an informal moment as I was about to leave). It was almost as if something I had said was *new*, that he even felt that I was a *new* experience. He must have thought the same of everyone, but I felt his personal delight, I was *new*. Everyone else was *new*, every dream and dream association was *new*. After the years and years of patient, plodding research, it was all *new*."

The loved and beloved relationship that emerged between Hilda Doolittle and Freud was typical of the many intense, close relationships he had with women—his mother, daughters, and many female colleagues. Freud received libidinal gratification from H. D.'s admiration, love and creativity—so much so, he sometimes became oblivious to his professional task. On occasion he turned the analytic session into a party, like using moments of H.D.'s hour to "celebrate" an insight by lighting his cigar.

One of Freud's limitations as a therapist, which emerged particularly in his work with Helene Deutsch, was his reluctance to help patients feel the hatred which they transferred toward him from their parents. H. D. had experienced considerable disappointment and anger with the men in her life—a brother and father had died prematurely, and her marriage had been unsuccessful. If Freud had been more neutral and less seductive and stimulating with H. D., some of the hatred she had bottled up toward her father, brother, and husbands would have been directed at him. Instead, he concentrated on her relationship with her mother—extremely important to analyze but, in this case, made to serve as a defense for analyst and patient. By concentrating on her feelings toward her mother, H. D. could avoid experiencing hatred toward Freud in the transference, and he could avoid experiencing it directed at him.

It may be that H. D.'s unconscious disappointment with Freud came through in her descriptions of him. Finding it difficult to feel anger toward him or to blame him for an incomplete analysis, she called him a "*blameless* physician." Again, perhaps to protect herself from noting Freud's imperfections, it appears as if the lady doth protest too much when she says "there is no evil" in Freud.

Both Freud and H. D. shared a zest for life. Just as he maintained his courage and enthusiasm until the very end, so she did not give up her quest for understanding more and ever more about her conflicts. She had received a measure of love from Freud, and he in turn enjoyed her poetic approach to life, her imaginative spirit, and the satisfaction of seeing her grow emotionally stronger, although he was not successful in freeing her to express the hatred that should have been part of her analytic experience.

A Wearer of the Ring

Dr. Ruth Mack Brunswick has a place in Freud's life which few if any of his biographers have noted, certainly not Jones. Only in Roazen's *Freud and His Followers* is there substantial material concerning Freud's relationship with her. Yet many have regarded her as one of Freud's favorites in Vienna. She saw him often, went to dinner at his house, visited him during summer vacations, and was close to his children. In many ways she was like an adopted daughter, considered a rival by Anna.

Ruth Brunswick, who became a psychoanalyst, was twenty-five when she sought out Freud, after graduating from Radcliffe College and Tufts Medical School. She had a chaotic relationship with her father, Judge Julian Mack, a distinguished American jurist and well-known Jewish philanthropist. Virtually nothing is known of her relationship with her mother.

Ruth played an important role as mediator between American analysts and Freud's inner circle in Vienna. A confidante of Freud's, she smoothed over many of the inevitable disharmonies between the two groups. She was also the intermediary through whom Americans went to Freud as patients, and Freud referred to her many American patients. Unknown to most people she was one of the few recipients of Freud's cherished "ring," copies of which he gave loyal adherents whom he felt would continue to bring psychoanalytic findings to the world and keep his "first love" alive. Among others who received a ring were Lou Andreas-Salomé, Marie Bonaparte, his daughter Anna, and Dorothy Burlingham. The ring was of antique beauty, used as a seal in the days of old Rome, bearing the head of a man with a delicately carved beard. (Sachs, who received one, thinks it was Jupiter.)

Roazen describes Ruth Brunswick as a woman with "charm and intelligence," who was "demonstrative and explosive, outgoing, effusive and warm. She was also an elegant person with cultivated manners, as well as vivacious and possessed of a lively intellect. . . . She was literate and verbal, well read, and one of the few Americans not stigmatized as an American in Freud's eyes." Freud used her as a screen for his ideas, as he did many of the women in his life —Minna, Marie Bonaparte, Lou Andreas-Salomé, and Martha, when she was his fiancée. But Roazen describes Ruth Brunswick as the domineering type, not just compliant recipient of Freud's ideas.

For some time Freud was closer to her than to his own daughter Anna. He gave Ruth part of the manuscript of his book on Woodrow Wilson, while Anna saw it only in print. Freud's attention to the American woman psychoanalyst aroused the envy of many of his followers, both male and female. They particularly resented any part she played in supervising Freud's health; it was she who arranged in 1931 for a special prosthesis for his mouth.

When Ruth, born in 1897, arrived in Vienna in 1922, she was married to Dr. Herman Blumgart, who accompanied her. The marriage was rocky and, in addition to psychoanalytic training, she wanted help with personal problems. Blumgart, who had been a heart specialist in the United States, wanted her to return to America and tried to persuade Freud to end her analysis. Freud refused, the marriage broke up. Blumgart went home alone and pursued an outstanding career in his profession.

His wife apparently already had in mind another man as husband, Mark Brunswick, a musician, five years younger. He was in love with her, and Freud encouraged the relationship. He also took young Brunswick as a patient, something later analysts would never do because of the close relationship between Ruth and Mark. Freud clearly tried to be matchmaker, in control of the arrangement, something he was not able to be with his own children. Though Freud hated weddings, as he did all ceremonies, he did attend the Mack-Brunswick ceremony.

While he consistently wrote that an analyst should model himself after the surgeon who concentrates solely on professional tasks and does not relate any personal feelings to the patient—the analyst should be "an opaque mirror" to his patient—perhaps more than any other well-known Freudian analyst, Freud himself violated this precept. As noted in his relationships with Marie Bonaparte, Lou Andreas-Salomé, Helene Deutsch and H. D., in addition to being their analyst, he was at times companion, mentor, father-figure and rescuer. So too with the Brunswicks; Ruth and Mark saw much of Freud in his family surroundings, and there were many social gatherings.

Freud's strong Oedipal rivalry seriously handicapped his working relationship with the Brunswicks. He violated another cardinal rule of psychoanalysis by discussing Mark's analysis with Ruth. She grew closer to Freud, more alienated from her husband, and when Mark asked Freud whether it would be a good idea to divorce Ruth, his advice was, "Yes."

As was to be expected, Mark and Freud also became estranged, and Ruth and the Professor grew more intimate. By the time the Nazis took over Austria, Ruth had made her mark in analysis in Vienna, where she settled, largely because of Freud's active assistance. In addition to the coveted "ring," Freud gave her another gift—the referral of the Wolf-Man, one of his favorite patients, the one who took Helene Deutsch's place on the couch.

Freud gave unstintingly to Ruth—ideas, patients, warmth, and admiration. He was also much influenced by her. Though he always doubted whether psychoses could be treated by the psychoanalytic method, he took seriously her conviction that it could. She was also able to influence his thinking with regard to the importance of the mother in the development of the child. Freud was an Oedipal man, and his positive Oedipal countertransference to Ruth helped him accept her idea that pre-Oedipal life was crucial in the cause of psychic conflict. He wrote that she was "the first to describe a case of neurosis which went back to a fixation of the pre-Oedipal stage and had never reached the Oedipus situation at all."

In Vienna, she was almost continuously in analysis with Freud. Despite her scientific productivity and her able functioning as a psychoanalyst, she tended to convert her troubled feelings into gastrointestinal illnesses. As a physician, she was able to prescribe medicine for herself and habitually took sleeping pills and pain killers. By 1933, she had gradually slipped into a serious drug addiction.

Obviously an addiction would be a difficult neurotic symptom for Freud to treat, given his own addiction to nicotine, and he seemed to react to Ruth's plight with anger and intolerance. Just as he never tied his own smoking to oral problems but viewed it as identification with his father, he may have been insensitive to Ruth's pre-Oedipal problems which contributed to her addiction. It is of interest that she ascribed many psychological difficulties of her patients to their pre-Oedipal life.

Adversaries of psychoanalysis have been critical of the many years a patient is on the couch; some have referred to the treatment as a form of addiction. They do not understand the dynamics of a long analysis, or what takes place between analyst and patient. If pre-Oedipal dependency wishes are analyzed, not gratified, if the patient's infantile fantasies are examined, not indulged, then the ego will be strengthened and the dependency wishes of the patient will diminish. But if the analyst becomes a supportive parental figure, the patient will not develop the necessary frustration tolerance and other ego controls necessary for mature living. This was the case with Sigmund Freud and Ruth Mack Brunswick. The analysis was too mutually gratifying and, therefore, did not liberate the patient. Ruth enjoyed her dependency on Freud—it was not viewed by either analyst or analysand as a problem to be studied in the treatment.

Freud's disappointment with her lack of analytic progress began to show as he himself grew sicker. She then became more demanding of Freud and jealous of Anna Freud's role in caring for her father. By 1937, Freud had trouble controlling his irritation with Ruth, though outwardly he still treated her as one of his favorites.

She stayed in Vienna all through the war. Her death on January 25, 1946, came as a shock to everyone. The cause of death was given as "a heart attack induced by pneumonia." But she actually died of too many opiates, combined with a crippling fall in the bathroom.

Despite her importance to Freud and analysis, no obituary appeared in the *International Journal of Psychoanalysis*. Perhaps no one wanted to face her sad ending. Or perhaps there was too much bitterness toward her for her special relationship to Freud. Or perhaps her death raised questions about Freud's analysis of her. Dr. Herman Nunberg, a distinguished psychoanalyst, wrote in an American journal of psychoanalysis only that she had "a sudden tragic death."

Of Other Women

A number of women were helped by Freud to become psychoanalysts, including Joan Riviere, one of England's leading analysts. She met Freud at The Hague Congress in 1920, and two years later went to Vienna to study with him. After his death, she wrote of his "indomitable tenacity," of his "vivid, eager mind seizing on every detail with astonishing interest and attention . . . there was in him a conjunction of the hunter on an endless trail and the persistent immovable watcher who checks and revises . . ."

Once they talked of a young scientist interested in psychoanalysis, and Freud said, "mournfully," she recalled, "But I can't regard it as normal, you know, that he has married a woman old enough to be his mother!" She could not help laughing at the discoverer of the Oedipus complex making such a remark, she said, and reported, "he met the laugh with a twinkle . . ."

Freud was also involved with women in other areas besides psychoanalysis. When Maria Montessori, the noted Italian physician and educator, asked him to support her progressive school, he wrote on December 20, 1917: "It gave me great pleasure to receive a letter from you. Since I have been preoccupied for years with the study of the child's psyche, I am in deep sympathy with your humanitarian and understanding endeavors, and my daughter, who is an analytical pedagogue, considers herself one of your disciples.

"I would be very pleased to sign my name beside yours on the appeal for the foundation of a little institute as planned by Frau Schaxel [later Mrs. Willy Hoffer, wife of a London psychoanalyst]. The resistance my name may arouse among the public will have to be conquered by the brilliance that radiates from yours."

Though Freud consistently wanted to be father-figure, big brother, mentor, and at times was chauvinistic toward his women colleagues, he helped many become successful psychoanalysts, as attested by the accounts of Princess Marie Bonaparte, Lou Andreas-Salomé, Dr. Helene Deutsch, Dr. Ruth Mack Brunswick, and Joan Riviere, among others. His letters show kindness and affection for these women. In their letters to him, the women reveal their gratitude for increased self-knowledge and decreased mental pain. He also was close to Yvette Guilbert, whose concert singing he admired, and Eva Rosenfeld, Anna's friend, whom he analyzed.

Freud, as we know, enjoyed being son to his mother, big brother to his sisters, father to his daughters, and companion to his wife. He thrived on the gratifications he derived from relationships with women colleagues. Though helpful, and almost always admired and loved by them, his narcissistic involvements and unresolved neurotic problems frequently interfered with his psychoanalytic treatment of them.

There was much rivalry among the women in Freud's life. Mother, sisters, wife, sister-in-law, daughters, students, colleagues, patients and friends, competed with each other for the Professor's love. Some were able to acknowledge their jealous feelings and thus could handle them without too much interference in their work and love relationships. But others were not able to face their feelings, became tormented by them, and their work and love relationships suffered.

Part Three

The Women in Freud's Cases

The First Lady of Psychoanalysis

Freud's first patients—all women—led to his discovery of what he called "the great secret." This was the power of the inner world of fantasy as contrasted to the outer world of reality ruled by parents and society. Often, in the conflicts of the human psyche, the two worlds clashed. If the inner world won, there were apt to be mental and emotional disturbances.

Originally, listening to the women, Freud was astounded to hear patient after patient insist that as a little girl she had been raped by her father. Freud wrote Fliess he could not believe so many fathers could inflict such indignities on so many daughters.

One day as he heard a woman describe how her father had seduced her, the thought came to him she might be expressing her wish as a little girl rather than relating a fact. In her world of "psychic reality" it was as though the wish had been carried out and now was accepted as truth. This would explain much of the irrational behavior of people who acted on the basis of their fantasies, rather than what most people called logical thought.

This discovery opened up for Freud a new understanding of human behavior. He *listened* to his patients, as no other physician had before, instead of hypnotizing them, suggesting they feel better, or giving them electrical shocks, or urging a rest cure, or greater will power. He *encouraged* the women to speak freely of what disturbed them.

One of his most endearing characteristics helped lead him to the truth—his own wish, expressed to Martha, that she tell him all her thoughts, all her feelings. This same wish applied to patients, or Freud would never have become the world's first psychoanalyst. It was a wish that went back to childhood when he wanted to know all his mother's thoughts and feelings. And out of his empathy, sympathy, and ever-deeper exploration of the mind, plus the conviction that physical ills with no organic cause masked mental turmoil, came the art-science of psychoanalysis.

Freud's formulations on psychoanalytic treatment went through several phases. After he gave up hypnosis, his efforts concentrated on making the "unconscious conscious." Through the use of free association and analysis of dreams, he helped the patient become aware of what caused the psychic pain. Freud later prescribed that "where there is id, there shall be ego." He discovered that if the person was to mature emotionally, work productively, and be as rational as possible, sexual and aggressive drives had to be mastered and controlled by the ego. Still later, Freud made his most outstanding therapeutic discovery—the transference. He showed that in every therapeutic relationship the analyst is perceived not only as he is but also as the patient wishes him to be and fears he might be (in the image of the parents of childhood). Freud showed that in any therapy the patient will distort the person of the therapist, and it is through the discovery of how and why the patient makes these distortions that recovery is immeasurably aided.

Freud also was able to show that "resistances are the heart of every therapy," which means that each patient, no matter how deeply he suffers, wants to preserve the status quo and unconsciously fights recovery. Near the end of his career, Freud realized that much of analysis consists of assisting the patient to know what impulses he refuses to face, what infantile fantasies still seek gratification, what defenses are used, and how and why the patient still seeks to punish himself. Finally, shortly before his death, Freud recognized the importance of ego psychology, which focuses on judgment, reality testing, interpersonal relations and the various defenses the ego musters in its battle with the primitive id.

A limitation in the examination of the women in Freud's cases is that, with the possible exception of Dora, all his written presentations consist only of trying to "make the unconscious conscious." In his reports we do not find much evidence of the interpretation of transference, countertransference, resistance, and other crucial aspects of psychoanalytic treatment. But from the very start of his work, Freud tried to listen carefully to his patient's words and detect unconscious meaning.

One other man had also "listened" to his patients but, not possessing Freud's "unrelenting love of truth," was unable to grasp the significance of what they were saying. This was Dr. Josef Breuer, Freud's close friend, and the man probably most responsible for the interest Freud took in treating hysterical women. For years they discussed their cases, stimulated each other to new theories of the mind.

The first woman to inspire Freud to grasp the concept of the unconscious and the entire "mental apparatus," as he named it, was a patient of Breuer's. Freud wrote Lou Andreas-Salomé on July 30, 1915, "I never saw his [Breuer's] famous first case, and only learned of it years later from Breuer's report." Freud may have been referring to "seeing" her as a patient, for she was a friend of Martha's and occasionally visited their apartment in the late afternoon when he was treating patients. The Jewish community of Vienna, whose ancestors came from Germany, was small and closely knit; Martha and Bertha Pappenheim, the young woman's name, may even have been distantly related through Heinrich Heine.

Bertha Pappenheim was born in Vienna in 1859 to a wealthy Orthodox Jewish family. She had two older sisters who had died and a younger brother, Wilhelm. He had achieved a university education, denied women at that time. Bertha's education, typical of an upper class girl of that era, included instruction in languages, so she spoke fluent French, Italian, and English by the age of twenty, though she deeply resented the discrimination against women by universities.

During her teen-age years she was an attractive, seemingly "normal," and sociable young woman, very interested in helping the sick and the poor. But, according to Breuer, she was "subject to excessive daydreaming due to an extremely monotonous existence in her puritanically minded family." He found her "markedly intelligent with an astonishingly quick grasp of things and penetrating intuition." Her powerful intellect could "digest solid mental pabulum . . . she had great poetic and imaginative gifts . . . under the control of a sharp and critical common sense."

Breuer told Freud of the case five months after he ended her treatment. Freud, then interning in the Viennese General Hospital, had just become engaged to Martha. He was twenty-six, Breuer, forty, and the patient, twenty-three.

Freud wrote Martha he was visiting Breuer on the night of November 18, 1882, when Breuer first described how he had helped a young woman recover from "hysteria." (The name given the malady reflects the belief that hysteria was peculiar to women; the term comes from the Greek "hystera" which means "uterus" or "womb.") A gentle, compassionate man, Breuer was known to his colleagues as "the doctor with the golden touch" because of his success with difficult patients and his ability to diagnose complicated illnesses.

Breuer had first seen Bertha a few days before Christmas in 1880. He had been called to her home by an anxious mother who was afraid her daughter might be dying of tuberculosis, for an epidemic was then raging through Vienna. In the next bedroom lay the daughter's father, dying of a lung infection. Bertha had been his night nurse for several months before falling ill.

Bertha (later named "Anna O." by Breuer in his case report, to protect her identity) was stretched out mute and silent as a corpse on her bed, both legs and arms paralyzed. She was almost blind, coughed constantly, suffered severe migraine headaches, and had hallucinations during which she saw snakes and the heads of skeletons. Her mother thought the persistent cough was a sign of tuberculosis.

But Breuer, on examining the young woman, decided she had "hysteria." As was the custom in those days, Breuer hypnotized Bertha, then suggested she "feel better." While in the trance, she started to mumble words. At first he could make no sense of them. Then he listened carefully, realized she was telling stories about a sick girl nursing her father. She spoke of memories that "tormented."

Intrigued by what she was telling him under hypnosis, Breuer went to see her every day, sometimes twice a day, over a period of eighteen months, a very unusual thing for a doctor to do with an hysterical woman. He discovered that, as she related experiences she had endured, most of them at her father's bedside as she nursed him, and released feelings of rage, shame and humiliation which had been repressed at the time, her physical symptoms vanished one by one.

For instance, one day he hypnotized her and asked in reference to her poor vision, whether she could remember the first time that it had become impaired.

She recalled sitting in the chair one night next to her father's bed, wondering if he were going to die, when her eyes suddenly clouded with tears. At that moment her father asked what time it was. She tried to blink back the tears, not wanting him to think she was crying because he might die. She picked up his watch from the table by the bed, and tried to tell the time through her tears. She brought the watch close to her eyes and squinted at it.

As she told Breuer this story, tears came to her eyes, as though reliving the scene. When he returned the next day, he noticed her eyes were clear and she could once more see perfectly.

Another time she described an evening when she was sitting beside her father's bed at their country home, exhausted from lack of sleep. She tried to keep awake in case he needed anything but fell asleep for a few minutes. When she woke with a start, she felt very guilty. Suddenly she thought she saw a large black snake slithering across the wall behind her father's bed, ready to attack him. She tried to raise her right arm to drive the snake away, but her arm was paralyzed. She looked again at the wall and the snake had disappeared. Ever since, she said, at times she saw snakes crawling around her room, though her nurse called her crazy.

Breuer thought it probable she had wanted to drive off the illusionary snake (or it may have been a real one, for there were snakes in the fields surrounding the house) with an arm partially paralyzed by the sensation of "pins and needles," caused by the unnatural position of her arm as she slept. The paralysis of her right arm, spreading later to her left arm and legs, became associated with the snake. After talking to Breuer about the experience and reliving the horror and guilt she had suppressed that evening, she no longer saw snakes in her room.

She called it "the talking cure," as one by one the mental and physical symptoms vanished, and after eighteen months of treatment she was normal once again. Then one day Breuer told her she was well and he would no longer visit her. He did not explain to Freud why he broke off the treatment so abruptly. But Freud speculated on what had troubled Breuer, in a letter to Stefan Zweig, on June 2, 1932: "What really happened with Breuer's patient I was able to guess later on, long after the break in our relations, when I suddenly remembered something Breuer had once told me in another context before we had begun to collaborate and which he never repeated. On the evening of the day when all her [Bertha's] symptoms had been disposed of, he was summoned to the patient again, found her confused and writhing in abdominal cramps. Asked what was wrong with her, she replied: 'Now Dr. B.'s child is coming!'

"At this moment he held in his hand the key that would have opened the 'doors to the Mothers,' but he let it drop. With all his great intellectual gifts there was nothing Faustian in his nature. Seized by conventional horror he took flight and abandoned the patient to a colleague . . ." Freud said he later heard from Breuer's youngest daughter, who had read Freud's reconstruction of this event, that she had asked her father if it were true and "he confirmed my version." Jones also reported that Mrs. Breuer became so jealous of her husband talking about Bertha at home that she asked him to stop seeing the young patient.

Evidently attracted to Bertha and frightened by her fantasy of pregnancy, Breuer had fled, feeling guilty and unable to cope with Bertha's and perhaps his own sexual fantasies. He did not understand, as Freud later did, that Bertha as a patient had transferred to him the repressed sexual feelings of childhood for her father. Freud theorized that repressed childhood sexual fantasies were the chief cause of neuroses (he later amended this to include repressed aggressive fantasies).

Breuer helped Bertha enough so her severe physical and emotional symptoms disappeared and she could function the rest of her life in the world of work. His experience with her helped pioneer psychoanalysts recognize the crucial importance of their ability to be the recipient of the patient's infantile sexual fantasies without feeling seduced or becoming seductive. The effective analyst, as Freud was to say, must behave "as if" he were a real sexual object without acting sexually with the patient or blocking the patient's expression of sexual fantasies.

Little is known about Bertha Pappenheim's life for the seven years after her treatment with Breuer prematurely terminated. She had a relapse, spent some time in a sanatorium. She then moved with her mother to Frankfurt, became interested in volunteer social work, and raised funds to build her own institution for the care of delinquent, retarded, and pregnant, unwed girls and their babies. She obviously identified with the pregnant, unwed girl—as she had experienced herself with Breuer.

She never resolved her mixed feelings toward her father—intense sexuality, anger and guilt—a conflict not helped by Breuer's inability to understand and work through her transference to him. Her anger toward men remained within. She never married. Later she said, "If there will be justice in the world to come, women will be law-givers and men will have to have babies." She stood steadfast against both abortion and the use of psychoanalysis to help disturbed young women. When asked if a troubled girl at the institution should be sent to a psychoanalyst, she said, "Never! Not as long as I am alive!" Undoubtedly this was a reaction to her rejection by Breuer.

Despite Breuer's inability to handle either her deep feelings for him or the feelings she had stirred in him and which he was unable to face, Freud realized he had undertaken a pioneer doctor-patient relationship. Though Bertha was far from cured, Breuer had helped ease, at least outwardly, some of her serious mental and physical symptoms, allowing her to function and become a feminist leader. Breuer brought into being what she named "the talking cure," which revealed the subtle but powerful connection between fantasies in the mind, repressed emotions, and bodily ailments without physical cause. Freud urged Breuer to write the case for a medical journal, insisting it was important for other doctors to know of his startling discovery about the nature and treatment of hysteria. But Breuer refused. Freud said he then put the case out of his mind until three years later when he studied in Paris and told Charcot about it, but "the great man" did not seem impressed.

When Freud returned to Vienna on April 4, 1886, the memory of "the talking cure" once again haunted him and, in his words, he "made" Breuer tell him "more about it." It took him six more years to persuade Breuer—nine years after he had left Bertha Pappenheim so abruptly—to co-author an article about "the talking cure" and its implications for hysteria. By then Freud had treated a number of his own women patients. The two authors called their article "On the Psychical Mechanisms of Hysterical Phenomena, Preliminary Communication."

In preparation for the article Freud wrote three "memoranda," the first in the form of a letter to Breuer on June 29, 1892. In it he says, "I am tormented by the problem of how it will be possible to give a two-dimensional picture of anything that is so much of a solid as our theory of neurosis." It is interesting he uses the word "tormented," the very word Bertha used to describe how she felt, the word that led Breuer to question her further. In the memoranda Freud refers to a "second state of consciousness." This is believed the first time in his published scientific work he mentions the unconscious.

The "Preliminary Communication" appeared in 1893, in two issues, January 1 and January 15, of the periodical *Neurologisches Centralblatt*, the principal German neurological journal, which came out fortnightly in Berlin. The article was immediately reprinted in the Vienna Medical Journal, where it set off a hostility against Freud's theories that, for the next few decades, was open and intense.

The authors referred to Bertha as "one of our patients" and reported that in this "highly complicated case of hysteria" all the symptoms "which sprang from separate causes, were separately removed" when they succeeded "in bringing clearly to light the memory of the event by which it was provoked and in arousing its accompanying affect [emotion], and when the patient had described that event in the greatest possible detail and had put the affect into words." They warned that "recollection without affect almost invariably produces no result." In other words, there had to be expression of emotion. Words alone—intellectualization—were of no help.

Freud's classic line appears in this article: "Hysterics suffer mainly from reminiscences." He explained that the reminiscences in question related to situations where an impulse to act was repressed. The symptoms appeared "in place of" the action and represented a "strangulated" emotion. Both the traumatic incident and the emotion were lost to the patient's memory as though they had never happened. But the symptoms persisted.

The authors explained that an injury that has been repaid, even if only in words, is recollected quite differently from one that has had to be accepted without an emotional outlet. The injured person's reaction to the trauma only has a completely "cathartic" effect if it represents an adequate reaction, "as, for instance, revenge."

Language may serve as a substitute for action, they maintained. With the help of language, an emotion can be discharged almost as effectively as if it had been expressed in action, when "it is a lamentation or giving utterance to a tormenting secret, e.g., a confession." If there is no reaction to a trauma in deeds or words, "or in the mildest cases in tears," any recollection of the event retains its emotional impact.

But if the original experience plus the emotions it produced can be brought into consciousness, the "imprisoned" emotions are discharged, the unconscious psychic mechanisms that maintain the symptom cease to operate, and the symptom disappears. This was the basis of the "cathartic method," which Freud was to develop into a far more complex treatment.

He also speaks in this article of a little girl who suffered for years from attacks of general convulsions, regarded as epileptic. When he hypnotized her, she promptly was seized by one of her attacks. He asked what she saw and she said in great fear, "The dog! The dog's coming!" It turned out she had the first of her attacks after being chased by a savage dog and had been reliving this dangerous moment through her convulsions.

The article became enlarged into a book, *Studies on Hysteria,* published three years later in 1895, which Freud also persuaded Breuer to co-author. In it Breuer discussed the case of Anna O. at great length, a discussion which Freud described as the "starting point of psychoanalysis." Freud also wrote about four of his women patients in great detail. Each author then presented his theories about the cause of hysteria.

This became the world's first book on psychoanalysis. It holds some of the basic concepts of psychoanalysis: repression, psychic determinism, unconscious mental activity, overdetermination, defenses, resistance, the sexual cause of neurosis, psychic trauma, conflict, conversion, rudiments of transference, and ambivalence. Another of Freud's classic lines is found here: "Much is won if we succeed in transforming hysterical misery into common unhappiness."

Breuer admitted in the book that he had "suppressed a large number of quite interesting details in the case of Anna O." But in comparison with the medical cases of that era, it was an impressive case history, the first of its kind.

Within the next few years Breuer found himself unable to accept Freud's sexual theories and withdrew from Freud professionally and personally. In a letter to the psychiatrist Dr. August Forel on November 21, 1907, Breuer wrote, "I confess that plunging into sexuality in theory and practice is not to my taste. But what have my taste and my feeling about what is seemly and what is unseemly to do with the question of what is true?" He was in effect supporting Freud but not openly, afraid it would damage his reputation.

Freud felt unhappy that this man, whom he addressed in a letter on May 3, 1889, as "Dearest Friend and best loved of men!" had rejected his theories and abandoned him. But he never forgot his gratitude for Breuer's original encouragement and financial help, which Freud eventually repaid. Thirty years later, when Hannah Breuer, widow of Breuer's oldest son Robert, asked Freud in London for help in emigrating to America, Freud got in touch with Dr. Abraham A. Brill of New York City, translator of Freud's work in America, to arrange the necessary papers. Hannah Breuer and her daughter Marie, before leaving for America, visited Freud, who by then spoke only with difficulty. He asked with deep solicitude about other members of the Breuer family and spoke of his early friendship with Robert's parents.

Freud made Breuer famous by persuading him to publish his research on Bertha Pappenheim. The case of Anna O. is historic in the annals of mental healing. Up to then no one knew the cause of mental illness. It was attributed to the devil, to evil spirits within, to heredity, and treated by hypnotism, electric shock, diet or drugs —all useless. Bertha Pappenheim dramatically demonstrated what mental illness was.

She proved that repressed emotions caused by traumatic experiences could affect the functioning of the mind. Through the "talking cure" that she unconsciously initiated, she indicated the potential healing power of what Freud later called "free association." She, Breuer, and Freud all played a part in this drama of mental healing.

To Keep from Screaming

The first case in which Freud used Breuer's cathartic method, he said, was with a woman he identified as Frau Emmy von N. He had started to use hypnosis eighteen months before. After he put Frau Emmy in a trance, he questioned her about experiences in her life that might be related to her many ailments.

She was so ill he had to visit her at her home where, on May 1, 1889, he found "a still youthful woman with fine, characteristically delineated features lying on a divan, a leather pillow under her neck . . ." She played restlessly with her fingers. She spoke with effort in a low voice, sometimes with a slight stutter, and frequently her face became contorted in tic-like movements. She also interrupted her remarks with a curious "clacking" sound that reminded Freud of a "ticking ending with a pop and a hiss."

She was perfectly coherent, highly educated, and intelligent and had a "full command over her store of memories," Freud said. But every two or three minutes she suddenly broke off her words, contorted her face into an expression of disgust and horror, stretched out her hands toward Freud, spreading and crooking her fingers, and exclaimed in a voice charged with anxiety, "Keep still! Don't say anything!—Don't touch me!"

He speculated she was probably under the influence of a recurrent hallucination of a "horrifying kind" and "was keeping the intruding material at bay with this formula." The interpolations ended with equal suddenness, and she took up what she had been saying without explaining or apologizing for her behavior, "probably, therefore, without herself having noticed the interpolation," Freud said.

Her family originally came from central Germany but had settled for two generations in the Baltic Provinces of Russia, where it possessed large estates. She was the thirteenth of fourteen children, of whom only four survived. She had been brought up by an overenergetic and severe mother. When she was twenty-three, she had married a much older man, a wealthy industrialist, who died of a stroke and left her to bring up their two daughters, one now sixteen, the other fourteen. They were often ailing and suffered anxiety, which she attributed to her illnesses. Since her husband's death fourteen years before, she had been ill constantly with a variety of bodily pains, accompanied by depression and insomnia. She was being treated by a physician in Vienna for the physical pain.

Freud reported that even during her "worst states" she remained capable of playing her part in the management of a large industrial business, of keeping "a constant eye on the education of her children, of carrying on her correspondence with prominent people in the intellectual world—in short, of fulfilling her obligations well enough for the fact of her illness to remain concealed."

Freud suggested she separate from her two daughters, who had a governess, and temporarily live in a nursing home, where he would visit her twice a day. She agreed to do this at once. The next evening he visited her in the nursing home and set a schedule for him to hypnotize her, once in the morning and again in the evening, as Breuer had done with Anna O. Under hypnosis, he would ask her to speak of her past.

One week after he had first seen her, on the morning of May 8, before being hypnotized, she told Freud she had an abnormal fear of animals—worms, mice, and especially toads. She said she had read in the *Frankfurter Zeitung*, which lay on the table before her, a news story about a boy who was tied up and a white mouse put into his mouth, whereupon the boy died of fright. As she related the incident, her face held an expression of horror, and she clenched and unclenched her hands, saying, "Keep still!—Don't say anything! Don't touch me!—Supposing a creature like that was in the bed!"

During the hypnosis Freud tried to disperse these animal fantasies, for he had picked up the newspaper, and while he found an item about a boy who was maltreated, it did not mention a mouse, so he realized this was her hallucination. He asked why she had been so frightened by the story of the boy. She said, "It has to do with memories of my earliest youth."

She explained, "When I was five years old, my brothers and sisters often threw dead animals at me. That was when I had my first fainting fit and spasms . . . Then I was frightened again when I was seven and I unexpectedly saw my sister in her coffin; and again when I was eight and my brother terrified me so often by dressing up in sheets like a ghost; and again when I was nine and I saw my aunt in her coffin and her jaw suddenly dropped."

This woman, as a child, had lived with death, seeing nine of her siblings die. A brother once threw a dead toad at her and she became hysterical. Her fear of worms, she explained under hypnosis, appeared after she was given a present of a pretty pincushion and found the next morning that small worms had crawled out of it—it had been filled with bran that was still wet.

Freud also learned her mother had been so depressed she was sent to what, in those days, was called a lunatic asylum. A governess told Frau Emmy, then a little girl, that patients in these places were tortured, spun around and around on a wheel, frozen to death in ice packs, and beaten.

Freud traced Frau Emmy's disgust with food to childhood, when she was forced by her mother to eat her allotted portion of meat. If she refused, she had to sit at the table, sometimes for hours, until she had eaten all of it. Often the fat became hard and congealed and she swallowed with disgust, wanting to vomit.

The words "Don't touch me!" mirrored her unconscious reaction to fear, originating from several threatening experiences. A brother who had become suicidal after taking morphine, at times would seize her in a death grasp, and she would go cold with fright. One time a friend of the family had gone mad in the house and caught her arm in a tight grip, as though he wanted to kill her. When she was twenty-eight and her daughter was ill, the child in her delirium had caught her mother's neck so forcibly she almost choked her mother to death.

Freud concluded Frau Emmy played so restlessly with her fingers, or kept rubbing her hands one against the other almost continuously, "so as to prevent herself from screaming." He explained, "We are all of us accustomed, when we are affected by painful stimuli, to replace screaming by other sorts of motor innervations." He cites the person who avoids screaming in fear while sitting in the dentist's chair by tapping his feet nervously.

He found as the cause of her "clacking" sound an experience with her daughter when the child had been so sick. Frau Emmy recalled an evening when she was exhausted by worry and long hours of watching by her daughter's bedside, and the child had at last fallen asleep. She had cautioned herself, "Now you must be perfectly still so as not to awaken the child." But this wish had been matched by an antithetical one: the wish for the child's death which she of course repressed in the fear that by being woken from needed sleep the little girl might die. "Similar antithetic ideas arise in us in a marked manner when we feel uncertain whether we can carry out some important intention," Freud said.

Frau Emmy told Freud she had hated this daughter the first three years of her life because her birth had kept Frau Emmy in bed several weeks, making it impossible for her to nurse her ill husband back to health. Her death wishes for this child were strong, causing the deep conflict that ended in the "clacking" sound. In her later state of exhaustion at the child's bedside, she lost control and produced a strange noise that was a compromise between keeping quiet and waking her daughter. "It was her horror at the noise produced against her will that made the moment a traumatic one," Freud said.

The "clacking" originated at a moment of violent fright—that she might wake her daughter and cause her death—and appeared at any future feeling of fright, Freud explained. These moments of fear were so many that they perpetually interrupted her speech causing the "meaningless *tic*," Freud said.

This case furthered his exploration of the effect on the psyche of the repression of emotions and thoughts connected with harrowing experiences in childhood (not only in adulthood, as Breuer had discovered). Frau Emmy had suffered far more than her share of terror as a little girl, a terror that produced what she called "storms in my head"—a seriously disturbed mother, the death of nine young brothers and sisters, a cruel brother who became a morphine addict, a thoughtless nurse who told her horror stories about her mother's treatment in the lunatic asylum. No mention is made of the father, but we can assume he offered his daughter little love or protection.

Her adult life had also been filled with pain—her husband dying when her second child was born, his relatives fighting to take a large estate from her, and the recurring illnesses of her two daughters.

The treatment lasted seven weeks, during which her nervous symptoms disappeared. She returned a year later for eight weeks. As a result of these two separate sessions of treatment, Freud was able to help her, he said, but only temporarily. The therapeutic success "was considerable," but "it was not a lasting one" in that it was not "thorough," and some of her symptoms returned. This was but the start of psychoanalysis, when he was slowly constructing new theories and techniques. In a footnote to the case added in 1924 he comments, "I am aware that no analyst can read this case history today without a smile of pity. But it should be borne in mind that this was the first case in which I employed the cathartic procedure to a large extent."

Though Freud was able to help Frau Emmy rid herself of her symptoms temporarily, in her treatment he could not apply what he was later to discover. In patients like Frau Emmy, who obviously had repressed strong reservoirs of hatred, there was a powerful wish to spite the analyst. Such patients experience the analyst as the hated parents and siblings of the past and express their revenge by trying to prevent the analyst from receiving the pleasure of curing them. Furthermore, a patient like Frau Emmy had so many pressing internal and external conflicts that to treat her for several weeks, or even several months, and expect improvement was unrealistic. She brought into every one of her relationships a basic distrust and, like Bertha Pappenheim, this distrust was felt in the transference relationship with the analyst and needed to be verbally expressed and analyzed.

A contemporary psychoanalyst would consider the treatments of both Bertha Pappenheim and Frau Emmy as superficial, for they concentrated almost exclusively on the removal of the patient's symptoms—something akin to modern behavior modification therapy. Freud was later to discover that for therapy to be really effective—for the patient to mature, to feel self-esteem, and get pleasure from living, loving, working, and playing—many conflicts related to dependency, hatred, spite, and competition had to be understood and faced.

Patients similar to Frau Emmy, who have a strong, punitive superego and deep unconscious wishes for revenge, reveal what Freud was to describe as "a negative therapeutic reaction." Though they may become aware of some of their unconscious wishes, both their need to prevent themselves from getting better and their unconscious desire to defeat the analyst's efforts, create therapeutic stalemates, which these patients end by breaking off the therapy.

As we know, Freud had difficulty being the recipient of a patient's negative transference feelings. Frau Emmy later expressed her anger at him by asking for a referral to another doctor who used hypnosis. Freud eventually learned she had seen a number of doctors, none of whom were able to help. He once met her oldest daughter, who told him her mother had broken off relations with both children, refusing to aid them financially. Frequently, when a patient who needs therapeutic assistance is not substantially helped, he refuses to help others. This seems to have occurred with Frau Emmy.

In his chapters in *Studies on Hysteria* Freud included several brief reports of other women he had treated. One was Cecilia M., from whom he had "collected the most numerous and convincing proofs," he said, that the emotions connected to a forgotten traumatic experience, along with its remembrance, could regularly be brought to consciousness and the nervous symptoms removed.

He described Cecilia as an unusual person, of a special artistic temperament, who wrote poetry. She was possessed by an "hysterical annihilation psychosis." She complained her life was "as if broken up into fragments." Then one day on the couch there suddenly emerged "an old reminiscence of the most plastic delineation with fresh and new feelings, and from then on she lived through anew for almost three years all the traumas of her life which were believed to have been long forgotten," he said.

All the memories were accompanied by "a frightful expenditure of suffering and with a return of all the old symptoms she ever had." This "wiping out of old debts" spanned a period of thirty-three years and enabled Freud, he said, to recognize "all her states" (conscious and unconscious). She found relief from anxiety only if she had the chance to "talk away" the reminiscences with all the feelings "and bodily manifestations that tortured her at the time."

After this catharsis, "she was quite well and in touch with everything for a number of hours," but became irritable, anxious or despairing until the next session. Once she said to him, "Am I not an outcast; is this not a sign of the deepest degradation that I told you this yesterday?" referring to a humiliating experience she had recounted. Freud assured her that what she had said was not in any way reason to condemn herself. The next hypnotic session brought to the surface a memory for which she had deeply reproached herself twelve years before (he does not describe it).

She also suffered from a violent facial neuralgia that suddenly appeared two or three times a year and persisted for five to ten days. One day during an attack she recalled a scene with her husband where a remark he made had humiliated her. She suddenly cried aloud in pain and said to Freud, "It was like a slap in the face." At this, both the facial neuralgia and the pain disappeared. Freud noted, "She had felt as if she really received a slap in the face, and her neuralgia in the cheeks had symbolized this slap."

Sometimes she became victim to a violent pain in her right heel, felt stinging sensations at every step, which made walking temporarily impossible. She remembered a time she was sent to a sanitarium in a foreign country. For eight days she would not leave her room, afraid of joining the other patients. Then the house physician appeared to take her to the dining room. The pain in her foot suddenly struck as she took his arm on leaving her room. It vanished after she told Freud of this scene, saying she had feared she would not make the "proper impression" on the strangers in the dining room. Freud connected "proper impression" to the "impression" a foot would leave.

Celia would occasionally get migraine headaches. She recalled that at the age of fifteen she once lay ill in bed, watched by her austere grandmother. She suddenly cried out, complaining to her grandmother of a pain in her forehead between the eyes. Speaking of this pain thirty years later, she told Freud her grandmother had gazed at her so "piercingly" that it seemed as if her look penetrated deeply into the brain. As Cecilia said this, she burst into laughter and the pain vanished.

The study of this woman gave Freud the opportunity to gather a collection of such symbolizations. A series of physical sensations considered organically determined were of psychic origin, he noted. A number of her traumatic experiences were accompanied by a stabbing sensation in the heart, as she said, "I felt a stitch in my heart." About a mental pain she would say, "Something sticks in my head." Once she commented about an unpleasant scene, "I have to swallow that."

There was a time early in treatment, Freud said, during which her every thought was a hallucination whose solution sometimes offered "great humor." Once she had the fantasy that both her physicians (Breuer, who evidently had referred her to Freud, was still treating her for her physical pains) were hanged in the garden on two adjacent trees. The image disappeared when she told Freud that the evening before, Breuer had refused her request for a certain drug, after which she placed her hopes in Freud, but found him just as inflexible. She had been furious at both and thought: "They are worthy of each other, the one is a *pendant* of the other!" Freud interpreted this as her wishing Breuer and he had died together, and she had translated this into the literal image.

She had "a remarkable form of premonition," he said. When she was in her "best state of mind" she would say, "It is now a long time since I have feared witches at night," or "How pleased I am that my eye-aches have stayed away so long." Freud was then quite sure, he said, that the next session would begin with her expressing a fear of witches or the dreaded pains in the eyes.

"It was always as if that which was already lying fully completed in the unconscious was shining through and the unsuspecting 'official' consciousness (to use Charcot's expression) elaborated the idea which came to the surface as a sudden thought into an expression of gratification, a lie which was quickly and surely enough punished."

Freud's treatment of Cecilia was reasonably successful. There are two important reasons for this. By not condemning or criticizing her, by "giving her permission" to talk about her hatreds, jealousies, and other emotions that humiliated her, Freud emerged as a "benign superego" for Cecilia. Psychoanalytic treatment is most effective when the patient slowly identifies with the analyst's nonpunitive, noncensuring attitude toward the patient's words and acts. Cecilia met in Freud a man who ostensibly cared about her, respected her, and tried to understand her, no matter what she reported. Psychoanalysts have learned this therapeutic attitude is necessary for treatment to be effective. What particularly helped Cecilia is that Freud, in this instance, was able to accept her criticisms of Breuer and himself. Psychoanalysts have also learned that when a patient can feel free to criticize an analyst and realize there will be no retribution, the prognosis for recovery is usually favorable.

In describing another case, Freud said he had gained deep insight into the interesting contrasts between the "most far-reaching somnambulistic obedience and the most stubborn persistence of the morbid symptoms because the latter were deeply founded and inaccessible to analysis." He treated unsuccessfully for five months "a young, active, and talented girl, afflicted with a severe disturbance of her gait." She "showed analgesias and painful spots in both legs, a rapid tremor of the hands, she walked with heavy, short steps, her body bent forward swaying like a cerebellar affliction, so that now and then she fell to the floor." She had been labelled as suffering from multiple sclerosis.

But her state was "strikingly cheerful" when Freud saw her at the suggestion of a specialist who thought she might be suffering from hysteria. Freud said he made an attempt to improve her gait by hypnotic suggestions but without success.

One day as she entered his office, swaying, supporting one arm on her father and the other on an umbrella, the tip of which was markedly worn off, Freud became impatient and screamed at her during the hypnosis, "You have carried on long enough, before noon tomorrow that umbrella you are holding in your hand will break and you will have to go home without an umbrella." He added in his report of the case, "I do not know how I came to the stupidity of directing any suggestion to an umbrella. Later, I was ashamed of it, but had no inkling that my smart patient would take it upon herself to save my reputation from her father who was a physician and witnessed this procedure."

The next day the father told Freud, "Do you know what she did yesterday? We were walking in the Ringstrasse. Suddenly she became exaggeratingly cheerful and began to sing in the street: 'We lead a free life . . .' and accompanied herself by beating time with the umbrella, and striking it against the sidewalk until it broke."

Freud commented in his report, "She naturally had no idea that through her own good sense she transformed a foolish suggestion into a very successful one."

When her condition showed no improvement after "assurance, command, and treatment in the hypnosis," he turned to what he called "psychic analysis" and asked about the emotional state that preceded the outbreak of her ailment. She then related in the hypnosis, but without any show of emotion, that shortly before her illness a young relative died, to whom she had considered herself engaged for many years. When her condition still did not improve, Freud told her in the next session he was quite convinced the death of her cousin had nothing to do with her bodily affliction, that something else must have happened which she did not mention. She started to speak but hardly uttered one word when she turned mute, and her elderly father, sitting behind her, began to sob bitterly.

"I naturally did not press the patient any further, nor have I ever seen her again," Freud said.

One wonders what caused the father to "sob bitterly," whether it was a case of incest or some other trauma related to him or her mother. We must remember psychoanalysis was in its infancy and in the years to come, it would be unthinkable for an analyst to permit a parent in the room during a session—nothing would be more likely to cause a patient's reluctance to speak of what troubled him.

Freud frequently learned from his failures. From cases like the young girl just described, he realized that resistances whose overcoming lies at the heart of therapy cannot be bypassed. The girl needed something from Freud that she did not get. She needed him to verbalize her fear and embarrassment in talking about a disturbing memory and then, sensing his empathy and understanding, she may have been able, in his presence, to face the memory and the emotions it aroused. From such cases, Freud also became keenly aware of the crucial importance of countertransference. When an analyst cannot maintain his empathy in the face of the patient's resistances and latent hatred, he may depart from his analytic stance and give orders, commands, or advice. But he then reduces the patient to a submissive child, and the patient's self-mastery and personal growth will be further impeded.

Another way an analyst can act out his negative feelings is by assigning a diagnostic label that implies gloom and doom. Freud's negative reaction to the young girl's psychological difficulties may have caused him to place too much emphasis on her multiple sclerosis so that he did not devote enough attention to her fears, anxieties, and painful memories.

He eventually pointed out that when childish wishes or pathological character traits in the patient are similar to those in the analyst, and the analyst is unable to recognize this, he may say the patient is untreatable. During most of his lifetime Freud, interestingly enough, felt "narcissistic neuroses" were untreatable (they are being treated today). Using Freud's own formulation, we can infer there were dimensions of his own narcissism he did not wish to face.

He also spoke of treating an eighteen-year-old girl "whose complicated neurosis showed its good share of hysteria." She complained of attacks of despair that had "a twofold content; in the one, she felt a pulling and pricking on the lower part of the face, from the cheeks down towards the mouth, in the other the toes of both feet were spasmodically extended and moved restlessly to and fro."

In treating this patient he asked her to tell him whatever thoughts came to mind, not to be ashamed to speak of anything, that there must be an explanation for her two symptoms—the facial and the foot distortions.

She at once became "red in the face," then, without needing hypnosis, told him that since the beginning of menstruation she had suffered from severe headaches which made any continuous occupation impossible and interrupted her education. Finally, when the headaches disappeared, she worked hard to catch up with her sisters and friends. In doing this she exerted herself "beyond measure," Freud said, and usually ended up depressed. Comparing herself physically with the other girls, she felt inferior. Her front teeth protruded, and she practiced for fifteen minutes at a time hiding them by pulling her upper lip down over her teeth. The failure of this "childish effort" led to more depression, "and from that time on, she gave as the content of one kind of her attacks a pulling and prickling from the cheeks downward," Freud explained, as though she were still trying to fulfill the wish to hide the offending teeth.

He called "no less transparent" the cause of the trouble with her toes. She told him the first of these attacks appeared after a mountain-climbing party in a resort. She recalled that one of the standing themes of mutual teasing among her brothers and sisters was to call attention to her undeniably large feet. So she forced her feet into very narrow shoes, until her father would not permit this and insisted she wear comfortable footwear. She got into the habit of playing with her toes in her shoes, "like one does when one wishes to measure by how much the shoe is too big," Freud said.

During the mountain-climbing party, which she did not find in any way tiring, her family continued to make remarks about her shoes, visible under her short skirt. One of her sisters said, "But you are wearing particularly large shoes today." She kept thinking obsessively about her humiliating feet and on her return home, had the first attack of toes cramping and moving involuntarily as "a memory symbol," in Freud's words, for the whole depressive trend of thought.

When she reported memories connected to both her cheeks and her toes, only the nervous gestures associated with her cheeks disappeared. Freud commented, "There must surely have been a fragment connected with it [the foot] which has not been confessed." He added a postscript: "I also discovered later that this simpleminded girl worked overzealously to beautify herself because she wanted to please a young cousin." This was the "fragment." It told of her frustrated sexual wishes, which then appeared in symbolic way in the offensive feet.

In terms of today's lengthy psychoanalytic treatment, these are "primer" illustrations. But for Freud at this time they were nuggets of psychic gold upon which he built his later complex theory and technique.

The Smell of Burnt Pudding

Not satisfied with the ephemeral results of treatment with patients such as Frau Emmy, Freud gave up hypnotism. His early cases convinced him hypnosis was useless because, unless the patient's observing ego was involved, there was no permanent easing of mental pain. The patient had to be conscious, not lying in a trance, to become aware of the connection between past and present. Freud learned that if the psychotherapy was to be effective, the patient had to be an active participant.

As he explained in *An Autobiographical Study,* hypnotism "screened from view an interplay of forces which now came into sight and the understanding of which gave a solid foundation" to his theory that buried memories were the cause of mental illness and many concomitant physical ailments.

He concluded patients really knew everything that related psychologically to their illness and would eventually communicate it to the doctor if they spoke freely of each thought that entered their minds, unveiling the traumatic memories. He was primarily interested in learning *why* the memory had originally been banned from consciousness.

Between the years 1892 and 1895 he developed the technique of free association without hypnotism. He found that as patients were able to speak their thoughts, particularly the ones they believed dangerous to self-esteem or in some way frightening, they became aware of the fears and wishes that had caused their emotional and physical distress.

This was for him a hard-wŏñ victory, for doctors were not supposed to listen to patients. They were the ones, traditionally, to do the talking and tell patients what to do. Freud had to try his best to overcome his medical training and any tendency in himself to be the dictator.

It was not only what a patient said but the tone of his voice and his gestures that Freud used in exploring the unconscious. He once remarked, "Betrayal oozes from every pore," meaning the patient betrayed the repressed feelings with each word, movement, sigh, or silence.

In 1892, in the case of "Miss Lucy R.," Freud was able to dramatically illustrate repression as the "foundation-stone on which the whole structure of psychoanalysis rests." Lucy R. was a young English governess employed in the house of a widower, the managing director of a factory in outer Vienna. A colleague of Freud's who was treating Lucy R. for sinus had referred her to him. She had completely lost her sense of smell and almost continuously imagined olfactory sensations she found disturbing. She was also "in low spirits and fatigued," suffering a loss of appetite and "efficiency" in caring for the widower's two little daughters.

Freud diagnosed her condition as mild hysteria. Though he tried to hypnotize her she would not fall into the trance, so, he says, "I therefore did without somnambulism."

In her first session she spoke of being tormented by the sensations of strong odors that occurred without warning every so often. Freud decided, he said, to make this the starting point of the analysis. He asked what odor she smelled most frequently, and she replied, "A smell of burnt pudding."

During the sessions that followed, he uncovered the fact that she had first smelled burnt pudding two months before, two days prior to her birthday. She had been cooking for her two small charges when she received a letter from her mother in Glasgow. She wanted to read it at once but the children tore it out of her hands and insisted she wait until her birthday. While they were teasing her, she became aware of a strong odor: the pudding she had been cooking was burning. Ever since she had been pursued by the smell, which became stronger when she felt upset.

Then she confessed to Freud that, though fond of the children, she had given notice to her employer. She thought other members of the household were talking against her to both her employer and his father, who also lived in the house, and she was not getting the support from the two men she had expected. Her employer had advised her, in a friendly way, to think it over for a few weeks. Because of mixed feelings, for she liked the children, she had stayed on. She also felt if she left she would be breaking a deathbed promise made to the children's mother, a distant relative, to take her place and devote herself to the children.

Freud connected the hallucination of the smell of burnt pudding to what he called "the little scene" that had wakened a conflict when she received the letter from her mother. The sensation of smell associated with the conflict—to leave or not to leave the children—persisted as symbolic of the repressed painful emotions.

It was still necessary, he believed, to explain why, out of all the feelings and thoughts evoked by the experience in the kitchen, she had chosen the smell of burnt pudding to symbolize it. He asked what she was feeling at the time. She said she had been suffering from such a bad cold that she was unable to smell anything. Yet the odor of burnt pudding had affected her nose.

Freud wanted to know why the conflict had led to her loss of the sense of smell, and to fatigue. What lay behind the memory of the burnt pudding? What did Lucy R. *not* want to recall?

From analyzing similar cases, he knew that in hysteria one essential condition had to be fulfilled: an idea had to be *intentionally repressed from consciousness*. This intentional repression was the basis for the conversion of the psychic energy into a physical symptom, as the energy cut off from the psychic association found an outlet, stimulating some allied part of the body.

Freud believed the basis for Lucy R.'s repression could only stem from some deep feeling of displeasure or discomfort—a basic incompatibility between the idea that had to be repressed and the maintenance of her self-respect. Only one conclusion could be reached, Freud speculated, and he was "bold enough" to mention it. He told Lucy R. he believed she was in love with her employer, though she was perhaps unaware of her feelings, and that she harbored a secret wish to take the dead wife's place, as the wife herself had suggested on her deathbed. Freud told Lucy R. he thought she was afraid the servants had guessed her wish and were making fun of her.

She admitted, in what Freud called "her usual laconic fashion," that this was true. He asked, in astonishment, why, if she knew she loved her employer, she had not mentioned it.

She replied, "I didn't know—or rather I didn't want to know. I wanted to drive it out of my head and not think of it again; and I believe latterly I have succeeded."

Freud later commented he had "never managed to give a better description than this of the strange state in which one knows and does not know a thing at the same time."

He then asked why she was unwilling to admit her feelings, was she ashamed of loving a man? She said it was distressing to her because the man was her employer and rich, whereas she was poor. She thought people would laugh if they knew her daydream.

She told Freud that for the first few years she had lived happily in the house, until one day her employer, whose behavior had always been very reserved, started to discuss the children's welfare, seeming far more friendly than usual. He told her how much he depended on her for the care of his motherless daughters. As he said this, she thought that he looked at her "meaningfully," and at that moment she dared allow herself to feel her hidden love. She hoped they would grow closer, but he had never again tried to talk to her on a personal level, so she made up her mind to banish all thoughts of love between them. She admitted to Freud there was little likelihood he would ever choose her as a second wife.

Freud expected that, with new awareness, Lucy R. would feel better, but she continued to complain about fatigue and depression. The imaginary smell of burnt pudding did not disappear completely, though it became weaker and occurred only when she felt anxious. The persistence of this smell led Freud to believe it was associated with deeper conflicts she had repressed. He kept asking her to talk of anything that might have to do with burnt pudding. She mentioned her employer's father and quarrels between him and his son that had taken place in her presence.

About this time her nasal disorder became so severe that she canceled her treatment temporarily. It was the Christmas season, and on her return she told Freud she had received many gifts from her employer and his father, and even from the servants, as though they all wanted her to stay. But, she said, she was not impressed with the presents.

Freud asked again about the smell of burnt pudding. She said it had disappeared but now she was troubled by the smell of cigar smoke. She recalled it too had been present earlier but overwhelmed by the stronger odor of the burnt pudding.

Here Freud noted, in his description of the case, that one symptom might be removed only to have another take its place, and that symptom after symptom had to be pursued until the underlying cause became known. He began questioning her about the new symptom, hoping to uncover its origin. He asked if there were any occasions she remembered smelling cigar smoke.

Speaking hesitantly, she recalled a scene in the dining room. There had been a guest, the chief accountant at her employer's factory, an old man who was very fond of the children and often came to lunch. As the children said goodbye after eating, the accountant tried to kiss them, and their father became furious, shouting at him not to touch them. Lucy R. told Freud that at this moment she felt a stab in her heart. The men were smoking cigars, she recalled.

Thus, there was a second experience which, like the first, was painful and had been banished from her mind. But why was this experience so painful, Freud wondered. He asked which of the two scenes had occurred first, and she said the one associated with the cigar smoke had happened two months earlier.

Freud asked why, when her employer reprimanded the old man, she felt "the stab" in her heart, since she herself was not involved. She replied she did not feel it was good manners for him to shout at an old man who was a friend and a guest in the house.

She then remarked that her employer had never liked anyone to kiss his children, at which there emerged the memory of a third and even earlier experience. She recalled that a few months before this scene, a woman friend of her employer's came to visit and as she was leaving had kissed the two little girls on the lips. Their father said nothing to the woman, but after she left he turned in fury on Lucy R. He shouted that he would hold her responsible if anyone kissed his children on the lips, that it was her duty not to allow it, and if it ever happened again, he would fire her and get someone else to take care of his daughters.

She had felt deep shock. This happened at a time she still felt it possible he might care for her, and his threat to fire her now smashed all her hopes. She told herself she must have made a mistake, he could never have any feeling of love for her or he would never have uttered these words. This had been the true trauma—the utter rejection of her love—and the experience with the chief accountant had revived memories of the devastating first scene.

Two days after she told Freud of the earliest scene, Lucy R. came to his office smiling and cheerful. Thinking perhaps her employer had proposed marriage, Freud asked if anything had happened. She said nothing had occurred, it was just that Freud had never known her free of depression and fatigue. When he asked about her marital prospects in the house, she said she had none but did not intend to make herself miserable over it, that she had given up all thought of her employer loving her.

Freud examined her nose, found its state now almost normal. She said she could distinguish between smells if they were strong. It was the end of nine weeks of treatment and she decided to stop. Freud reported he met her by chance four months later at a summer resort, and she seemed in good spirits and reassured him her nose did not trouble her any longer.

Through this case Freud gave a clear and simple example of the purpose of repression, as well as how repressed emotions and thoughts may return in disguised forms as "the body speaks." Lucy R.'s psychic distress was transposed into physical symptoms, a process Freud named "conversion," in which body and mind interact in the interest of emotional and mental survival so self-esteem may be preserved.

Freud's analytic behavior with Lucy R. provided her with what his followers have termed a "corrective emotional experience." He helped her express her sexual yearnings in an atmosphere of benign neutrality. This gave her an experience with a cigar-smoking father-figure who understood but did not admonish, accepted but did not advise, interpreted but did not seduce. Freud's acceptance of Lucy's forbidden wishes helped her eventually to tolerate them. As he later pointed out, when sexual and aggressive wishes become more acceptable to the conscious ego (that is, where id was, there is now ego), self-esteem rises and the punitive superego loses much of its power.

If Lucy R. were in analysis today, the roots of her anxiety would be traced to childhood conflicts. In all likelihood her father smoked cigars and her mother may at times have cooked puddings that burned, bringing forth wrath from her father. To have this powerful an effect in adult life, a symptom must be tied in memory to dangerous or frightening childhood experiences.

"This Proud Girl with Her Longing for Love"

Also about this time, in the autumn of 1892, Freud was asked by a colleague to examine Fräulein Elisabeth von R., a young lady who had suffered for two years from such pains in her legs she had difficulty walking. The colleague (probably Breuer) told Freud he thought she had hysteria, that he knew the family slightly, and during the last few years its members had suffered many misfortunes.

First, the patient's father had died, then her mother had to undergo a serious eye operation, and soon after, one of the patient's sisters died of a heart ailment of long standing. The largest share of the nursing of these sick relatives had fallen to Elisabeth.

In his first interview with her, Freud found a young woman of twenty-four who seemed very intelligent, mentally normal, and had a cheerful air, "the *belle indifférence* of a hysteric," he thought, attributing this phrase to Charcot. She walked with the upper part of her body bent forward, without any support. In her walk she was almost normal. But she complained of great pain in walking and of being quickly overcome by fatigue. She said that after a short walk or after standing for a while, she had to rest, which lessened the pain but did not do away with it.

Her case represented "the first full-length analysis of a hysteria undertaken by me," Freud noted. With Elisabeth von R. he experimented with a procedure he later developed into the regular method of free association. He described it: "This procedure was one of clearing away the pathogenic psychical material layer by layer, and we liked to compare it with the technique of excavating a buried city."

He would begin by asking her to tell him what she knew about a certain event and then carefully note the points at which some train of thought remained obscure "or some link in the causal chain seemed to be missing." Then he would "penetrate into deeper layers of her memories at these points."

He did not hypnotize Elisabeth but asked her to "lie down and keep her eyes shut," though he did not object if she occasionally opened them, changed her position on the couch, or sat up. When she seemed deeply moved by what she was saying, she seemed to "fall into a state more or less resembling hypnosis. She would then lie motionless and keep her eyes tightly shut."

She told him she was the youngest of three daughters and was "tenderly attached to her parents" on whose estate in Hungary she had spent her youth. Her mother suffered frequent infections of the eyes as well as nervous states. Elisabeth had been drawn "into especially intimate contact with her father, a vivacious man of the world, who used to say that this daughter of his took the place of a son and a friend with whom he could exchange thoughts," Freud said. The description of the father-daughter relationship sounds quite similar to the one Freud had with his daughter Anna.

He described her as "greatly discontented with being a girl." She wanted to study or become musically trained rather than marry and raise children as her mother had done. But she put her mother and older sisters first when they needed her, as they did when her father fell ill of a chronic disorder of the heart and died. She had nursed him for eighteen months, sleeping in his room, ready to wake if he called her at night, and she also looked after him during the day, forcing herself to appear cheerful. The start of her illness was connected with this period of nursing (like Anna O.), and she had taken to her bed when her legs, chiefly her thighs, hurt her. But it was not until two years after her father's death that she became incapable of walking naturally because of the pain.

When the year of mourning ended, her oldest sister married, and soon after, her second sister. Then this second sister died of a heart condition aggravated by a second pregnancy. Elisabeth's widowed brother-in-law seemed inconsolable and withdrew from his wife's family to whom he had been close, especially Elisabeth.

"Here, then, was the unhappy story of this proud girl with her longing for love," said Freud. "Unreconciled to her fate, embittered by the failure of all her little schemes for re-establishing the family's former glories, with those she loved dead or gone away or estranged, unready to take refuge in the love of some unknown man—she had lived for eighteen months in almost complete seclusion, with nothing to occupy her but the care of her mother and her own pains."

Freud now took another step forward in the process of psychoanalysis. He gave up asking her about a specific symptom. Instead, he asked her to report "faithfully" whatever appeared before her "inner eye" or passed through memory. She recalled she had been attracted to a young man who had taken her home after a party, but on returning to her father's sickbed found him worse, felt guilty, and saw the young man only a few times before giving him up completely.

In remembering the first time her legs hurt, she told Freud of the day she and her brother-in-law had taken a walk. When Freud asked what happened on the walk that might have brought on the pain, she gave "the somewhat obscure reply that the contrast between her own loneliness and her sick sister's married happiness (which her brother-in-law's behavior kept constantly before her eyes) had been painful to her."

Another time, when she thought of how lonely she was and had "a burning wish" to be as happy as her sister, she suffered violent pains in her legs. On still another occasion she thought of herself as strong enough to be able to live without the help of a man, but then was overcome by a sense of her weakness as a woman, by a longing for love, and the desire to have a husband like her dead sister's. In Freud's words, her "frozen nature began to melt." It had inevitably become clear to him what troubled her, though she did not seem aware of her conflict.

One day she spoke of the time just after her sister died when she and her mother stood at the bed looking at her sister's corpse. Elisabeth told Freud that at that very moment the thought shot through her mind like a flash of lightning in the dark, "Now he is free again and I can be his wife."

Freud said this moment brought before his eyes "in concrete form" the cause of her pains: "This girl felt towards her brother-in-law a tenderness whose acceptance into consciousness was resisted by her whole moral being. She succeeded in sparing herself the painful conviction that she loved her sister's husband by inducing physical pains instead, her thoughts relating to this love had been separated from her awareness, forcing the 'incompatible idea' out of her associations."

The recovery of this repressed idea had a "shattering effect" on her, Freud said, when "I put the situation drily before her with the words: 'So for a long time you had been in love with your brother-in-law.'" She complained at once of the most "frightful pains," denied his explanation was true, insisted she was incapable of such wickedness. It was a long time, he said, before his "two pieces of consolation—that people are not responsible for their feelings and the fact she had fallen ill under the circumstances was sufficient evidence of her moral character—made any impression on her."

He kept probing into the original impressions made by her brother-in-law, the start of the feelings she had repressed. On his first visit to the house, her brother-in-law had mistaken her for the sister he was to marry before her older, "somewhat insignificant-looking sister," arrived, Freud said. One evening while she and her prospective brother-in-law were carrying on a lively conversation, her sister interrupted with the remark, "The truth is, you two would have suited each other splendidly." Another time at a party where the guests did not know of his engagement, one woman criticized a defect in his figure, and while her sister, his fiancée, sat in silence, Elisabeth flared up to defend him with a zeal she could not understand.

Freud said he was anxious to learn what chance there was that her wish, of which she was now conscious, would come true. Her mother told Freud she had long ago guessed Elisabeth's fondness for her late sister's husband, that no one seeing the two of them together could doubt her wish to please him. The mother also said neither she nor the family advisers were in favor of his marrying Elisabeth, for both his physical and mental health had been affected by his wife's death and he did not seem sufficiently recovered to take on a new marriage.

Freud told Elisabeth what her mother had said (under today's psychoanalytic conventions it would be a rare event for an analyst to see the parent of a patient, much less tell the patient what the parent said) and encouraged her to face with calmness the uncertainty of her future. Her physical condition had improved, she no longer talked of pains in her legs and walked naturally. He regarded her as "cured," and she left Vienna with her mother to spend the summer with her oldest sister and her family.

In the spring of 1894 Freud said he saw her at a private ball he attended. He reported, "I did not allow the opportunity to escape me of seeing my former patient whirl past in a lively dance." Later he heard she had married someone unknown to him.

The treatment of Elisabeth was successful, again, not only because the unconscious was made conscious and certain id material became mastered by the ego, but because she experienced Freud as a benign superego that did not judge, censor, advise, or reinforce. In this type of safe atmosphere, where Elisabeth could be her own person, she could begin to accept herself as she felt accepted by Freud.

In his article on Elisabeth von R., Freud mentioned briefly two former cases where he had used hypnotism. One was the treatment of Miss Rosalia H., twenty-three years old, who for a number of years made great efforts to educate herself as a singer. She came to him complaining that "her beautiful voice did not obey her in certain notes." She felt choking and tightening sensations in her throat so that the tones sounded strained, and her teacher would not permit her to appear in public. The slightest excitement, seemingly without provocation, evoked the choking sensation and prevented free expansion of her voice.

It was not difficult to recognize in this "annoying sensation" an hysterical conversion, Freud said. He undertook a "hypnotic analysis" and learned the young woman had become an orphan at an early age and was brought up by her aunt, who had many children, so she had to share the life "of a most unfortunate family existence." The husband of her aunt, "seemingly a pathological personality, abused his wife and children in the most brutal manner, and what especially pained her was his unconcealed sexual preference for the servant and nurse girls in the house . . ."

When her aunt died, Rosalia became the protectress of the younger children, who were harassed by their father. She tried to suppress the anger and contempt she felt for him. It was then the choking sensation in her throat originated. Whenever she was "compelled to swallow an affront, whenever she had to remain silent on hearing a provoking accusation," she suffered a scratching in her throat and the tightening and failure of her voice, Freud said.

A music teacher showed a deep interest in her, assuring her that her voice entitled her to choose the profession of singing. She secretly took lessons from this man, and because she often went to his studio with the choking sensation in her throat following some violent scene in the house, a connection was formed between the singing and the hysterical reaction, Freud said: "The apparatus of which she should have had free control was filled with the remnants of innervation [nerve stimulation] from those numerous scenes of suppressed excitement."

She finally left her uncle's house, moved to Vienna to escape this family, but found her throat still affected. "Unfortunately, she had no luck with her relatives," as Freud put it, for she went to live with another uncle who treated her with such friendliness that his wife objected, jealous not only of her husband's attentions to Rosalia, but also of her musical talent, for the wife had wished to become a singer. Rosalia felt so uncomfortable in the house that she did not dare sing or play the piano when her aunt was within hearing distance.

One day she brought a new symptom to Freud. She complained of a disagreeable prickling sensation in her fingertips, which forced her to make peculiar, jerky movements. Under hypnosis she recalled without hesitation a series of scenes starting in early childhood in which she had suffered some injustice without defending herself, some act that made her fingers jerk. She had been forced to hold out her hand in school while the teacher struck it with a ruler. She recalled a sexual scene in which her cruel uncle, who suffered from rheumatism, one day asked her to massage his back. She did not dare refuse and massaged him as he lay in bed. Suddenly he threw off the covers, jumped up, tried to grab her and throw her down on the bed. She managed to escape and locked herself in her room. Freud asked if she had seen her uncle nude and she said she did not remember. Freud speculated that the current unpleasant jerky sensations in her fingers could be explained either as due to the suppressed impulse to punish him or originated from the fact she was massaging him prior to the assault.

She told Freud that the uncle with whom she now lived one day begged her to play the piano, and she sat down at it, accompanying herself as she sang, believing her aunt was away from the house. Suddenly the aunt appeared in the door. Rosalia jumped up, closed the piano, and flung away the sheet of music. Freud commented, "The movement of the fingers which I saw during the reproduction of this scene resembled a continuous jerking, as if one literally and figuratively would reject something like throwing a sheet of music or rejecting an unreasonable demand."

A second case involved Miss Matilda H., a pretty girl of nineteen who suffered an incomplete paralysis of the legs when Freud first saw her. Months later he was called again to help her by her mother, who said Matilda was depressed and suicidal, irritable, and unapproachable, lacking consideration for anyone.

Under hypnosis she informed Freud the reason for her depression was the breaking of her engagement many months before. She had become engaged in a hurry, and on closer acquaintance with the man, things displeasing in him to her and to her mother became more and more evident. On the other hand, the material advantages of the engagement were too important to make the decision of a rupture easy, and thus both she and her mother hesitated to end it. Finally her mother took action and delivered a decisive "no" to the man.

Shortly after, the patient said, she woke as from a dream and began to think of the broken betrothal, weighing once again the various pros and cons. She lived in a time of doubt, angry at her mother, finding her own life unbearable. During her sessions with Freud, she did little but burst into tears.

One day, near the anniversary of her broken engagement, her depression suddenly disappeared. Freud said, in writing of the case, "This was attributed to the success of my great hypnotic cure." He was later to write about "anniversary mourning" or "joy" and how someone might be either overelated or very depressed on the occasion of the anniversary of a birth, engagement, marriage, desertion, or death. These two cases mark the end of his use of hypnotism whose results, at best, were always temporary. He does not reveal whether Matilda's paralysis of the legs cleared up.

Incest on a Mountain

Though he does not give the exact year but places it only in the 1890s, Freud unexpectedly treated a very brief case while on vacation in the Hohe Tauern, one of the highest ranges in the Eastern Alps. Wanting to "forget medicine and more particularly the neuroses," one day he climbed a mountain renowned for its panoramic views and for its refuge hut.

Feeling refreshed after food at the hut, he sat outside enjoying the magnificence before his eyes, so lost in thought that at first he did not connect the words with himself as he heard, "Are you a doctor, sir?"

He looked up to observe "the rather sulky-looking girl of perhaps eighteen who had served my meal and had been spoken to by the landlady as 'Katharina.'" He judged by her dress and bearing that she was no servant but daughter or some relative of the landlady.

He replied, he said, after "coming to myself," by admitting he was a doctor. Then he asked, "But how did you know that?"

"You wrote your name in the Visitors' Book, sir," she said. "And I thought if you had a few moments to spare . . . The truth is, sir, my nerves are bad. I went to see a doctor in L——— about them and he gave me something for them; but I'm not well yet." Freud did not want to identify where the doctor lived for this might place the locale of the girl's home.

There he was, once again with the neuroses, Freud reflected, then thought humorously that he was interested to find neuroses "could flourish in this way at a height of over 6,000 feet." He decided to give some of his vacation time to this troubled girl, making it clear in his later report of the case he did not consider this brief exploration of a young woman's fears and memories a "psychoanalysis" in any sense of the word but only a "simple talk."

He asked Katharina to tell him what symptoms she suffered. She said she sometimes felt so out of breath she feared she would suffocate. He wanted to know if other physical symptoms accompanied the difficulty in breathing. She mentioned a pressure on her eyes, a heaviness in her head, a buzzing sound in her ears, and the fear

166

someone would seize her suddenly. At such moments, she said, she thought she was "going to die."

Freud asked if she thought of anything, or saw any image in her mind during the attacks. She said she always saw a frightening face that looked at her angrily. Freud asked if she recognized the face and she said she did not. Then he asked when she had first suffered an attack, and she placed the time as two years earlier.

He tried what he called "a lucky guess," based on his findings, he said, that young girls often experienced anxiety when they faced the world of sensuality for the first time, especially if they had a mother or father who in some way terrorized them. He suggested to Katharina she might have seen or heard something that embarrassed her very much, which she would rather not have known about.

She said frankly, "Heavens, yes! That was when I caught my father with the girl, with Franziska, my cousin."

In his original report on the case, Freud disguised her father as her uncle, then in a footnote added in 1924 said he ventured after the passage of so many years "to lift the veil of discretion" and reveal she was not the niece but the daughter of the landlady and had suffered anxiety as a result of sexual assaults not on the part of her uncle but her father.

Freud asked her to describe the time she saw her father and her cousin together. This took place at a refuge hut in another part of the valley that her mother was then operating. Several men who had climbed the mountain asked for food. Her mother had gone for the day and Katharina looked all over for Franziska, who did the cooking, but could not find her. She then searched for her father but he too seemed missing. Her little brother Alois suggested Franziska might be in their father's room, and they went there but found the door locked, which seemed strange to her.

She told Freud, "Then Alois said: 'There's a window in the passage where you can look into the room.' We went into the passage; but Alois wouldn't go to the window and said he was afraid. So I said: 'You silly boy! I'll go. I'm not a bit afraid.' And I had nothing bad in my mind. I looked in. The room was rather dark, but I saw my uncle [father] and Franziska; he was lying on her.'"

She stopped. Freud said, "Well?"

"I came away from the window at once, and leant up against the wall and couldn't get my breath—just what happens to me since. Everything went blank, my eyelids were forced together, and there was a hammering and buzzing in my head."

"Did you tell your mother that very same day?"

"Oh no, I said nothing."

Freud asked why she had been so frightened—had she understood what she saw? She said she had not, that she was then only sixteen and did not know why she felt frightened, that she had forgotten all her thoughts while experiencing the attack. Freud asked if the face she saw when she lost her breath was Franziska's, and she said it was the face of a man. Freud then asked if it could be her father's as he lay in bed with Franziska. She said she did not see his face clearly, that it had been too dark in the room.

Her memory seemed blocked but Freud hoped she might reveal more if she kept talking. He asked what happened next. She said her father and Franziska must have heard something because they came out of the room shortly. All that day she felt sick when she thought about what she had seen. Several days later she stayed in bed, vomiting constantly.

Freud recalled research with Breuer on anxiety attacks and their finding that "being sick means disgust." He asked Katharina if she had felt disgust when she looked into the room.

"Yes, I'm sure I felt disgusted," she said, then added reflectively, "but disgusted at what?"

"Perhaps you saw something naked? What sort of state were they in?"

"It was too dark to see anything; besides they both of them had their clothes on. Oh, if only I knew what it was I felt disgusted at!"

Freud said he had no idea, either. But he told her to talk on and say whatever came to mind, "in the confident expectation that she would think of precisely what I needed to explain the case."

She told him her mother had sensed she was hiding a secret, and she finally revealed to her mother what she had seen. There were angry scenes between her mother and father, and finally her mother took her and her brother to the other place she was managing (the refuge hut at which Freud had dined), leaving her father with Franziska, who had become pregnant.

Freud said, to his "astonishment," she then dropped these

"threads" and began to talk of events occurring two or three years earlier. The first set of memories related to times her father had made sexual advances to her, when she was fourteen. She described how she had spent a winter's night with him at an inn in the valley. He sat in the bar drinking and playing cards while she went upstairs to bed in the room they were to share. She was not quite asleep when he came into the room. She fell asleep, only to wake suddenly "feeling his body" in the bed.

She jumped out of the bed, remonstrated, "What are you up to, Uncle [Father]? Why don't you stay in your own bed?"

He tried to pacify her, saying, "Go on, you silly girl, keep still. You don't know how nice it is."

"I don't like your 'nice' things," she said. "You don't even let one sleep in peace."

She stood by the door, ready to flee into the hall, until finally "he gave up and went to sleep."

Freud asked if she had known what her father was trying to do. She said she had not known at the time, though it had later become clear. She had resisted him, she said, because it was "unpleasant to be disturbed in one's sleep and because it wasn't nice."

She then spoke of further sexual experiences with her father. Another evening in an inn she had to defend herself when once again he became very drunk. Still another night, when Franziska, her father, and she were staying at an inn, her father sleeping in one room and Franziska and she in another, she had wakened to see her father standing at the door, turning its handle. When she asked what he was doing, he said, "Keep quiet. I was only looking for something."

Another time the whole family had spent the night in their clothes in a hay loft during a trip, and she was wakened by a noise. She thought she noticed her father, who had been lying between Franziska and her, turning away from Franziska.

Freud asked whether she had felt frightened on these occasions. She said she thought she had, but was not sure. He noticed as she concluded her "two sets of memories," she became "like someone transformed." The sulky, unhappy face "had grown lively, her eyes were bright, she was lightened and exalted."

The understanding of the case now became clear to him, he said.

What she had told him, in apparently aimless fashion, had provided an "admirable explanation" for her behavior when she discovered her father and cousin in sexual intercourse. She had established a connection between this image and what had happened previously in her life. Her vomiting had been caused not by disgust at the sight of her father and Franziska in bed, but by the memories it had stirred within her of the attempted seductions by her father.

Freud said to her, "I know now what it was you thought when you looked into the room. You thought: 'Now he's doing with her what he wanted to do with me that night and those other times.' That was what you were disgusted at, because you remembered the feeling when you woke up in the night and felt his body."

"It may well be," she said "that that was what I was disgusted at and that that was what I thought."

Freud then asked, "What part of his body was it that you felt that night?"

But she would not answer, merely smiled "in an embarrassed way, as though she had been found out . . ." He said he could imagine the tactile sensation she had later learned to interpret. Her facial expression seemed to say she supposed he was correct in his conjecture. But he felt he could not explore further, and "in any case I owed her a debt of gratitude for having made it so much easier for me to talk to her than to the prudish ladies of my city practice, who regard whatever is natural as shameful."

But he had one more thing to clear up, the hallucination of the angry face that appeared during her attacks and struck terror into her. Where did it come from? He asked her about this. He says that as though *her* knowledge also had been extended by their conversation, she promptly replied, "Yes, I know now. The head is my father's head—I recognize it now—but not from *that* time."

She recalled that when her mother and father started to fight, her father raged at her. He kept saying it was all her fault, if she had not told on him her mother would not be insisting on a divorce. He threatened to get even. If he saw her alone, even at a distance, she said, "his face would get distorted with rage and he would make for me with his hand raised." She always escaped but felt terrified he would catch her by surprise and kill her.

Freud said he could understand that it should have been precisely this last period, when there were scenes of fury in the house, that "left her the legacy" of the hallucinated face as symbol of the trauma. He concluded his article by saying, "I hope this girl, whose sexual sensibility had been injured at such an early age, derived some benefit from our conversation."

Katharina was one of the patients Freud thought seduced by a parent. Later he realized Oedipal fantasies often lay behind a woman's conviction that she had been seduced as a girl. Whether Katharina was actually seduced or her whole story was a fantasy, is difficult to know. Her story seems plausible as she tells it.

What Freud discovered about both real and fantasied seduction is that the victim feels an inner excitement that he usually represses. When Katharina referred to her "disgust," she was also talking about her own sexual wishes, repellent to her.

After reviewing many cases like Katharina's, Freud concluded such patients possessed strong wishes for revenge following the excessive sexual stimulation and the accompanying rage and guilt, which often became transformed into helplessness. In talking to Freud, Katharina was probably using the interview as a chance to gratify some of her revengeful wishes as well as to gain some mastery and control over the incestuous desires.

While Freud attributed her unawareness of sexuality to simple ignorance, he was later to write that adolescents and children possess greater sexual knowledge than is believed. He became convinced a lack of sexual awareness was due to "rejection" of sexual knowledge already acquired. He concluded that teen-agers possess "sexual knowledge far oftener than is supposed or than they themselves believe."

Afraid to Pick a Mushroom

All these cases appeared in *Studies on Hysteria,* the book Freud wrote with Breuer. After his break with Breuer, Freud went his lonely way until Fliess became his friend and father-confessor. As Freud wrote Fliess on January 17, 1897, "You obviously enjoy the turmoil in my head," adding he would continue to describe his new discoveries.

In a letter dated April 28, 1897, Freud reported he had started treatment the day before of a young woman who had a brother who "died insane." Her chief symptom, Freud said, "dates from the time she heard the carriage driving away from the house taking him to the asylum." Since then, she had been terrified of driving in a carriage, convinced she would be hurt or die in an accident. Once, summoning momentary courage when she dared ride in a carriage, the horses shied and she jumped out, breaking a leg.

In this letter Freud said, "Today she came and said she had been thinking over the treatment and had found an obstacle." When he asked what it was, she replied, "I can paint myself as black as necessary, but I must spare other people. You must allow me to mention no names."

"Names don't matter," he said. "What you mean is your relationship with the people concerned. We can't draw a veil over that."

She still objected, saying, "The criminal nature of certain things has become clear to me, and I can't make up my mind to talk about them."

Freud reassured her, knowing what she was worried about, "On the contrary, I should say that a mature woman becomes more tolerant in sexual matters."

She agreed, then said, "When I consider that the most excellent and high-principled men are guilty of these things, I'm compelled to think it's an illness, a kind of madness, and I have to excuse them."

"Then let us speak plainly," Freud said. "In my analyses I find it's the closest relatives, fathers or brothers, who are the guilty ones."

"It has nothing to do with my brother," she protested.

"So it was your father, then," Freud said.

She admitted that when she was between eight and twelve, her "allegedly otherwise admirable and high-principled father" would regularly take her into his bed "and practice external ejaculation (making wet) with her." Even at that time she felt anxiety. A sister six years older, to whom she later confessed her father's acts, admitted she had the same experience. A cousin told her that at the age of fifteen she had resisted the advances of her grandfather.

Freud commented in his letter to Fliess, "Naturally she did not find it incredible when I told her that similar and worse things must have happened to her in infancy."

Almost a month later, on May 2, 1897, Freud wrote Fliess he found that wealthy women often identify with women of "low morals," who are connected sexually with fathers or brothers. Fear of becoming a prostitute like the women associated with this father or brothers, derives from this identification, Freud said. He commented, "There is tragic justice in the fact that the action of the head of the family in stooping to relations with a servant-girl is atoned for by his daughter's self-abasement." Such a father would also be seductive to his daughter, arousing her sexual desires, her wish to be identified with his mistress, and her guilt.

In this letter Freud mentions as patient "a girl last summer who was afraid to pick a flower or *even* a mushroom, because it was against the will of God; for He forbids the destruction of any germs of life."

Freud explained her fear as arising "from a memory of religious talks with her mother, who inveighed against taking precautions during intercourse because they meant the destruction of living germs." The girl also insisted all objects handed to her must be "wrapped up," and Freud interpreted this as relating to a condom.

He concluded that "identification with the patient's mother was the chief content of her neurosis." As early as 1897 he sensed the many ways a young girl identified with her mother's feelings, thoughts, acts, and attitudes had much to do with her emotional development. Psychoanalysts since have devoted much time to exploring this concept more deeply.

Slowly Freud was perceiving how adult sexuality was related to childhood sexual development, how the sexual fantasies and experiences of the child could influence emotional, sexual and mental development. He was also beginning to recognize that people are their own worst enemies. They "don't dare to eat a peach" or "pick a mushroom" because they feel so guilt-ridden. He realized that all the reassurance in the world was of no avail. They needed to discover in themselves the sexual and aggressive wishes they were repressing, wishes that aroused anxiety and contributed toward their strong need for punishment, their inability to enjoy pleasure, and their lack of emotional freedom.

A Daughter's Revenge

Freud wrote five long case histories, each more than one hundred pages, that have become classics in psychoanalytic literature. The first, "A Fragment of an Analysis of a Case of Hysteria," is known as the case of "Dora," the name he gave a young woman to protect her identity. According to Jones, she was the sister of a Socialist leader whose name, he said, could not be disclosed.

Mention of Dora probably appeared for the first time in a letter Freud wrote Fliess on October 14, 1900, reporting a "new patient, an eighteen-year-old girl." He finished writing the case on January 24, 1901, but Jones says that "for motives of professional discretion [perhaps to protect her brother or father]" he did not publish it until April 1905. Freud told Fliess the title of the case was to be "Dreams and Hysteria," that it was "a fragment of an alalysis of a case of hysteria," the phrase he ultimately chose as its title.

Dora was part of two triangles. The first included her mother and father, the second, her father's mistress, a married woman, and her husband. Dora also had a brother, one and a half years older. Her father, a man in his late forties, was a wealthy manufacturer in Vienna. Dora had always been very attached to him and concerned about his health, for since she was six, he had suffered severe illnesses. First he had tuberculosis and the family moved from Vienna to a small town in the southern part of Austria which had better climate. There his lungs improved and he was soon able to resume travel to his many factories, and eventually to live once again in Vienna.

When Dora was twelve, her father had what Freud called a "confusional attack," followed by symptoms of paralysis and slight mental disturbances. A friend persuaded him to see Freud who, on learning the father had syphilis before his marriage, prescribed "an energetic course of treatment" for syphilis and all the disturbances disappeared. Four years later the father brought his daughter to Freud because she was "neurotic" and suffered from a nervous cough and hoarseness of the throat.

Freud said he never met Dora's mother but from the accounts

given by Dora and her father. "I was led to imagine her as an uncultivated woman and above all as a foolish one, who had concentrated all her interests upon domestic affairs . . . She presented the picture, in fact, of what might be called the 'housewife's psychosis.' She had no understanding for her children's more active interests, and was occupied all day long in cleaning the house with its furniture and utensils and in keeping them clean—to such an extent as to make it almost impossible to use or enjoy them." He said relations between Dora and her mother had been "unfriendly" for years and "the daughter looked down on her mother and used to criticize her mercilessly . . ."

Dora's father told Freud that he, his wife, and daughter had formed "an intimate friendship" with a married couple named Herr and Frau K., that Frau K. had nursed him during his long illness, earning his "undying gratitude." He also reported that Herr K. had always been kind to Dora, taking her on long walks, giving her small presents, and Dora had taken care of the K.'s two little children, at times becoming almost a mother to them.

Dora, an intelligent, attractive girl, spoke freely to Freud. She told him that Herr K., who had flirted with her over the years, one day when she was eighteen, as they walked along the shores of a lake at a summer resort where the two families gathered, kissed her ardently. He avowed his love and spoke of his hope of marrying her when he got a divorce. He intimated he had not had sexual relations with his wife in a long time.

At this, Dora struck him in the face and fled. When she told her parents what had happened, they did not believe her. They asked Herr K. if the story were true. He denied it and claimed it was her imagination, that his wife had told him Dora read only books about sex and was interested in nothing else. Dora felt outraged, tried to get her parents to give up seeing the K.'s but they refused.

Dora then told Freud of an earlier episode with Herr K., "which was even better calculated to act as a sexual trauma," Freud said. When she was fourteen, one day Herr K. arranged with her and his wife to meet at his office for a view of a street church festival. Then he persuaded his wife to stay home, dismissed all his clerks and was alone when Dora arrived.

As the time for the festival approached, he asked Dora to wait for him at the door that opened on the staircase leading to the top floor from which they were to see the festival. He then closed the outer shutters of the room, darkening it. He walked over to Dora, who expected him to lead her to the roof, suddenly clasped her in his arms and kissed her on the lips, whereupon Dora felt "a violent feeling of disgust." She tore herself out of his arms, hurried past him down the stairs and out into the street. Neither of them mentioned what happened until Dora confessed it to Freud. She said that for some time after she avoided being alone with Herr K., cancelled plans to join the K.'s on a short trip.

The office scene—second in order of mention to the one at the lake but first in order of time—showed, according to Freud, that something unnatural was troubling Dora, that "the behavior of this child of fourteen was already entirely and completely hysterical." He said he would, without question, consider a young woman hysterical whose own sexual responses and fantasies disgusted her, arousing feelings that were "preponderantly or exclusively unpleasurable." He called this psychic mechanism *reversal of affect* (emotion) and said it was one of the most important and yet one of the most difficult problems in the psychology of the neuroses and as yet he did not fully understand it.

The scene had left behind one consequence, he said, "in the shape of a sensory hallucination" that occurred from time to time, even while Dora was telling him her story. She said she would still feel on the upper part of her body the pressure of Herr K.'s embrace. Freud said he reconstructed in his mind the scene during which, on Herr K.'s "passionate embrace," Dora no doubt felt not merely his kiss on her lips but the pressure of "his erect member" against her body. This perception had been "revolting" to her, dismissed from memory, repressed and replaced by the "innocent sensation" of pressure on her thorax. This in turn derived an excessive intensity from its repressed source and caused the nervous cough and hoarse throat. Freud added, "Once more, therefore, we find a displacement from the lower part of the body to the upper." This had occurred in several other cases.

He also theorized that feelings of disgust may arise when the sight or thought of the genitals "act as a reminder" of the excremental functions, especially when the genitals are male, since that organ performs the function of micturition as well as the sexual function. The function of micturition is the earlier known of the two functions and the only one known during the presexual period, he pointed out, and thus disgust may become one of the means of emotional expression in the sphere of the adult sexual life. He added that the words of one of the early Christian Fathers, "between urine and faeces we are born," cling to sexual life and cannot be detached from it "in spite of every effort at idealization."

Freud did not hesitate to speak frankly of sexual subjects to Dora. He explained, in writing the case, "There is never any danger of corrupting an inexperienced girl. For where there is no knowledge of sexual processes even in the unconscious, no hysterical symptom will arise; and where hysteria is found there can no longer be any question of 'innocence of mind' in the sense in which parents and educators use the phrase."

In his remark about the "innocence of mind," Freud was trying to answer some of his critics. To this day, opponents of psychoanalysis aver that making the unconscious conscious, particularly making the unconscious sexual wishes conscious, corrupts the mind of the innocent. Freud was able to demonstrate, over and over, through the analytic method that children, adolescents, and adults all have strong sexual desires and sexual memories lodged in the psyche. He was further able to clearly show that when people are anxious, frightened, and have disabling neurotic symptoms, their emotional health can be restored as they are helped to become aware of what they really feel.

In an attempt to discover more of the reasons for Dora's nervous cough and sore throat, Freud pointed out to her there was a contradiction between insisting her father's relation with Frau K. was a love affair and, on the other hand, maintaining her father was impotent, as she had done. She told Freud she knew there was more than one way of obtaining sexual gratification. He asked whether she was referring to the use of organs other than the genitals for the purpose of sexual intercourse and she admitted she was. Freud then told her she must be thinking precisely of those parts of the body which, in her case, were in a state of irritation—the throat and oral cavity.

She strongly denied this. But Freud said, in reporting the case,

". . . the conclusion was inevitable that with her spasmodic cough, which, as is usual, was referred for its exciting cause to a tickling in her throat, she pictured to herself a sense of sexual gratification *per os* [oral intercourse] between the two people whose love-affair occupied her mind so incessantly." A short time after she "tacitly" accepted this explanation, her cough vanished.

This case had many ramifications; Freud was now probing more deeply into all the passions that might complicate a life. He uncovered Dora's deep anger at her father not only for the years of his sexual involvement with Frau K. but for "handing her [Dora] over to Herr K. as the price of his tolerating the relations between her father and Frau K." She was unconsciously driven by a desire for revenge for the many hurts her father inflicted on her as he made her part of his promiscuous sexual life.

Freud also discovered Dora had become furious at Herr K. after a young governess at the K. home told her one day that Herr K. had made advances to her, she had become sexually involved with him, but then he quickly discarded her. She said she was staying on hoping he would change his mind and resume the affair. Freud now interpreted Dora's slap, when Herr K. kissed her, as also due to disgust at his duplicity.

Freud pointed out to Dora she also felt a love for Frau K., with whom she had lived for years on a footing of closest intimacy, sharing her bedroom whenever Dora stayed at the K. home. When Dora spoke of Frau K. to Freud, he said she would praise her "adorable white body" in a tone "more appropriate to a lover than to a defeated rival." Freud commented, "When, in a hysterical woman or girl, the sexual libido which is directed towards men has been energetically suppressed, it will regularly be found that the libido which is directed towards women has become vicariously reinforced and even to some extent conscious." Dora flew into a rage when Frau K., to whom she had been so devoted, refused to believe her accusations against Herr K., and even turned traitor by saying Dora was only interested in books about sex. Dora felt as though everyone she loved had betrayed her.

Freud interpreted at length two of Dora's dreams. The first was a recurrent one. He related the dream in Dora's words: "A house was on fire. My father was standing beside my bed and woke me up. I dressed myself quickly. Mother wanted to stop and save her jewel-case; but Father said: 'I refuse to let myself and my two children be burnt for the sake of your jewel-case.' We hurried downstairs, and as soon as I was outside I woke up."

She told Freud she first had the dream three nights in a row while staying at the lake resort where the scene with Herr K. had taken place. She had dreamed it the fourth time a few nights before she told it to Freud.

He deduced the dream was a reaction to her experience with Herr K. at the lake. But to reach the childhood conflicts that formed the underlying interpretation of the dream, he had to know more of her thoughts about the various images in the dream. He asked what occurred to her in connection with the dream.

"Something occurs to me but it cannot belong to the dream, for it is quite recent," she said, "whereas I have certainly had the dream before."

"That makes no difference," he replied. "Start away! It will simply turn out to be the most recent thing that fits in with the dream."

She told him her mother and father had been quarreling because her mother locked the dining room door at night and her brother's room could be reached only through the dining room. Her father did not want her brother "locked in" because "something might happen in the night so that it might be necessary to leave the room," she said.

She then remembered, when the family first arrived at the lake resort, a violent thunderstorm was raging and her father said he was afraid of fire when he saw their small wooden cottage that lacked a lightning conductor. Freud at once made the connection between her father's fear of fire and her words that "something might happen in the night so that it might be necessary to leave the room."

Whereupon she confessed that, as a little girl, she had been a persistent bed-wetter. Since water was needed to put out a fire, Freud associated her bed-wetting to the "fire" in her dream. He also commented that "wetting" was associated with the sexual act.

Also in connection with the dream, Dora spoke of returning at midday from the traumatic walk with Herr K. and lying down as usual for a short nap on the *chaise longue* in Frau K.'s bedroom. She woke suddenly to see Herr K. standing beside her. She asked sharply what he wanted. He said he was not going to be prevented from coming into his own bedroom when he felt like it. She later asked Frau K. if there were a key to the bedroom door and was given one. The next morning while dressing hurriedly she locked the door, fearing Herr K. might burst in.

Freud pointed out that in her dream she had seen her father standing by her bed just as Herr K. had done in reality. He interpreted the dream as containing a resolution to one of her conflicts as in it she assured herself, in effect, "I shall have no rest and I can get no quiet sleep until I am out of this house." The dream, Freud said, turned this the other way and said, "As soon as I was outside, I woke up." Dreams often present the reverse, for opposing thoughts exist peacefully in the unconscious, Freud discovered.

Her mother's "jewel-case," appearing in the dream, was a symbol of the female genitals, Freud said. The meaning of the dream was becoming clearer, he noted in his report: "She was saying to herself, 'This man [Herr K., and, earlier in her life, her father] is persecuting me; he wants to force his way into my room. My 'jewel-case' is in danger, and if anything happens it will be Father's fault.' "

Then, a most important fact, Dora told Freud that when she was a child her father would come into her room at night to wake her, just as Herr K. had done, but for another purpose—not to seduce her but to keep her from wetting her bed.

With this confession, Freud said, the "essence of the dream might perhaps be translated into words such as these: 'The temptation is so strong, Dear Father, protect me again as you used to in my childhood, and prevent my bed from being wetted!' " The dream mirrored the infantile conflict—to wet or not wet the bed—and the adult conflict that stirred memories of it—the temptation to yield to Herr K.'s seductive behavior and "get wet sexually" opposed by several forces rebelling against the temptation—feelings of morality, of jealousy, and wounded pride because of the young governess' disclosure, and Dora's strong tendency to repudiate her own sexuality.

The second dream occurred a few weeks later. She described it in these words: "I was walking about in a town which I did not know. I saw streets and squares which were strange to me. Then I came into a house where I lived, went to my room, and found a letter from Mother lying there. She wrote saying that as I had left home without my parents' knowledge she had not wished to write to me to say that Father was ill. 'Now he is dead, and if you like you can come.' I then went to the station and asked about a hundred times: 'Where is the station?' I always got the answer: 'Five minutes.' I then saw a thick wood before me which I went into, and there I asked a man whom I met. He said to me: 'Two and a half hours more.' He offered to accompany me. But I refused and went alone. I saw the station in front of me and could not reach it. At the same time I had the usual feeling of anxiety that one has in dreams when one cannot move forward. Then I was at home. I must have been traveling in the meantime, but I know nothing about that. I walked into the porter's lodge, and inquired for our flat. The maidservant opened the door to me and replied that Mother and the others were already at the cemetery."

A prominent motive of the dream, Freud said, was a fantasy of revenge against her father, "which stands out like a facade in front of the rest." He referred to the dream as picturing her father as ill, then dead, then in his grave. Behind this fantasy lay concealed thoughts of revenge against Herr K. because of his affair with the young governess, Freud added.

Her refusal in the dream to let a man help her reach the station, Freud said, might be translated into her feelings about Herr K.: "Since you have treated me like a maidservant, I shall take no more notice of you, I shall go my own way by myself, and not marry."

At the end of the analysis of this dream, Dora broke off the treatment, Freud said, as though she were also taking revenge on him for the eleven weeks they had talked about her conflicts in his office. She paid him a visit a year and a quarter later on April 1, 1902, after reading in the newspapers that he had been given the title of Professor by the University of Vienna. She told him that for four or five weeks after she stopped seeing him she had been "all in a muddle," but now she was absorbed in work (he did not say what kind). She also informed him she had no thoughts of marrying.

But she did marry, he eventually learned, and died years later in New York City. He had been able to help her enough so she could accept an intimate relationship with a man. He claimed his analysis of Dora was "imperfect" because it had only started when she broke off the treatment. It was also "imperfect," he added, because Dora did not seek him on her own but consented to see him at her father's request—always a handicap for both patient and analyst when the patient is present because someone has asked him to undertake treatment. There are those who question whether Freud's treatment of Dora ever began.

Freud remarked in a footnote to the case that the longer the interval of time that separated him from the end of her analysis, the more probable it seemed to him that the fault in his technique lay in an omission. He said he failed to discover in time, and to inform Dora, that her homosexual love for Frau K. was the strongest unconscious current in her mental life. He reported he ought to have guessed this because Dora had chosen Frau K. as her sexual confidant, which young girls often do older women. Frau K., ironically, had been the one who later accused Dora of being interested only in sex.

Freud said he should have "attacked this riddle" and looked for the motive of such an extraordinary piece of repression. If he had done this, the second dream would have given him the answer, for "the remorseless craving for revenge expressed in that dream was suited as nothing else to conceal the current of feeling that ran contrary to it— the magnanimity with which she forgave the treachery of the friend she loved [Frau K.] and concealed from every one the fact that it was this friend who herself had revealed to her the knowledge which had later been the ground of the accusations against her." Freud admitted, "Before I had learnt the importance of the homosexual current of feeling in psychoneurotics, I was often brought to a standstill in the treatment of my cases or found myself in complete perplexity." He also said he had made some mistakes with Dora in not dealing adequately with the "transference," which as yet remained to be studied in all its complexities.

But this was 1900, when Freud was only at the beginning of his exploration of neurosis and the symptoms he called the result of "bodily compliance," the physical expression of a repressed emotion, as well as the relation between neurotic symptoms and repressed sexual perversions. At the time he treated Dora, he was preoccupied with the connection between dreams and neurotic symptoms. He had just published *The Interpretation of Dreams* and the Dora analysis is a continuation of the theories in that revolutionary volume.

His purpose in writing the Dora case, he said, was to show the value that interpretation of dreams had in analytic treatment. Unless the therapist had learned the art of interpreting a dream, he could not hope to penetrate the structure of a neurosis, Freud asserted.

Dora's dreams revealed her sexual love for Herr K., for Frau K., and for Dora's father, all of which had been so repressed as to be totally unknown to her. The only figure in her life—and the most important—whom Freud neglected to mention as the object of her love was her mother. In his later studies he realized that the deep attachment of a daughter to her mother was her original love, and the daughter had to be able to move away emotionally from her mother before she could love her father.

Dora's dreams disclosed the extraordinarily complicated interplay in a life of emotions such as love, hate, jealousy, disgust. This case history of Freud's has for years served as a model for students of psychoanalysis.

As students and teachers of psychoanalysis have reviewed the case of Dora, they have enriched the understanding of Dora's psychodynamics, her transference to Freud and Freud's countertransference.

Dr. Jules Glenn in *Freud and His Patients* has updated the analytic findings on Dora. He points out that Freud did not take into sufficient consideration that Dora was a teen-ager. As an adolescent, her wish for autonomy led to a struggle against the analyst who was foisted on her by her father and experienced by her as the agent of the parents from whom she was trying to break away. Glenn also states Freud seemed to overlook the fact that the intimacy of the analytic situation is often a threatening one for a young woman who is consciously trying to avoid attachments to older men. Dora, like many adolescents, made liberal use of the defense mechanism of reversal and changed her love (toward parents, Herr K., and Frau K., Freud and others) to hatred. Freud did not fully appreciate these complexities when working with her since, at the time, he was not yet aware of the many vicissitudes of transference. Nor did Freud fully appreciate Dora's adolescent seductiveness, possibly because her antagonism masked it.

Glenn also suggests Freud was premature in his interpretations of Dora's symptomatic acts and, as a result, she might have experienced him as a dangerous sexual adult who was trying to seduce her.

As psychoanalysts have become more aware of how the ego copes with the anxiety activated by impulses, the superego, and the outside world—a major contribution to psychoanalytic thought by Anna Freud—they have noted how much a patient like Dora feared betrayal and desertion and did to Freud what she feared would be done unto her. She betrayed him by deserting him.

Glenn lists as the major determinants of Dora's leaving the analysis: 1. Distrust of Freud, who reminded her of her lying father— the transference was intensified by the reality that Freud had accepted her as a patient on the request of her father, thus Freud was experienced by her as joining the family in a conspiracy against her. 2. Fear of Freud as a seducer who discussed sexual matters openly and prematurely. 3. Fear of her own sexual wishes toward Freud. 4. A need to turn from passivity to activity when threatened with desertion and betrayal. 5. Revenge against Freud, transferred from her desire to get even with other adults.

What also should be considered in a further understanding of Dora's analysis is that Freud did not encourage her to continue. When she announced she had come for her last session, he accepted it as a *fait accompli.*

Like many analysts, Freud may not have found Dora's rejection of him easy to cope with and therefore rejected her by viewing her as untreatable. He seemed to underestimate her affection and desire for help, while overestimating her desires for revenge. There is the impression he used her hostile feelings as a justification for terminating treatment.

Finally, we know young women can induce erotic fantasies in the male analyst, which Freud might have felt impelled to deny. This may happen when the analyst rejects a patient prematurely. Dr. Alexander Grinstein in *Freud and His Dreams* has suggested Freud had erotic interests in female patients whom he unconsciously considered Oedipal substitutes. In Grinstein's discussion of Freud's Irma dream, in which Irma, a patient, is given an injection, Grinstein suggests the injection symbolized Freud's unconscious wish to inject Irma sexually.

For Love of Another Woman

Twenty years later, in 1920, Freud wrote of another eighteen-year-old girl in a twenty-nine-page article titled, "The Psychogenesis of a Case of Homosexuality in a Woman." It began with the sentence: "Homosexuality in women, which is certainly not less common than in men, although much less glaring, has not only been ignored by the law, but has also been neglected by psychoanalytic research."

He ventured the opinion that the narration of a single case "not too pronounced in type," in which it was possible to trace the origin and development of homosexuality in the mind "with complete certainty and almost without a gap may, therefore, have a certain claim to attention."

He did not give his patient a name but described her as "a beautiful and clever girl of eighteen, belonging to a family of good standing." She had "aroused displeasure and concern in her parents by the devoted adoration with which she pursued a certain lady 'in society' who was about ten years older than herself." The parents charged that this lady, in spite of her distinguished name, was a promiscuous, bisexual woman. She lived with a married woman, with whom it was alleged she was sexually involved, while at the same time carrying on affairs with men.

The patient "did not contradict these evil reports," Freud said, "but neither did she allow them to interfere with her worship of the lady, although she herself was by no means lacking in a sense of decency and propriety." When the girl made a serious attempt at suicide, the alarmed parents persuaded her to consult Freud.

She did not regard herself as needing help, but consented to see him for her parents' sake. Freud said he was skeptical in general about the prospects of bringing about change in a homosexual. He had succeeded in only a few favorable cases, where heterosexual urges were strong, or where the person had strong motives for wanting a change in sexual choice.

Though both Freud and the patient understood intellectually the causes of her homosexual behavior, only slight progress was made beyond this. Freud perceived that "a powerful motive" maintaining

186

the homosexuality was her wish to be avenged on her father for his opposition to her behavior. Freud broke off the treatment, he said, advising the girl to be analyzed by a woman because her hatred of a man was so strong. There is no record as to whether she ever followed his advice, or how long treatment lasted. We must remember Freud was being discreet in these reports to protect the patient.

He discovered she had passed through a pronounced Oedipal phase in childhood but had not emerged from it, as most women do by transferring their love from their father to another man and working through their rivalry with and hostility toward their mother. Instead, she responded in exaggerated fashion "in the way many people do when disappointed in love: namely, by identification with the lost object." He explained this was one way of regressing toward infantile narcissism, thus avoiding conflict with the mother and trying to ensure her love. Not enough attention had been paid to this motive of "evasion," Freud added, one that plays an important part in the cause of homosexuality.

With the young woman's help, as she talked freely, he unraveled some of the motives that had impelled her to try to kill herself, stating in the article that suicide is the wish to murder turned on the self. He suggested, "Perhaps no one can find the psychical energy to kill himself unless in the first place he is thereby killing at the same time someone with whom he has identified himself, and is directing against himself a death wish which had previously been directed against the other person."

Until the young woman tried to commit "the desperate act" of killing herself, neither she nor her parents had any idea of the strength of her passions for the other woman, Freud said. He commented on how often people are unaware of the power of their love until some relatively slight incident (in this case, the father passing his daughter and the older woman on the street) reveals such love by the intensity of their response (her suicidal attempt).

The parents had told Freud their daughter waited hours for the older woman, standing outside her door, and that she often sent her bouquets. "It was evident this one interest had obliterated all others in the girl's mind," Freud said. The parents did not know to what extent their daughter had gone in her sexual relations with this questionable lady, but only that she never showed any interest in a man.

Also, she had previous attachments to members of her sex, which had originally aroused her father's suspicion and anger.

The suicide attempt followed immediately after the father met his daughter on the street walking with the older woman. He passed them by with a glance of fury. The girl promptly confessed to the older woman that the man who had given them such an angry look was her father and that he had forbidden her to continue the friendship. Whereupon her companion turned on her, ordered her to leave her side and never again wait for her outside her door, or speak to her.

In despair at this sudden rejection of her love, the girl rushed to the side of the road, flung herself over a wall and down the side of a hill that bordered a railroad track. She hurt herself so severely she spent several weeks in bed. After her recovery, her father did not dare oppose her with so much determination, and the older woman, up to then rather cool to her, was now moved by this unmistakable proof of the girl's passion and became more friendly.

About six months after the girl's attempted suicide, the parents appealed to Freud for help. The father, Freud said, was "an earnest, worthy man, at bottom very tender-hearted, but he had to some extent estranged his children by the sternness he had adopted towards them." When he first learned of his daughter's homosexual tendencies, he flared up in rage and threatened her. Freud comments, "There was something about his daughter's homosexuality that aroused the deepest bitterness in him, and he was determined to combat it with all the means in his power." Then Freud added cynically, "the low estimation in which psycho-analysis is so generally held in Vienna did not prevent him from turning to it for help."

The mother was "a still youngish woman, who was evidently unwilling to relinquish her own claim to find favour by means of her beauty." She did not take her daughter's passion for a woman as tragically as did the father, nor was she so incensed at it, Freud said. She was however "decidedly harsh towards her daughter and overindulgent to her three sons, the youngest of whom had been born after a long interval and was then not yet three years old."

Freud learned from the young woman that with the older women she had adored, she had enjoyed only a few kisses and embraces. "Her genital chastity, if one may use such a phrase, had remained intact," he put it. The older woman she now loved had never allowed her to do more than kiss her hand, and kept urging her to give up bestowing her affection on women. The girl herself insisted to Freud her love was pure and she was physically revolted at the idea of sexual intercourse.

Freud learned that at the age of thirteen she had displayed a sudden tender, exaggeratedly strong affection for a three-year-old boy she saw regularly in a playground at a Vienna park. She took to the child so warmly that a permanent friendship developed between his parents and her. Freud commented that one could infer at that time she wished to have a child herself. Then, after a while, she grew indifferent to the boy and started to take an interest in mature but still youthful women, which was when her father administered "a mortifying chastisement" (Freud does not go into detail as to its content).

Freud noted that her change from love of a small boy to older women occurred at the age of sixteen, simultaneously with the pregnancy of her mother and the birth of a third brother. The analysis revealed "beyond all shadow of doubt that the beloved lady was a substitute for—the mother," Freud said. He pointed out the first women she loved were mothers, women between thirty and thirty-five she had met with their children on summer holidays or in the family circle of acquaintances. Motherhood as a "condition of love" was given up only in her latest love, for this older woman had no children. But her slender figure, beauty, and offhand manner reminded the girl of her older brother, so that her latest choice "corresponded not only with her feminine but her masculine ideal, combining the homosexual tendency with heterosexual," Freud said.

He suggested that this girl, whose mother favored her three sons at her daughter's expense, as well as jealously keeping strict watch against any close relation between her daughter and her husband, obviously yearned "for a kinder mother." The mother did not help her daughter through the difficult period at puberty of the revival of the infantile Oedipus complex.

At the birth of her third brother, the daughter became keenly aware of her wish to have a child, and a male one, but "that it was her father's child and his image that she desired, her consciousness was not allowed to know," Freud said. "And then—it was not she who bore the child, but her unconsciously hated rival, her mother. Furiously resentful and embittered, she turned away from her father and from men altogether. After this first great reverse, she forswore her womanhood and sought another goal for her libido."

She entirely repudiated her wish for a child and the desire for the love of a man. Freud said that what happened was "the most extreme [development] possible of all open to her." She "changed into a man, and took her mother in place of her father as her love-object." Her relation to her mother had been ambivalent from the start "and it proved easy to revive her earlier love for her mother and with its help to bring about an overcompensation for her current hostility toward her. Since there was little to be done with the real mother, there arose from the conversion of feeling described the search for a mother-substitute to whom she could become passionately attached."

Freud mentions another reason for the girl's homosexuality, which he described as "advantage through illness." The mother still attached great value to the attentions and the admiration of men. If the daughter became homosexual and left men to her mother, in other words "retired in favour of" the mother, she believed she would remove something that had been partly responsible for her mother's dislike. She would no longer be a rival and her mother would love her.

She also remained homosexual to defy her father, Freud said, in that her behavior followed the principle of the talion (punishment in kind): "Since you have betrayed me [by giving mother a baby instead of me], you must put up with my betraying you." She *wanted* her father to know of her passion for the older woman, otherwise she would be deprived of satisfaction of her keenest desire—revenge. The father's reaction, one of fury, was as though he realized the deliberate revenge was directed against him, Freud says. He adds that the wish for revenge was also directed against her mother for becoming pregnant and producing another masculine rival to compete for her love.

Freud mentioned another important factor in female homosexuality. He said the analysis showed the girl had suffered in childhood from a strong "masculinity complex." She was " a spirited girl, always ready to fight, she was not at all prepared to be second to her slightly older brother; after inspecting his genital organs she had developed a pronounced envy of the penis, and the thoughts derived from this envy still continued to fill her mind." He added, "She was in fact a feminist; she felt it to be unjust that girls should not enjoy the same freedom as boys, and rebelled against the lot of women in general."

Her homosexuality was exactly what would follow from the combined effect in a person "with a strong mother-fixation of the two influences of her mother's indifference and of her comparison of her genital organs with her brother's," Freud claimed.

He concluded that two fundamental facts about homosexuality had been revealed by psychoanalytic investigation: homosexual men experience a specially strong fixation in regard to the mother and in all "normal" people there can be detected a very considerable measure of latent or unconscious homosexuality.

He urged further research on the causes of homosexuality, which contemporary psychoanalysts are conducting, but essentially Freud's conclusion, stated in this paper, still holds true, that there is a strong "mother fixation" on the part of both male and female homosexuals. This fixation arises from several factors, including the mother's seductive behavior to the child, not necessarily overt but subtle, and her attitude that men are no good, dangerous, and untrustworthy.

By this time Freud was even more aware of the resistance in a patient who did not want to be helped, as in this case he said she was "determined to retain the sole symptom" for which she was being analyzed. He recorded the case from the point of view of the causes of her unhappiness but he had been unable to help her resolve her conflicts because of her "lack of cooperation," which was when he referred her to a woman analyst.

In light of the further understanding of "resistance," a contemporary psychoanalyst would spend far more time helping such a patient overcome her resistance, not giving up so early in treatment or advising a change of analyst, as Freud did. The patient would be helped to gain a sense of trust in the analyst, and the analysis would become more of an emotional experience, rather than the brief intellectual one of earlier days, when theories were being developed.

It would appear the premature termination of the analysis of the "homosexual woman," as in the case of Dora, was due in part to Freud's failure to recognize her adolescent status, the defense mechanisms typical of that period, and his own countertransference responses (perhaps unconsciously thinking of the young patients as his rival sisters when they were adolescents). It may be conjectured that once again Freud had difficulty coping with his patient's hostile transference, a displacement of her hatred toward her father. Instead of helping her discharge this hatred by expressing it toward him and understanding it arose in childhood, he discharged her prematurely.

From this case Freud was later able to point out that what parents abhor in their children, they really hate in themselves. No doubt the homosexual woman's father despised his own homosexual wishes and, therefore, could not be very understanding of his daughter's plight.

One of the reasons not just parents, but people in general, are so punitive toward homosexuality is because anger and the wish to punish are convenient means of dispelling their own guilt about homosexual wishes. Psychoanalysis has reaffirmed many times that we despise in others what we cannot tolerate in ourselves.

From this case Freud showed that homosexuality is much more than just a "life style." The homosexual man or woman is a human being who is suffering. He or she must avoid sustained intimacy with the opposite sex to prevent devastating anxiety and rage. Contemporary psychoanalysts and therapists from other disciplines have proved the homosexual is an unfulfilled, unhappy person, unaware of the many unconscious wishes and fears causing his constricted sexual life.

Over the years Freud reported briefly on other cases. In 1913, he wrote of a woman patient who was sensitive about anyone gazing at her feet. This he traced to an experience in childhood when she had tried to imitate her brother's way of urinating, wet her feet, and her brother made fun of her. Freud concluded that unexpected self-criticism by a patient comes either from identification with someone the person dares not criticize or from covering a wish to boast about some other quality, such as when a woman, secretly proud of her charms, deplores a supposed lack of intelligence. He made the important statement that when symptoms are alluded to in a dream, the rest of the dream will contain some explanation of the cause of the symptoms.

Thus Freud kept learning from his women patients as he explored the many complexities of the primitive, or unconscious part, of the mind and its effect on the person trying to become civilized. He treated women patients from eighteen to fifty, some very rich, some moderately wealthy, some poor. They represented all degrees of emotional disturbances: some suffered episodes of psychotic behavior, some had been in institutions, some had mere anxiety attacks, others were only occasionally depressed.

Out of each case he learned something, perhaps only one insight, perhaps several, perhaps further proof of a theory he had just originated. He tried to ease the woman's suffering if he could. Even in his failures he observed why he had failed and this in turn led to deeper exploration.

Part Four

Freud's Contributions to Understanding Women

The "Riddle" of Femininity

Theories of great men are often colored by subjective influences and Freud's conclusions about the nature of femininity are no exception. We have shown that his wish to restore and preserve an idealized image of his mother influenced his choice of occupation, his important adult relationships, his concepts of personality development, his interactions with his children, and his activities as a therapist.

In particular, we have tried to show that the defensive operations he used to protect and preserve his idealized vision of his mother from invasion by his deep hostility toward her left their mark on his theories of psychosexual development and on his notions of womankind. He loved and cherished women, but he also feared them and, at times, showed contempt for them. Thus his unresolved ambivalence toward women affects several of his psychoanalytic theories of femininity.

Experts on the study of human behavior, with their biases, limitations and vulnerabilities, are influenced, as are all of us, by their own subjectivity, and some of their conclusions about intrapsychic and interpersonal behavior do not rest exclusively on impartial reflection. It has been well documented that theories of personality often tell us as much about the theorist and his personal problems as they do about the dynamics of others. Dr. Alfred Adler, discoverer of the "inferiority complex," was a sickly child, who hated his older brother and other children for being "superior." Dr. Harry Stack Sullivan, who theorized that "every child needs a chum," acknowledged he was a lonely boy who always yearned for a friend. Dr. Karen Horney, who never forgave her father, a seafaring man, for being away from home a great deal of the time, argued that the major cause of neurosis is the "feeling of loneliness in a hostile world."

The eminent psychologist Dr. Henry A. Murray, in *Explorations in Personality*, says, "Man—the object of concern—is the ever-varying cloud and psychologists are like people seeing faces in it . . . for each perceiver every sector of the cloud has a different function, name and value—fixed by his initial bias of perception."

197

In addition to Freud's personal biases, his attitudes toward women also have to be evaluated in the light of his time. The type of male narcissism found in his theories about women also appears in the writings of many authors of that era. Western culture at the turn of the twentieth century looked down on women as existing only to gratify a man's needs, bear children, and run the home. Later psychoanalysts showed the saliency of the patient's social context in the development of conflicts.

In terms of the thinking of his era, Freud's women colleagues could be considered emancipated as far as his attitude toward them was concerned. While he believed women belonged in the home, he took it "as a gross inconsistency to exclude women from practicing psychoanalysis" and encouraged them to study and train in that field. Yet he really could not tolerate women as competitors and equals, though he admired their intelligence and loyalty, doing his part, with much success, in keeping them dependent on him.

Jones says of Freud's attitude toward women that, if judged by today's standards, it would be called "rather old-fashioned." But, Jones adds, whatever Freud's intellectual opinions may have been, there were indications in his writings and correspondence of a different "emotional attitude." Jones concludes it might be fair to describe Freud's view of the female sex "as having as their main function to be ministering angels to needs and comforts of men." His letters and choice of wife showed he had only one type of sexual object in mind, a gentle, feminine one, Jones says. Yet, while believing women might belong to the weaker sex, Freud regarded them in some respects as finer and ethically nobler than men, according to Jones, in that "there are indications that he wished to absorb some of these qualities from them."

Freud clearly took into account how society held back women. Toward the end of his career he acknowledged, "We must beware . . . of underestimating the influence of social customs, which . . . force women into passive situations."

During most of his life he appears to have regarded women as less sexual than men. His fear of his sexual feeling for his mother was the main reason for his fairly frequent inability to see women as full sexual human beings. He wrote of children that they "behave in the same kind of way as an average uncultivated woman in whom the same polymorphously perverse disposition persists." One source of his hatred of America was that the women there were less sub-servient to men, and he did not like the shift away from the Old World views.

Freud was undoubtedly also frightened of the "woman" in him-self. He had strong yearnings to submit and be given to by a father-figure, but fought these wishes. He feared his own passivity, hated so to lose control that he refrained from drinking whiskey or taking aspirin. In clinical practice he often linked femininity and creativity. He once remarked to a highly artistic male patient, "You are so feminine you can't let it out." He meant this interpretation to be a compliment.

In discussing femininity, Freud explained that psychoanalysis "does not try to describe what a woman is—that would be a task it could scarcely perform—but sets about enquiring how she comes into being." He meant how a woman develops emotionally and men-tally into a mature, fulfilled human being. He did not claim to know all the answers to what he called "the riddle of the nature of femininity," which "throughout history people have knocked their heads against."

What he found out about femininity was "certainly incomplete and fragmentary," he admitted in 1933 but, he explained, he was only describing women "insofar as their nature is determined by their sexual function . . . It is true that that influence extends very far; but we do not overlook the fact that an individual woman may be a human being in other respects as well."

Freud's seeming humility in the last quotation is betrayed by his contempt when he says an individual woman "may be a human being." But he did acknowledge he was not an omniscient god when he added, "If you want to know more about femininity, en-quire from your own experiences of life, or turn to the poets, or wait until science can give you deeper and more coherent information." Psychoanalysts who followed Freud have added important research to the knowledge of the psychosexual development of a woman.

In assessing Freud's contributions to understanding women it is important to recognize those monumental discoveries that have weathered many a controversial storm and withstood the passage of time. Some of his theories, though validated clinically, are repudiated even today by some professionals and lay people. Concepts such as infantile sexuality, penis envy, and the Oedipus complex are revolutionary and stir anxiety in the souls of even the most intrepid and mature.

The most-quoted, supposedly hostile remark of Freud's concerning women is one he made to Marie Bonaparte: "The great question that has never been answered and which I have not yet been able to answer despite my thirty years of research into the feminine soul, is 'What does a woman want?' "

Though many have interpreted this remark as an accusatory complaint, it was meant to be a philosophical query about the frustrated women of the Victorian era in Vienna, who reflected their unhappiness at the constricted, puritanical times in which they lived. Even the daughters of the wealthy were not allowed to attend the universities, compete in any way with men, admit sexual desire or feelings of anger. Those women who dared suggest they be given the right to vote were ridiculed.

If Freud today had to answer his own question, "What does a woman want?" undoubtedly he would change his answer to a more appropriate one, in line with how times have changed since those overromantic, sexually inhibited days of the early twentieth century.

The second most quoted "anti-feminist" remark of Freud's is "anatomy is destiny." With his strong tendency toward a defensive type of male chauvinism, he may have been implying that only men, who possess the mighty phallus, can be achievers and free souls. But contemporary psychoanalysts have interpreted this statement as suggesting that how one views his or her anatomy will influence their destiny. Clinicians note every day that when patients feel self-hatred the root of the conflict most often lies in their self-demeaning attitude toward their sexuality.

Among Freud's outstanding contributions to the study of femininity are his pointing out the inadequacy of contraceptive methods, his condemnation of the practice of *coitus interruptus* as causing anxiety in women, and his deploring the inhibiting and destructive effects of feminine sexual ignorance.

With the rise of feminism, first in England, then in Germany and France, Freud's views on women changed. It is interesting that the woman who translated from English into German the first book on women's fight for equality, A *Vindication of the Rights of Women* by Mary Wollstonecraft, published in 1792, was the woman whose treatment Freud described as the starting-point of psychoanalysis—Bertha Pappenheim. She did the translation (she also paid to have it printed) at night, after working all day as administrator of the Jewish Girls Orphanage in Frankfurt.

It is true Freud did not use his energy to fight for the freedom of women politically, economically, or socially—only emotionally and sexually. But perhaps that is enough for one man. It is only fair that his statements about the role of women in society be kept separate from his scientific discoveries of female sexual development, even though the two are related. He recognized that the psychological slavery that social customs imposed on women affected their mental and sexual well-being.

For Freud to embark on freeing women from sexual bondage took Herculean effort in the backward 1890s. The importance of sexuality in causing hysteria first struck him, he said, when studying with Charcot in Paris. He overheard Charcot tell a colleague, as they discussed a nervous young married woman, that her husband was either "impotent" or "very awkward." The colleague expressed astonishment this should have anything to do with the wife's condition. Whereupon Charcot said, "In such instances, it is always a matter of sex—always—always—always."

This was 1895, just before Freud married Martha. Up to then, being a very moral young man, he may not have thought much about sex consciously, though he was very romantic when it came to "love," as his letters to his fiancée showed. One of the earliest written expressions of his feelings about women occurred in these letters. In the first flurry of love, he wrote Martha on June 18, 1882, "What sorceresses you women are!" A year later, on September 16, 1883, he expressed the opinion, ". . . no girl in love for the first time knows whether it is the real thing," intimating that he, a man, did, and was superior in that respect (one way to deal with a sorceress).

In this letter, Freud defends a woman. He writes of the death of a friend, a doctor heading a department at the Vienna General Hospital, who hung himself in a public bath only ten days after his return from a honeymoon. Freud says, "And why did he do it? . . .

the details that drove him to his death are unknown to us, but that they are linked up with his marriage is beyond doubt . . . As an explanation, the world is ready to hurl the most ghastly accusations at the unfortunate widow [who had led him a merry chase before accepting his proposal]. I don't believe in them . . . He died from the sum total of his qualities, his pathological self-love coupled with the claims he made for the higher things of life." Freud in 1883 was already on his way to understanding the deeper levels of the mind.

That October 6 he wrote Martha of feeling inferior to a woman: ". . . you are so good and—between ourselves—you write so intelligently and to the point that I am just a little afraid of you. I think it all goes to show once more how quickly women outdistance men." Here he was paying tribute to the higher intelligence of women.

But he made up for the compliment in a letter on October 23, writing, "I will let you rule the house as much as you wish, and you will reward me with your sweet love and by rising above all those weaknesses for which women are so often despised." He did not state what they were.

In a letter dated November 15, 1883, he discussed at length the work of John Stuart Mill, English economist and philosopher (1806–1873), whose essay on the emancipation of women Freud was translating into German. The fact Freud chose this essay to translate suggests an unconscious choice in favor of women, even though he did not fully agree with Mill's theories.

He described Mill to Martha as "perhaps the man of the century who best managed to free himself from the domination of customary prejudices." Then he said, ". . . I recollect that in the essay I translated a prominent argument was that a married woman could earn as much as her husband. We [Martha and he] surely agree that the management of a house, the care and bringing up of children, demands the whole of a human being and almost excludes any earning, even if a simplified household relieve her of dusting, cleaning, cooking, etc. He [Mill] had simply forgotten all that, like everything else concerning the relationship between the sexes."

He went on to say that in Mill's presentation "it never emerges that women are different beings—we will not say lesser, rather the opposite—from men. He finds the suppression of women an analogy to that of Negroes. Any girl, even without a suffrage or legal competence, whose hand a man kisses and for whose love he is prepared to dare all, could have set him right." Here Freud, twenty-nine years old and still in medical college, sounds naive, though he then says that "law and custom have much to give women that has been withheld from them."

On one occasion he and his friend Schönberg met a woman who talked "in a very affected, unintelligent fashion," and Schönberg spoke contemptuously of her, but Freud told Martha he defended the woman on the grounds she was a good wife and mother. He added, "If a woman feels orthographically, so to speak, it doesn't matter so much whether she writes and speaks orthographically or not." He thus dismisses this woman's ability to talk or spell correctly as far less important than her ability to feel the "correct" impulses.

Five months before they married, he wrote Martha on April 19, 1884: ". . . it is a general rule that brides are happier than bridegrooms," and we may wonder if this mirrors his ambivalent feelings about marrying after four and a quarter years of courtship and opposition by Martha's mother. Then he added he looked forward to "sharing life with a beloved who is going to be not only a housekeeper and cook but a precious friend and a cherished sweetheart as well."

As he became a psychoanalyst, he started discovering the many facets of the sexual life of young girls and women. By 1938, when forced to leave Vienna because of the Nazi invasion, he was still working out his theories of female sexuality. He wrote his son Ernst on May 13, 1938, in discussing a sudden inhibition in himself to the writing of letters: "Perhaps you will remember that I once traced back the so-called 'physiological feeble-mindedness of women' (Möbius) [Dr. Paul Möbius, a German neurologist] to the fact that women were forbidden to think about sex. As a result they acquired a dislike for thinking altogether." He may have been unconsciously referring to his inability to feel sexual desire in this year before he died.

In a letter to Dr. Max Eitingon a month later, June 6, 1938, Freud announced that for the first time in his life he found himself temporarily cared for by a woman physician. On the flight to London, his regular doctor, Dr. Max Schur, scheduled to accompany him, suddenly had an emergency appendix operation. Anna substituted Dr. Josefine Stross, "a lady pediatrician," Freud said, "who looked after me very well."

By then he had drastically changed his views about women's place as being in the home. All around him women were rising to the top of professions, not only in the medical and psychoanalytic worlds but in business and politics. For this progress Freud certainly deserves some credit, for liberation in one area of life sparks liberation in other areas.

Original Views

The rough beginnings of Freud's theories on female sexuality appear in his letters to Fliess. He states one of his first views on neurosis in women in an undated letter at the end of 1892. At that time he believed impotent husbands caused hysteria in women, in line with Charcot's thinking.

He told Fliess that no nervous disease can exist "without a disturbance of the sexual function" and that in men, neurosis is accompanied by "relative impotence," while in women it is "a direct consequence of neurasthenia in men, through the agency of the reduction in potency in the latter."

Perhaps only teasing, he wrote Fliess on February 8, 1893, amplifying this theory, asking him to keep the letter from the eyes of his young wife (Fliess had just married Ida Bondy of Vienna). Freud explains, "The contention I am putting forward and desire to test by observation is that neurasthenia is always *only* a sexual neurosis." He spoke also of a "precondition" to neurosis, saying, "A great deal will of course depend on the predispositions of the two partners: whether (1) she herself was neurasthenic before marriage, and (2) whether she was made hystero-neurasthenic during the time when intercourse occurred without preventives."

He had been able to trace as due to "coitus interruptus," he said, anxiety neurosis in women, including symptoms of hypochondria, claustrophobia, agoraphobia, obsessive brooding, and depression. In one woman with "tormenting hypochondria" which started at puberty, he discovered she had been sexually assaulted when eight years old. He commented that the "terror of sexuality" could in itself produce neurosis.

Eight months later in a letter dated October 6, 1893, he told Fliess that when he asked women patients about their sexual feelings "they all go away impressed and convinced, after exclaiming: 'No one has ever asked me that before!' " Freud was daring to test his theory and finding that the women were accepting his very personal questions about the most tabooed topic in their lives, in spite of strong resistance, natural in the face of society's prohibitions.

205

He was studying frigidity in women when he wrote Fliess in January 1895 that he wondered why anaesthesia was so predominantly a characteristic of women. He speculated, "This arises from the passive part played by women. An anaesthetic man will soon cease to undertake sexual intercourse, but women have no choice."

He maintained women become frigid more easily for two reasons: "Their whole upbringing aims at not arousing somatic sexual excitation (S. S.) but at translating into psychical stimuli any excitations which might otherwise have that effect," and, "Women very frequently approach the sexual act (or get married) without any love —that is, with only slight somatic sexual excitation (S. S.) and tension of the terminal organ [the vagina]. In that case they are frigid and remain so." Marriages of convenience arranged by parents were the rule in those days, especially in wealthy Jewish families, who did not want their fortunes dissipated by poor in-laws.

Freud, who was also studying the nature of depression, found in most of his women patients that a low level of tension "in the terminal organ seems to constitute the main disposition to melancholia." Where this low level is present, any neurosis "easily takes on a melancholic stamp." He speculated that "potent individuals easily acquired anxiety neurosis" while the impotent "inclined to melancholia." This theory he soon abandoned as invalid.

In the same letter, to illustrate the cause of paranoia, he cited the case of a woman patient sent by Breuer. Declaring that people became paranoiac "about things that they cannot tolerate," he told of this unmarried woman of thirty who shared a home with her brother and older sister. She belonged to the "superior working-class" and her brother was a minor executive of a small manufacturing business. They had, for a time, rented one of the rooms in the house to an acquaintance, "a much-traveled, rather mysterious man, very clever and intelligent."

This man lived with them for a year "on the most companionable terms," then abruptly left. He returned in six months, stayed a short while, then vanished for good. The two sisters lamented his absence, spoke of him only in praise. Then one day the younger sister (Freud's patient) told the older one that once when she was dusting the boarder's room, as he lay in bed, he had "called her up to the bed, and, when she had unsuspectingly obeyed, put his penis in her hand," Freud said. Soon after, the boarder left the house forever.

Within the next few years the younger sister began to have delusions of persecution, believing women neighbors were castigating her for her behavior and saying untrue things about the former boarder. Freud diagnosed this to Fliess as "a neurosis which can easily be interpreted as a sexual one." He said she fell into a "fresh bout of paranoia" every so often.

The older sister told Freud that whenever conversation turned to the attempted seduction, her sister denied anything had occurred. Freud tried in his treatment to overcome the paranoia by bringing her memory back to the scene but was unsuccessful.

One day, he wrote Fliess, he "got her to tell me all about the lodger in a state of 'concentration hypnosis.'" In reply to his searching enquiries, he said, "as to whether anything 'embarrassing' had happened, I was met by the most decided negative and—I saw her no more. She sent me a message [even though she was still in therapy] to say that it upset her too much." He added in frustration, "Defence! That was obvious. She *wished* not to be reminded of it and consequently deliberately repressed it."

Then Freud asked, "What was the peculiar nature of the paranoiac defence?" And answered: "She was sparing herself something; something was repressed. And we can guess what that was. She had probably in fact been excited by what she had seen and by recollecting it. So what she was sparing herself was the self-reproach of being 'a bad woman.' And the same reproach was what reached her ears from outside . . . To start with it had been an internal reproach; now it was an imputation coming from outside. The judgment about her had been transposed outwards . . ."

She gained something by this, however, Freud noted: "She would have had to accept the judgment from inside; but she could reject the one from outside." He added, in italics, "*In this way, the judgment, the reproach, was kept away from her ego.*" The purpose of the paranoia was to "fend off an idea that was intolerable to her ego by projecting its subject-matter into the external world . . . paranoia is a misuse of the mechanism of projection for purposes of defence."

The alcoholic, for instance, will blame the woman with whom he is impotent, never admitting he has become impotent through drink, Freud pointed out: "However much alcohol he can tolerate, he cannot tolerate this knowledge." Whatever the delusional idea, people "love their delusion as they love themselves. Herein lies the secret," he said.

He cited the case of "an old maid" who keeps a dog, or an old bachelor who collects snuffboxes, as illustrating "something quite analogous" that takes place with obsessions, which deal with the psychic mechanism of *substitution*. The old maid uses the dog as substitute for "a companion in marriage," Freud said, while the bachelor uses his collection of snuffboxes for "a multitude of conquests."

Freud put it: "Every collector is a substitute for Don Juan Tenorio—so too the mountain climber, the sportsman, and so on. These things are erotic equivalents. Women are familiar with them as well." He said gynecological treatment falls into this category in that there are two kinds of women patients: one category is loyal to their physicians, as they are to their husbands, whereas the other change their physicians "as though they were lovers."

Through his first women patients Freud was working out his theories of the various psychic mechanisms that govern daily life, telling them to Fliess, a companion willing to listen, encourage, and occasionally propose theories of his own that proved not very pertinent. But he served a purpose in that he admired and believed in Freud at a time when almost everyone else was condemning him.

The "Shocking" Book

The one book of Freud's that more than any other temporarily brought down the wrath of the world on his head, including most of the medical profession, was what he called "the little work" that completely revolutionized the scientific views on sexuality, leading to the so-called sexual revolution. The book contained what Freud described as "the 'red rag' of the sexual factor," and it shocked and antagonized even his close friend Breuer. Today the antagonism has somewhat abated, as more and more people accept Freud's belief that "from the beginning everything sexual should be treated like everything else worth knowing."

The book, appearing in 1905, was called *Three Essays on the Theory of Sexuality.* He started off by defining what he called "libido." He described it as sexual hunger, a natural part of the sexual instinct possessed by human beings and animals. For centuries, he said, "popular opinion" believed it absent in childhood, starting to emerge at puberty and revealed in maturity as "an irresistible attraction exercised by one sex upon the other," with its aim "presumed to be sexual union, or at all events actions leading in that direction."

But, he said, he had every reason to believe such views gave "a very false picture" of the true situation and, if studied closely, would reveal a number of errors, inaccuracies, and hasty conclusions.

He introduced two technical terms used throughout the book. One was "the sexual object," the person to whom the sexual attraction flowed. The second was "the sexual aim," the act toward which "the instinct tends." Numerous deviations, such as the "perversions," occur in respect to both the sexual object (in fetishism, a fur hat or lacy lingerie, for instance), and the sexual aim (not procreation but mere release of tension through masturbation or other means), so that the relation between the deviations and what is assumed to be "normal" requires "thorough investigation," he said.

He explained the perversions—homosexuality, voyeurism, fetishism, exhibitionism, sadomasochistic practices, the sexual use of the anal orifice, and sexual use of the mucuous membrane of the lips and mouth—were "fixations" occurring in the process of natural psychosexual development. If, for any reason, the process was impeded in its natural expression, the person would suffer either from a failure to reach mature sexual desire or from a need to keep expressing one of its earlier manifestations which society called "perversion."

The symptoms shown by the neurotic are formed in part, he said, at the cost of "abnormal" sexuality," so that *neuroses are, so to say, the negative of perversions.* (Italics are his.) He explained that the contents of the conscious fantasies of "perverts," of the delusional fears of paranoiacs that are projected in a hostile sense onto other people, and of the unconscious fantasies of neurotics, which psychoanalysis reveals lies behind their symptoms, all coincide "even down to their details."

An unconscious tendency to perversion "is never absent and is of particular value in throwing light upon hysteria," he asserted. The mental life of all neurotics, without exception, he insisted, "shows inverted impulses, fixation of their libido upon persons of their own sex."

In this book Freud theorizes that sexual feelings do not spring full-blown at adolescence, as people took for granted, but start at the day of birth when a baby grasps his mother's breast and begins to drink milk, thereafter associating the feel of warm flesh with being fed and nourished.

He said, "There seems no doubt that germs of sexual impulses are already present in the new-born child and that these continue to develop . . . the sexual life of children usually emerges in a form accessible to observation about the third or fourth year of life." He added he believed the processes connected to sexual development actually start "as early as during intrauterine life."

What he called "infantile amnesia, which turns everyone's childhood into something like a prehistoric epoch and conceals from him the beginnings of his own sexual life," is responsible for the fact that, in general, no importance is attached to the development of sexual life in childhood, he said. The amnesia is used to deny periods of masturbation, of sexual interest in and curiosity about the genitals of parents, brothers, and sisters.

Children, far from being the little innocents everyone believes them to be, Freud said, are very aware of their sexual feelings. He used the word "sexual" in this instance, he explained, to mean awareness of the "functions" and "appearance" of their bodies, the bodies of parents, as well as how babies are conceived and born, and about the act of sex. If they sleep in their parents' bedroom and happen to see or overhear their parents in sexual intercourse, they may become overstimulated, Freud said, and need to defend themselves in various ways against this too-early, excessive arousal of the sexual drive. He speculated that victims of St. Vitus dance, whose movements are so symbolic of the sexual act, may have been sexually overstimulated as children and are expressing both their frustration and repressed excitement of an earlier time.

One of Freud's earliest and most lasting convictions was that of bisexuality. He was convinced the infant was at first bisexual and that the difference between the sexes did not emerge in full form until puberty. He placed emphasis on two crucial expressions of bisexuality in mental life—penis envy and the castration complex.

He thought of children as "polymorphous perverse," which meant their sexual drive in its infantile aspects could fasten on man or woman, took pleasure in sexual looking, exhibitionism, and sadistic impulses, all part of preparation for mature sexuality.

He laid the foundation in this book of the stages of psychosexual development on which all future psychoanalytic thought was based. He theorized that the development took place in four main stages— oral, anal, phallic, and genital. The stages were not complete in themselves but overlapped and pervaded the others. The degree of "pathological perversion" depends, he said, on what he called the "vicissitudes" that the sexual drive encounters along the way to mature expression. He also described three stages of love—self-love, love of someone with the same body, and love of someone of the opposite sex.

For a little girl (and boy) the oral stage starts at birth and is followed by the anal stage, which begins whenever her parents decide to train her and focus her attention and concern on controlling her urine and feces. This usually occurs between one and a half and two years of age. It is followed by the phallic stage, which starts about four. The little girl now becomes aware of the contrast between the bodily cavity that appears as her genitals and the appendage on the body of the boy out of which flows his urine. The final, or genital stage, starts in the mid-teens, at the end of puberty which begins at eleven. Young women are then ready to menstruate, to mate, and to procreate.

Once he had established the existence of infantile sexuality, Freud did not hesitate to say he believed all children's questions about sex should be answered honestly. He thought sexual enlightenment should be offered children as soon as they asked questions, saying they would absorb only as much as they could accept at their age.

In *Three Essays on the Theory of Sexuality* Freud discusses at length the relationship of a daughter to her parents as crucial to natural psychosexual growth. Both girls and boys display from an early age a dependency on the people looking after them which is "in the nature of sexual love," he said. Anxiety in children is "originally nothing other than an expression of the fact they are feeling the loss of the person they love." It is for this reason they are frightened of every stranger and afraid of the dark where they cannot see the one they love, he added.

The excessive affection of a parent for a little girl may awaken her sexual desire "prematurely," before her body is adult enough to handle it, he warned. The postponing of sexual maturation gives the little girl time to build, among other restraints upon sexuality, the barrier against incest and to learn to observe "the moral precepts" that preclude from her choice of objects of love, those she has loved in childhood.

He explained the difference between "romantic love" and "sexual love" as a fantasy of adolescence due to the "infantile fixation of the libido." He said girls with an acute need for affection "and an equally exaggerated horror of the real demands made by sexual life have an irresistible temptation on the one hand to realize the ideal of sexual love in their lives and on the other hand to conceal their libido behind an affection which they can express without self-reproaches, by holding fast throughout their lives to their infantile fondness, revived at puberty, for their parents or brothers or sisters."

Psychoanalysis has no difficulty, he said, showing these women they are *"in love"* (italics his), in the everyday sense of the word, with these blood relations. He adds that in instances where someone who previously has shown no sign of neurosis, falls ill after an unhappy experience in love, it is possible to show with certainty that "the mechanism of his illness consists in a turning-back of his libido upon those whom he preferred in infancy."

He warned that in view of the importance of a child's relations to his mother and father in determining his later choice of a sexual object, any disturbance of those relations will produce "the gravest effects upon his adult sexual life." For instance, jealousy in a lover is never without an infantile root, he said. If there are quarrels between parents, or if their marriage is unhappy, the ground is prepared in their children "for the severest predisposition to a disturbance of sexual development or to a neurotic illness."

Freud changed his views on many theories during his life but he remained constant to the end on infantile sexual development. It formed one of the strongest pillars of both the theory and technique of psychoanalysis.

Phallic Envy

One of Freud's outstanding discoveries is the notion of "infantile omnipotence." He was able to demonstrate that every child aspires to be a god and "wants what he wants when he wants it." Children yearn to possess everything in sight; place them in a toy store, and they want to own it immediately. Every boy goes through a stage of wanting to possess breasts, have babies, and enjoy all the privileges and pleasures he believes girls and women enjoy, reasoning, "The grass is always greener on the other side." Just as every boy flirts with the idea of becoming a girl, every girl flirts with the idea of becoming a boy. Every patient in a successful analysis, at one point or other, finds himself wishing to be a member of the other sex.

Perhaps the most difficult theory of Freud's for women to accept is that of "penis envy." Most women deny they entertain any such thought. But the childish wish to have a penis, repressed by most women, emerges in their dreams and in the fantasies of those who lie on the psychoanalyst's couch, as they recall to consciousness early repressions. The wish is also at times freely expressed by little girls to their parents and peers.

Freud maintained that every girl, as she enters the phallic stage at the age of four, becomes aware that a boy possesses an appendage she lacks. She imagines that originally her clitoris, from which she derives sexual pleasure, was a penis before it was "cut off." She has the fantasy, according to Freud, that she has been castrated. She blames her mother, who, she believes, has a penis hidden in her body, for this castration and "does not forgive her for thus being put at a disadvantage," Freud says.

The little girl then turns to her father, expecting he will give her his penis. When she is disappointed here too, she slowly accepts the fact she will never have a penis, and her wish then becomes one of having a baby by her father, later transferring this wish to a man outside the family. This is her sexual preparation for femininity, according to Freud.

In talking of "the little creature without a penis," as Freud once described a girl, he said she imagines her penis was taken away because she was "bad." She feels inferior because of the evident superiority of the penis to her own "small and inconspicuous clitoris." From then on, Freud said, she "falls a victim to envy for the penis," which may cause a strong wish to be a boy that is never resolved unless she faces her feelings.

Freud declared, "You may take it as an instance of male injustice if I assert that envy and jealousy play an even greater part in the mental life of women than of men." This envy, he said, is rooted in the wish for the penis. He claimed that feminine "shame" is connected with "concealment of genital deficiency."

Envy of the penis serves a psychological purpose, Freud maintained. A girl is "driven out" of her attachment to her mother by this envy and "enters the Oedipus situation as though into a haven of refuge," Freud puts it. He claimed girls face a handicap spared boys because they do not fear castration as intensely as boys do and thus lack "the chief motive to surmount the Oedipus complex," remaining in it an indeterminate length of time. "In these circumstances the formation of the super-ego must suffer; it cannot attain the strength and independence which give it its cultural significance, and feminists are not pleased when we point out to them the effects of this factor upon the average feminine character," he wrote in his paper "Femininity" in 1933.

Psychoanalysts who have followed Freud have noted the strong contribution of a male-dominated society to penis envy. Many a woman has understandably reasoned, "If men get more—power, status, money—why not act like a man?" This hides the wish to be a man.

Contemporary psychoanalysts have added another dimension to Freud's theory of penis envy. While they agree penis envy is a universal fantasy of little girls at the age of four, they say it need not remain a driving force in a woman's life, creating unhappiness. If a little girl's emotional needs are understood by a loving mother and protective father, the normal fantasies of penis envy that occur during her phallic stage of sexual development will be accepted, then suppressed (a conscious act, rather than the unconscious act of repression). She will then proceed to the final, or genital stage of sexual development with ease. Her affectionate, tender feelings will merge with her erotic ones, and she will be able to love a man not for the physical attribute which, as a little girl, she envied and unconsciously wished to possess, but out of her feelings for him as a total person. She will want him not as possessor of the desired phallus, but as mate and father of her child. In Freud's words, her original wish for a penis has changed into the wish for a baby.

Thus penis envy does not have to be a lifelong state to which women are doomed, current psychoanalysts say, though it may continue to be "a regressive fixation in those women who are afraid of femininity." The wish to be a man remains intense only if a woman is afraid of sexual penetration or of pregnancy.

A little girl's yearning for a penis may be particularly intense in a family where both mother and father favor a brother or in some way convey the feeling they wish she had been a boy. This may occur if the first-born is a girl, or if there have been several girls and the expected newcomer turns out to be another girl instead of the wished-for boy. Every little girl wants to be loved *as a little girl*. If she does not believe she is loved for herself, the only way of winning parents' affection is to try to be a boy.

A little girl also has the fantasy, Freud said, that if she had a penis she could, like a boy, urinate without "messing" herself. Later feelings of shame may be based on memories of her mother's scoldings when she toilet-trained her daughter and her mother's shame at being a woman, Freud said. A girl may also think that if she had been a boy, she would not have to endure what she believes the embarrassment and humiliation of menstruation (some call it "the curse"), which brings fantasies of bleeding to death as punishment for all her "sins"—masturbation, fantasies of marrying her father, of killing off her mother and rival brothers and sisters.

In her unconscious envy of the penis, many a woman adorns herself with feathers, sequins, furs, glistening silver and gold ornaments that "hang down"—what psychoanalysts call "representations" of the penis. In some primitive tribes women bind their breasts so they protrude stiffly in front, like an erect penis.

A woman who unconsciously is still seeking the coveted penis may think of a man as a penis rather than as a person. In her fantasy he becomes her appendage, enhancing her, making her feel strong and powerful. Dreams of many a woman after her first act of intercourse reveal "an unmistakable wish . . . to keep for herself the penis with which she had come in contact," Freud noted, adding that "these dreams indicated a temporary regression from the man to the penis as an object of desire."

But when emotional trauma in the oral, anal, and phallic stages has not been too frightening, a girl will not fear becoming a woman or look on "woman" as inferior to man. She will not be afraid of being tender or sexually receptive to a man, rather than submissive, an attitude that conceals rage. Rather than angrily competing with men, the woman who is secure in her femininity will continue to seek equality in more and more areas.

Clitoris to Vagina

Another controversial theory of Freud's is that a woman must transfer her sexual interest from her clitoris, the girlhood site of excitability, to the vagina in order to be fully feminine. Freud called the transfer "intimately related to the essence of femininity." If a woman is unable to achieve this transfer and still considers her clitoris the main sexual area, this will be "the chief determinant" of her neurosis, Freud said.

He was the first scientist to point out there was a difference between clitoral and vaginal response and to theorize about the effect on a woman's psychosexual development if she failed to transfer her sexual gratification from one to the other. "With the change to femininity, the clitoris should wholly or in part hand over its sensitivity, and at the same time its importance, to the vagina," he stated.

He did not say he believed the clitoris should be completely desensitized, that its stimulation could not play a preparatory role in sexual intercourse. But if all sexual interest remained focused on the clitoris, he warned that the vagina was likely to remain anesthetic and the "clitoridal" (his term) woman would have no wish to be penetrated by the penis. He was talking primarily about a psychological shift from the active wanting of her mother to the passive aim of wanting to be wanted by her father. Any arrest at the clitoridal state indicated inability to face the Oedipal conflict or achieve the wish for a man's penis to enter the vagina, Freud said.

He further explained that if a girl could not give up the clitoris as the chief site of sexual stimulation, this was a sign she still wished to be a boy. She was afraid to become passive and tender and still wanted to be in control of the sexual situation.

Transference from clitoris to vagina as the new erotic site aids and abets what Freud described as the second task a woman had to perform in the course of her development, as contrasted to the "more fortunate man," who only needed "to continue at the time of sexual maturity the activity that he has previously carried out at the period of the early efflorescence of his sexuality."

This second task facing a girl is the change in her love object from her mother to her father. Though the little boy is spared this task, he must alter the quality of his love for his mother from a possessive, infantile one to a more mature love that takes in her needs, as he resolves his ambivalent feelings. This is true too of the little girl, toward both her mother and father.

When she abandons clitoral masturbation the little girl renounces a certain amount of general activity and passivity gains the upper hand as she turns to her father, Freud said: "This wave of development clears the phallic activity out of the way and smooths the ground for femininity." He added, "If too much is not lost in the course of it through repression, this femininity may turn out to be normal."

Freud called "the essential feature in the development of femininity" the ability of a girl to turn her sexual love from the mother to the father. At the time the clitoris is the little girl's chief erotic zone, one of her fantasies is that if it grows into a penis she can become the lover of her mother—her first love—and take the place of her rival father, Freud said (this is one of the unconscious fantasies of the lesbian). But when her erotic feelings turn to the vagina, the feminine receptacle for a man's penis, her fantasies turn to love of her father, and she proceeds on her way to sexual maturity.

This is not an easy act for many women, Freud warned, because the duration of a daughter's attachment to her mother has been "greatly underestimated." Many a woman remains "arrested" at the original attachment and never "properly achieves the changeover to men." Freud attributes one of the reasons to the fantasy of seduction by the mother, held by most little girls. He says, ". . . the seducer is regularly the mother . . . Here, however, the phantasy touches the ground of reality, for it was really the mother who by her activities over the child's bodily hygiene inevitably stimulated, and perhaps even roused for the first time, pleasurable sensations in her genitals."

Little girls may also be sexually overstimulated by mothers or other women who care for them. Freud mentioned the case of a little girl who had been masturbated almost daily by an emotionally disturbed governess, as the result of which the little girl suffered many hysterical and bodily symptoms.

If a girl has deep, unresolved hostility to her mother, it will pervade her life with a husband, Freud predicted, and "the second half of a woman's life may be filled by the struggle against her husband, just as the shorter first half was filled by her rebellion against her mother." Also, under the influence of a woman's becoming a mother herself, an identification with her own mother may be revived, bringing with it a compulsion to repeat the unhappy marriage between her parents if one existed.

Freud postulated a number of fantasies held by little girls, saying, "Enough can be seen in the children if one knows how to look." Among these fantasies are: babies are born through the mouth (in the oral stage); babies, like feces, come out of the anus (in the anal stage); the penis is a weapon that can annihilate, and sexual intercourse is a brutal, assaultive act by the man upon the woman (in the phallic stage).

Freud also pointed out that as a girl transfers her erotic love from her mother to her father, there is a change in the nature of her aims. She wants a man not to nurse her, as her mother did, but to protect her, think her lovely and desirable. This is part of what Freud called "the transformation of the instinct."

While little girls do not as yet know the techniques of adult sex, they may be filled with erotic longing, a yearning to be held close, kissed, caressed by their fathers. If a father loves his little daughter in a tender, wise way, she will not be unduly aroused. She will be able to handle her erotic feelings and eventually transfer them to a man outside the family. But if fathers, or other male adults in her life, respond in excessively seductive ways to her, she may live at too high a level of sexual tension. Her fantasies will be highly stimulated at an age when emotionally she cannot handle either the fantasies or her aroused physical feelings.

If a little girl is actually seduced by a man she will, as a woman, have intense sexual fantasies and, as psychoanalytic case histories show, can become a prostitute or nymphomaniac. Girls whose fathers have been excessively seductive, just short of actual incest, are apt to engage in much sexual acting out later in life.

The growing girl must eventually accept that her desire for her father is destined to remain forever unfulfilled. Sometimes her parents do not make this easy. A mother may be jealous and punish her daughter for her natural attraction to her father. A daughter's normal sexual feelings may thus come into conflict with her dependency on her mother and father. She is caught between the passion of her feelings and her need to hide them so her mother and father will continue to love her.

Freud pointed out that the rape fantasy is a common one to girls and women, with their father unconsciously pictured as the brutal attacker, though usually in the disguise of a stranger. Many women seek as lovers sadists who torture, beat, or rape them. This beating fantasy has its origin in the woman's incestuous attachment to her father Freud reported in his paper, "A Child is Being Beaten." The beating is punishment for the incestuous wishes. The brutal stranger also serves to take total responsibility for the sexual act; the woman is forced to submit to his will and thereby erases all her guilt over incestuous wishes.

The original fantasy in "A Child is Being Beaten" Freud says, is "I am being beaten (i.e., I am loved) by my father." This is later replaced by a fantasy which disavows the little girl's wish for sexual intimacy with her father. The beating is then performed in her imagination upon someone she views with jealous rage (a brother or sister) as she becomes a spectator of an event that takes the place of the sexual act, which in her early childhood fantasy looks like a beating administered to her mother.

Beating fantasies represent not only punishment for forbidden incestuous relations but also the regressive substitute for it. Freud said, "Here for the first time we have the essence of masochism." He explained that masochism is the transformation of an active sadistic impulse into the passive wish to suffer, due to a sense of guilt. The guilt in women that follows masturbation, he said, is due not to the act itself but to the fantasy which lies at its root—the incestuous wish for the father.

A woman must overcome her "deference" for a man and come to terms with the idea of incest with father or brother, for "deference" conceals hatred, and this hatred will always interfere with future love, Freud said. Unless she can accept that her incestuous love for her father or brother was a natural one, she cannot transfer desire freely to another man.

The ability to express tenderness is very important in achieving mature, or genital love, Freud believed. If a woman has an irreconcilable split between her sensual and her tender feelings, her erotic life will remain, as Freud put it, "divided between two channels: the same two that are personified in art as heavenly and earthly (or animal) love."

Just as there are men who can express love only "for a harlot," to use Freud's phrase, so there are women who can love only promiscuous men, the Don Juans. Such women are in fantasy reliving the childhood Oedipal triangle. They are for the moment successful in mastering the conflict as they keep their promiscuous lover (the father) by their side and triumph over the rival (their mother). That their lover will inevitably discard them, as he has all his other women, they know unconsciously, even as they consciously may fight against all odds to keep him sexually faithful.

Psychoanalysts point out that some of the new sexual practices believed by many to exemplify greater freedom actually prove that incestuous fantasies are at work. Orgies, in which couples perform the sexual act in full view of each other, are a reenactment of the primal scene. The child, now an adult, is getting even with his parents for having had intercourse in front of him, or in the room next to him, where he overheard sounds and words. When three persons are involved in the sexual act, this too relates to the primal scene, with the child, now an adult, at last able to fulfill his wish to join his mother and father in the act of sex.

In his exploration of female sexuality, Freud became the first to liberate woman (man, too) from the prejudiced belief that masturbation was "evil" and would harm body and mind. An issue of the day was whether masturbation caused physical illness. Freud believed its deleterious effects are psychological, because of guilt over the incestuous fantasies, rather than organic. He theorized masturbation was a necessary act in infancy in the development of sexuality, and those girls too frightened to explore the pleasure of their own bodies would have difficulty in later years accepting the body of someone of the opposite sex.

Freud was also responsible for the concept, so important in today's psychoanalytic thinking about femininity, that a little girl's beliefs of what it means to be a woman are formed, both consciously and unconsciously, in large part as a result of her relationship to her mother, whether the little girl is primarily loved and wanted or hated and rejected. Also important is the mother's feelings about being a woman and her attitudes toward "men."

Freud was the first to note that depressed mothers tend to have depressed daughters. The very disturbed mother who has little self-esteem cannot pass on a sense of self-esteem to her daughter and, therefore, will cripple her psychosexual development, psychoanalysts have found out.

The mother may be unhappy at having given birth to a girl and this, too, the baby girl will feel, and then hate herself for being feminine. Every girl is somewhat ambivalent about her mother, but if the girl is excessively hostile, this will create the fear of her mother's revenge as she projects her own hostility onto her mother. This anger and fear leads little girls to curb their feminine identification, for angry little girls do not wish to be feminine but masculine, so they can have the power they attribute to the penis.

Psychoanalysts also stress the importance of the father in helping his little girl feel feminine, so she will wish to become a woman. Fathers who remain aloof out of fear of emotional or sexual contact with their daughters interfere with the little girl's desire to accept her incipient womanhood. Moreover, if a father downgrades the mother, the little girl will not wish to emulate her.

If a mother and father have remained children themselves, emotionally speaking, the little girl will not stand a very good chance of becoming a mature woman. Too much of her psychic energy has to go into defenses she must erect and maintain against her exploding sexual and angry feelings as her natural passions clash with the passions of her parents.

Female Aggression

Freud arrived late in life, for reasons of his own, at formulating the theory that there was a second instinct as powerful as the sexual—the aggressive. He had always spoken of sadism, hate, and the desire for revenge from the earliest cases on, usually in relation to the frustration of Oedipal sexual wishes. In 1909 he was still saying that "every act of hate issues from erotic tendencies."

Then in 1914 he introduced the concept of "narcissism" and said it included the self-preservative urges emerging as the feeling of hate, summoned to ward off threats to life, either psychological or physical—if someone attacks us we instinctively feel hate so we can defend ourselves. But Freud's new theory became "profoundly unsatisfying" to him, Jones says, because then the inner conflict was between the "narcissistic" and the "erotic" impulses—between two forms of the sexual instinct—and Freud had come to think in terms of "opposing forces in the mind that created conflict," as opposed to two forms of a single instinct. Formerly he labeled the force opposing the sexual drive as "self-interest."

In 1915 Freud wrote *Instincts and Their Vicissitudes,* in which he differentiated between "hate" and the sexual drive. This was the start of his concept of a nonsexual part of the ego that opposed the sexual instinct. In 1919, he wrote the controversial book *Beyond the Pleasure Principle,* in which he established two opposing forces in the mind—the life instincts and the death instincts, or destructive instincts. He said both were of equal status and "in constant struggle with each other," though the latter "inevitably won in the end." Many psychoanalysts have disagreed with the notion of a "death instinct," but accept the existence of an "aggressive instinct." They believe that normal, healthy aggression should be differentiated from defensive hatred, tinged with the desire for vengeance.

In writing of women and aggression, Freud said that "social customs" had suppressed woman's natural aggressiveness. As noted, he warned, ". . . we must beware . . . of underestimating the influence of social customs, which force women into passive situations." He added that the suppression of women's aggressiveness that has been "imposed on them socially favours the development of powerful masochistic impulses, which succeed, as we know, in binding erotically the destructive trends which have been diverted inwards."

The adding of an aggressive instinct led to a change in his theory of equating "femininity" with passivity, submissiveness, and masochism, and "masculinity" with activity and assertiveness. He stated in 1933 that he now believed women could display great activity in a variety of ways and no longer considered "active" synonymous with "masculine" or "passive" with "feminine." He pointed out, "Analysis of children's play has shown our women analysts that the aggressive impulses of little girls leave nothing to be desired in way of abundance and violence."

Aggression is the motor force of both work and love, he said, part of all activity, large and small. It is only when the aggressive impulse becomes excessively tinged with hate and desire for revenge that it overwhelms feelings of love and tenderness and may end in psychosis, suicide, or murder.

Discussing aggression in a letter to Marie Bonaparte dated May 27, 1937, Freud said he would try to answer her question about aggression, that the whole topic had not as yet been "treated carefully" and what he had to say about it in earlier writings "was so premature and casual as hardly to deserve consideration."

He told her he thought "the instinct of destruction" could be "sublimated into achievements," that all activities "that rearrange or effect changes are to a certain extent destructive and thus redirect a portion of the instinct from its original destructive goal." The sexual instinct cannot be expressed without some measure of aggression and "therefore in the regular combination of the two instincts there is a partial sublimation of the destructive instinct," he said.

Curiosity, "the impulse to investigate," may be regarded as a complete sublimation of the aggressive impulse, Freud said, and in the life of the intellect, the destructive instinct "attains a high significance as the motor of all discrimination, denial and condemnation." The turning inward of the aggressive impulse is naturally the counterpart of the turning outward of love when it passes from the self to others, he told Marie Bonaparte, adding, "One could imagine a pretty schematic idea of all libido being at the beginning of life directed inward and all aggression outward, and that this gradually changes in the course of life," adding, "But perhaps that is not correct."

He concluded the letter by saying the repression of aggression was the hardest part to understand: "As is well known, it is easy to establish the presence of 'latent' aggression, but whether it is then latent through repression or in some other way is not clear."

He asked, "Please do not overestimate my remarks about the destructive instinct. They were only tossed off and should be carefully thought over if you propose to use them publicly. Also there is so little new in them." He later said in his book *The Outline of Psychoanalysis*, published posthumously in 1941, "All of psychoanalysis has to be reformulated in terms of understanding the aggressive drive as separate from the libidinal drive." Contemporary psychoanalysts are starting to do this.

In commenting on the case of a woman who dreamed she would like to see her daughter, seventeen, dead before her eyes, Freud said he discovered in the mother's analysis that she had harbored this death-wish for a long time. The child was born of an unhappy marriage, soon dissolved. Once while the mother was pregnant, in a fit of rage after a violent scene with her husband she had beaten her fists against her abdomen in the hope of killing the fetus within.

Freud commented: "How many others, who love their children tenderly, perhaps over-tenderly, today, conceive them unwillingly and wished at that time that the living thing within them might not develop further! They may even have expressed that wish in various, fortunately harmless, actions. Thus their [the child's] death-wish against someone they love, which is later so mysterious, originates from the earliest days of their relationship to that person . . ."

In his later years Freud placed great emphasis on the destructiveness of the aggressive urge in humans when they cannot control it. In *Civilization and Its Discontents*, written in 1930, he said, ". . . men are . . . creatures among whose instinctual endowment is to be reckoned a powerful share of aggressiveness. As a result, their neighbor is for them not only a potential helper or sexual object, but also someone who tempts them to satisfy their aggressiveness on him, to exploit his capacity for work without compensation, to use him sexually without his consent, to seize his possessions, to humiliate him, to cause him pain, to torture and to kill him."

Because of Freud's awareness of the role hate plays in the life of the child and the adult, aggression and its many effects on emotional and mental development constitutes one of the main themes of psychoanalytic study today. The hatred of a parent for a child and the effect on the child, for instance, was mentioned by Freud in 1931 in "Female Sexuality." He described the early attachment of a daughter to her mother as closely connected to the cause of neurosis and speculated that excessive dependence on the mother contained "the germ of later paranoia in women." He described this "germ" as "the surprising, yet regular, dread of being killed (?) devoured by the mother" (question mark is Freud's). This dread of the mother may, he said, arise from an unconscious hostility on the mother's part which the child senses.

Contemporary psychoanalysts have been able to refute Freud's notions about women having a weaker super-ego than men and contend that the strength or weakness of the super-ego does not depend on the sex of a person but on how his parents helped him face the conflicts of early life associated with weaning, toilet-training, and sexuality.

Freud also speaks of women as more "narcissistic" than men and having a stronger need "to be loved" rather than "to love." This is debatable, for men can surely be as narcissistic as women (Narcissus, after all, was a man), depending on their emotional experiences as a child.

Freud also said of women they "must be regarded as having little sense of justice." He relates this "to the predominance of envy [penis envy] in their mental life." He describes the demand for justice as a modification of envy, a putting aside of envy. This also is debatable, for a sense of justice would appear to stem from how parents treat a child and the values they inculcate in the child, rather than to which sex it belongs.

Freud maintained women have less capacity for sublimating their instincts than men, are more psychically "rigid and unchangeable," and "have made few contributions to the discoveries and inventions in the history of civilization." He ignores the fact that men through the centuries have kept women subjugated, and limited or forbid opportunities for sublimation.

There are areas important to women that Freud did not deal with very deeply, including the maternal instinct, undoubtedly strong in most women. When thwarted by the trauma in a woman's early emotional life, the maternal feelings appear in sublimated ways. Bertha Pappenheim was an outstanding example. Never able to marry and give birth, though she wanted children desperately as shown in her fantasied pregnancy with Breuer, she embarked on a career to help unwed, pregnant mothers and the illegitimate children born to them and deserted by their fathers. Bertha Pappenheim provided food, shelter, education, and training in various types of work for these deprived mothers.

What of the mother who hates her children enough to kill them, or encourages a lover to kill them and get them out of the way? Freud explained that such intense feelings of jealousy and violence relate to brutal parents who stunted the woman's maternal instincts in infancy.

When Freud insists that a mother's relation to her son is "altogether the most perfect, the most free from ambivalence of all human relationships," we may doubt this, if only because of the number of murders committed by sons of their mothers, or of wives and mistresses to whom these sons have transferred early murderous hatred. Men commit far more murders than women and if, as Freud said, the seeds of murder are sown in the nursery, the relationship between many a mother and son cannot be too sublime.

We have repeatedly stressed that Freud can be charged with failing to understand his ambivalence toward women and have related this to the fact that he never faced the depth of his fury at his overidealized mother. This may also have delayed his understanding of the strength of the aggressive impulse. But he was the one to point out the power of ambivalence, the mixed feelings of love and hate that exist in all of us. In Freud's life, his love and respect for women far outweighed his repressed hatred.

What Freud Gave Women

Freud was not perfect, he too was entitled to hostilities that ran deep. But so did his love, and perhaps if both feelings had not been so intense, he never would have made his great discoveries. The demand that he be "perfect," that fifty years ago he be able to foresee the future of feminism and become active on all fronts of women's liberation, is part of the unrealistic, infantile wish that one's parents be omnipotent. This wish has to be given up by those who want to take an objective look not only at themselves but at others and society.

Freud had good reason to trust women because no woman ever deserted or deceived him. Freud's mother, wife, daughters, women friends, and colleagues were all faithful. He handled ambivalence toward women by denying his hatred. With men, he either over-idealized them or overtly rejected them. Women responded to his wish for continued contact by never abandoning him, but some men could not tolerate his intermittent arrogance and rejected him, including Jung, Adler, and Rank.

Though somewhat handicapped by unresolved pre-Oedipal and Oedipal conflicts, Freud slowly and thoughtfully worked out some of the practical ways a woman could become sexually free, rather than, as in the past, humiliated, subordinated, and frustrated. He refused to go along with the doctors of that era in considering a nervous, hysterical woman as untreatable. He listened to her "crazy" thoughts, made sense of her nonsense, and discovered the way to mental healing.

He established the fact that mental health is not merely an absence of severe emotional illness but includes the ability to use potentials to a far higher degree. He declared most people used only a small part of their intelligence and creativity.

He added a new "right" not only to the lives of women but to human rights in general—the right to mental health. The belief a mother and father may consider a child his possession, to abuse physically and psychically, is no longer sacrosanct, or even tolerable. Child abuse has become a crime. A mother who cannot stop herself from assaulting her child can no longer blame the child's behavior, but is urged to find out why she hates the child, why, for instance, she may be inflicting on him a repressed fury for a cruel husband or a violent mother or father.

Freud's findings also led to emphasis on the prevention of mental illness, the bringing of help to children before their conflicts escalate into adult unhappiness. Today many national organizations exist to aid emotionally disturbed children, including the mentally retarded and those suffering physical illnesses without organic cause.

Lastly, Freud helped women realize it was not a crime to become aware of the inner self so they could be happier, more effective in whatever they wished to undertake, whether motherhood, career, or both. He aided them in understanding their nature and all human nature. It is a nature that includes hate, fear, deceit, greed, tyranny, prejudice—signs of every savagery that lies in the human heart. By tracing destructive behavior of the self and others to its sources in the mind, Freud helped at least to make peace a possibility for the individual.

Freud's gifts to women are great—greater than women yet know how to use. Whatever he may have owed the few women he loved, admired, respected, and befriended, he has repaid womankind a thousandfold by showing the way to emotional and sexual freedom— the essence of emancipation.

Books Consulted

Alexander, Franz, ed., with Eisenstein, Samuel; and Grotjahn, Martin. *Psychoanalytic Pioneers*. Basic Books, 1966.

Bernays, Edward L. *Biography of An Idea*. Simon & Schuster, 1965.

Bonaparte, Marie. *Female Sexuality*. International University Press, 1953.

Breuer, Josef; and Freud, Sigmund. *Studies on Hysteria*. Basic Books, 1957.

Clark, Ronald W. *Freud: The Man and His Cause*. Random House, 1980.

Deutsch, Helene. *Confrontations with Myself* W. W. Norton, 1973.

Doolittle, Hilda. *Tribute to Freud*. McGraw-Hill (Paperback Edition), 1975.

Eissler, K. R. *Sigmund Freud: His Life in Pictures and Words*. Helen and Kurt Wolff Books, Harcourt Brace Jovanovich, 1978.

Fine, Reuben. *A History of Psychoanalysis*. Columbia University Press, 1979.

Freud, Ernst L., ed. *Letters of Sigmund Freud*. Basic Books, 1960.

Freud, Martin. *Glory Reflected*. Angus and Robertson. 1957.

Freud, Sigmund. *Three Essays on the Theory of Sexuality*. Standard Edition Complete Psychological Works, VII, 1920.

——. *Some Psychical Consequences of Anatomical Distinction Between the Sexes*. Ibid., XIX, 1920.

——. *Female Sexuality*. Ibid., XXII, 1920.

——. *Beyond the Pleasure Principle*. Ibid., XVIII, 1920.

——. *An Autobiographical Study*. W. W. Norton, 1952.

——. *Dora—An Analysis of a Case of Hysteria*. Collier Books, 1963.

——. *The Origins of Psycho-Analysis: Letters to Wilhelm Fliess*. Basic Books, 1954.

Grinstein, Alexander. *Freud and His Dreams*. International Universities Press, 1980.

Jones, Ernest. *The Life and Work of Sigmund Freud*. Basic Books, I, 1953, II, 1955 III, 1957.

Kanzer, Mark; and Glenn, Jules. *Freud and His Patients*. Jason Aronson, 1980.

Leavy, Stanley A. ed. *The Freud Journal of Lou Andreas-Salomé*. Basic Books, 1964.

Mitchell, Juliet. *Psychoanalysis and Feminism*. Vintage Books, 1975.

Murray, Henry. *Explorations in Personality*. Science Editions, 1962.

Peters, H. F. *My Sister, My Spouse.* W. W. Norton, 1962.
————. *Zarathustra's Sister.* Crown Publishers, 1977.
Pfeiffer, Ernst, ed. *Sigmund Freud and Lou Andreas-Salomé Letters.* Helen and Kurt Wolff Books, Harcourt Brace Jovanovich, 1966.
Puner, Helen. *Freud: His Life and Mind.* Howell, Soskin, 1947.
Roazen, Paul. *Freud and His Followers.* Alfred A. Knopf, 1975.
Sachs, Hanns. *Freud, Master and Friend.* Harvard University Press, 1946.
Stolorow, Robert, and Atwood, George E. *Faces in a Cloud: Subjectivity In Personality Theory.* Jason Aronson, 1979.
Strean, Herbert. *The Extramarital Affair.* The Free Press, 1980.

Index

233